Fundamentals of Layout Design for Electronic Circuits

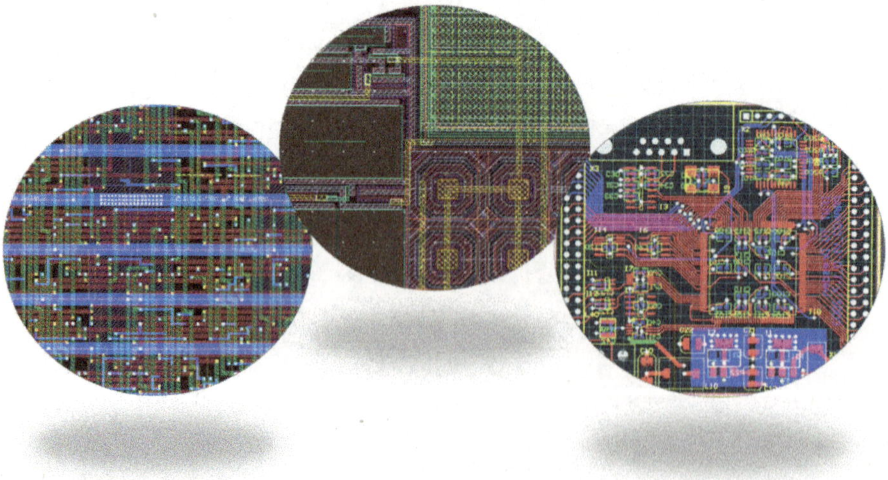

Jens Lienig · Juergen Scheible

Fundamentals of Layout Design for Electronic Circuits

Springer

Jens Lienig
Electrical and Computer Engineering
Dresden University of Technology
Dresden, Saxony, Germany

Juergen Scheible
Electronic Design Automation
Reutlingen University
Reutlingen, Baden-Wuerttemberg, Germany

ISBN 978-3-030-39283-3 ISBN 978-3-030-39284-0 (eBook)
https://doi.org/10.1007/978-3-030-39284-0

This Springer imprint is published by the registered company Springer Nature Switzerland AG
The registered company address is: Gewerbestrasse 11, 6330 Cham, Switzerland

Foreword

The advances in technology and the continuation of Moore's law mean that we can now make transistors that are smaller than human cells. We can also integrate trillions of these transistors in a single chip and expect that all these transistors turn on and off a few billion times a second synchronously. This engineering feat has been made possible by the ingenuity of computer scientists and mathematicians, who design the algorithms to enhance the performance of the computers, and the inventiveness of engineers who are able to build these complex and intricate systems. Generating the schematic network of a trillion-transistor circuit inside a CAD program is enormously difficult—getting it laid out during physical design so that the circuit works in real silicon flawlessly is, however, the real challenge we face today.

I have been teaching courses on physical design for almost two decades to computer science and electrical engineering students. I have always had to carefully walk the tight rope that separates the teaching of theory from practice. One of the most difficult parts has been finding a textbook that gives a balanced view between theory and actual design. On one hand, the current and future engineers need to know the design algorithms and how to deal with the ever-increasing number of transistors. On the other hand, they need to know how to fabricate ICs and what are the constraints that exist because of the ever-reducing transistor sizes. And here this book comes in: It covers the theoretical concepts and the technical know-how in a practical, application-oriented manner for every layout engineer. It starts with silicon material and IC fabrication and how the silicon material can be manipulated to make microelectronic devices and operate the circuit. Then, the book comes back to changes that happen in the silicon as a result of circuit operation. All of these topics are covered in a practical manner with lots of demonstrations to cement the concepts.

This book is able to connect the theoretical world of design automation to the practical world of the electronic-circuit layout generation. The text focuses on the physical/layout design of integrated circuits (ICs), but also covers printed circuit boards (PCBs) where needed. It takes the reader through a journey starting with how we transform silicon into reliable devices, discusses how we are able to perform such engineering feats, and the important practical considerations during this process. Then, the book bridges to how these vast and complicated physical structures can be best represented as data and how to turn this data back into a physical structure. It continues with the discussion of the models, styles and steps for physical design to give a big picture of how these designs are made, before going into special hands-on requirements for layout design of analog ICs. Finally, it ends by discussing practical considerations that could extend the reliability of the circuits, giving the designers and engineers a 360-degree point of view of the physical design process.

I have known Jens Lienig through his work and books for many years. In his books, he first captures the reader's attention by giving a big picture, with examples and analogies, that provides the reader with an intuitive understanding of the topics to come. Only then does he go into the details, providing the depth of knowledge needed to design high performing systems. Through this combination, his readers are able to understand the material, remember the details, and use them to create new ideas and concepts. This, along with his genuine care for his readers, vast knowledge of the field and practical experience, makes Prof. Lienig the ideal person to write such a book—and he has found the perfect match: Juergen Scheible, who has a wealth of theoretical and practical experience in designing commercial circuits. His extensive experiences as the Head of the IC Layout Department for Bosch means that he has been responsible for layout design of not only a whole slew of designs including smart power chips, sensing circuits and RF designs, but also creating new design flows to adapt to ever-changing technologies. When it comes to design, Prof. Scheible knows all the tricks that come from years of industrial experience—the multitude of rules and constraints one must consider when drawing a layout in a given technological framework. The combined experience and knowledge of these two authors have made a great tapestry of theory and practice, and hence, the resulting book is a must-read for every layout engineer.

I am delighted to write this foreword not only because I have the highest regard for both authors, but also because I cannot wait to use the book for teaching physical design. The combined expertise of the authors and the attention they have paid to theory and practice, big picture and detail, illustrative examples and written text, make this book the perfect go-to resource for students and engineers alike.

Calgary, Alberta, Canada

Prof. Laleh Behjat
Department of Electrical
and Computer Engineering
University of Calgary

Preface

When an engineer in a London Post Office got tired of sorting hundreds of tangled cables between their connectors, he filed a patent named "Improvements in or Connected with Electric Cables and the Jointing of the Same" in 1903—probably without foreseeing the broad implications of his "flat foil conductors laminated to an insulating board". Thus was born the *printed circuit board* (*PCB*), which became an engineering success. The first boards required extreme manufacturing skills—electronic devices were fixed between springs and electrically connected by rivets on Pertinax. Copper-laminated insulating layers were introduced in 1936, leading the way to reliable, mass-produced printed circuit boards. These boards enabled the manufacture of affordable electronic devices, such as radios, which have become indispensable items in everyone's home ever since.

The invention of miniature vacuum tubes in 1942 started the first generation of modern electronics. The earliest large-scale computing device, the Electronic Numerical Integrator and Computer (ENIAC), contained an impressive 20,000 vacuum tubes.

In 1948, the invention of the transistor kickstarted the second generation of computing devices. These transistors proved to be smaller, cooler and much more reliable than their predecessor vacuum tubes, enabling truly portable electronic gadgets, such as small transistor radios.

The 1960s saw the dawn of the third generation of electronics, ushered in by the development of *integrated circuits* (*ICs*). Together with semiconductor memories, they enabled increasingly complex and miniaturized system designs. Subsequently, we witnessed the first microprocessor in 1971, followed soon thereafter by a host of technical breakthroughs whose impact remains evident today. In 1973, Motorola developed the first prototype mobile phone, in 1976 Apple Computer introduced the *Apple I,* and in 1981 IBM introduced the *IBM PC*. These developments foreshadowed the *iPhones* and *iPads* that became ubiquitous at the turn of the twenty-first century, followed by intelligent, cloud-based electronics that complement, facilitate, and enhance our lives today. These days, even the cheapest smartphone contains more transistors than there are stars in the Milky Way!

This spectacular success in the art of engineering relies on a crucial step: transforming an abstract, yet increasingly complex circuit description into a detailed geometric layout that can subsequently be manufactured "for real" and without flaws. This step, referred to as *layout design*—or *physical design*, as it is also known in the industry and which is the term we use in this book—is the final stage in every electronic circuit design flow. All of the instructions necessary for fabricating PCBs and ICs must be generated in this step. Essentially, all components in the abstract circuit description, consisting of the device symbols and the wire connections between them, are translated into formats that describe geometric objects, such as footprints and drilling holes (for PCBs) or mask layout patterns comprising billions of rectangular shapes (for ICs). These blueprints are then used during fabrication to "magically create" the physical electrical network on the surface of a silicon chip in the case of an IC—which performs exactly the same functions as envisioned in the initial circuit description when electrons are sent through the system. Without this physical design stage, we would not have even the simplest radios, let alone laptops, smartphones or the myriad of electronic devices we take for granted nowadays.

Physical design was once a fairly straightforward process. Starting with a netlist that describes the logical circuit components and interconnections, a technology file, and a device library, a circuit designer would use floorplanning to determine where different circuit parts should be placed, and then the cells and devices would be arranged and connected in the so-called place and route step. Any circuit and timing problems would be solved by iteratively improving the layout locally.

Times have changed; if previous generations of circuit designs represented complexity analogous to towns and villages, current-generation designs cover entire countries. For example, if the wires in one of today's ICs, such as found in a smartphone, were to be laid out with regular street sizes, the area of the resulting chips would stretch over the continental U.S. and Canada combined, covered entirely with streets shoulder-to-shoulder! Hence, today's multi-billion transistor circuits and heterogeneous stacks of printed circuit boards require a far more complex physical design flow. Circuit descriptions are *partitioned* first in order to break down complexity and to allow parallel design. Once we have arranged the contents and interfaces of the partitions during *floorplanning*, these blocks can be handled independently. *Placement* of devices is the first step, followed by *routing* their connections. *Physical verification* checks and enforces timing and other constraints, and multiple measures are applied in a *post layout process* to ensure manufacturability of the IC and PCB layout.

The field of physical/layout design has grown well beyond the point where a single individual can handle everything. Constraints to be considered during layout generation have become extremely complex. The stakes are high: one missed reliability check can render a multi-million-dollar design useless. The fabrication facility to produce a single technology node can easily cost over a billion dollars. Research papers describe solutions to a myriad of these problems; their sheer volume, however, renders it impossible for engineers to keep pace with the latest developments.

Given the high stakes and the incredible complexity, there is a pressing need to temporarily step back from these rapid developments and to consider the *fundamentals* of this extremely broad and complex design stage. Students need to learn and understand the basics behind today's complicated layout steps—the "why" and "how", not just the "what". Engineers and professionals alike need to refresh their knowledge and broaden their scope as new technologies compete for application. With Moore's law and thus, continuous down-scaling being replaced by novel and heterogeneous technologies, new physical design methodologies enter the field. To successfully master these challenges requires sound knowledge of physical design's basic methods, constraints, interfaces and design steps. This is where this book comes in.

After a thorough grounding in general electronic design in Chap. 1, we introduce the basic technology know-how in Chap. 2. This knowledge lays the foundation for understanding the multiple constraints and requirements that make physical design such a complicated process today. Chapter 3 looks at layout generation "from the outside"—what are its interfaces, how and why do we need design rules and external libraries? Chapter 4 introduces physical design as a complete end-to-end process with its various methodologies and models. Chapter 5 then dives into the individual steps involved in generating a layout, including its multifarious verification methodologies. Chapter 6 introduces the reader to the unique layout techniques needed for analog design, before Chap. 7 elaborates on the increasingly critical topic of improving the reliability of generated layouts.

This book is the result of many years of teaching layout design, combined with industrial experience gained by both authors before entering academia. Chapters 1–7 are well structured for teaching a two-semester class of layout/physical design. For use in a one-semester class, Chap. 1 (introduction) and Chap. 2 (technology) can be assigned for self-study, with instruction starting with Chap. 3 (interfaces), followed by design methodologies (Chap. 4) and design steps (Chap. 5). Alternatively, Chap. 4 can also be used as an effective starting point, followed by the detailed design steps of Chap. 5, intermittently extended with material from the respective interfaces, design rules and libraries presented in Chap. 3. All figures of the book are available for download at www.springer.com/9783030392833.

A book of such extensive scope and depth requires the support of many. The authors wish to express their warm appreciation and thanks to all who have helped produce this publication. We would like to mention in particular Martin Forrestal for his key role in writing a proper English version of our manuscript. Special thanks go to Dr. Mike Alexander who greatly assisted in the preparation of the English text. His knowledge on the subject matter of this book has been appreciated. We thank Dr. Andreas Krinke, Kerstin Langner, Dr. Daniel Marolt, Dr. Frank Reifegerste, Matthias Schweikardt, Dr. Matthias Thiele, Yannick Uhlmann, and Tobias Wolfer for their many contributions. Our warm appreciation is also due to Petra Jantzen at Springer for being very supportive and going beyond her call of duty to help out with our requests.

Rapid progress will continue apace in layout design in the years to come, perhaps by some of the readers of this humble book. The authors are always grateful for any comments or ideas for the future development of the topic, and wish you good luck in your careers.

Dresden, Germany Jens Lienig
Reutlingen, Germany Juergen Scheible

Contents

1 Introduction 1
1.1 Electronics Technologies 2
1.1.1 Printed Circuit Board Technology 2
1.1.2 Hybrid Technology 5
1.1.3 Semiconductor Technology 7
1.2 Integrated Circuits 13
1.2.1 Importance and Characteristics 13
1.2.2 Analog, Digital and Mixed-Signal Circuits 14
1.2.3 Moore's Law and Design Gaps 17
1.3 Physical Design 21
1.3.1 Main Design Steps 21
1.3.2 Physical Design of Integrated Circuits 23
1.3.3 Physical Design of Printed Circuit Boards 26
1.4 Motivation and Structure of This Book 27
References 29

2 Technology Know-How: From Silicon to Devices 31
2.1 Fundamentals of IC Fabrication 32
2.2 Base Material Silicon 33
2.3 Photolithography 34
2.3.1 Fundamentals 34
2.3.2 Photoresist 35
2.3.3 Photomasks and Exposure 36
2.3.4 Alignment and Alignment Marks 38
2.3.5 Reference to Physical Design 39
2.4 Imaging Errors 40
2.4.1 Overlay Errors 40
2.4.2 Edge Shifts 42
2.4.3 Diffraction Effects 42
2.4.4 Reference to Physical Design 44

2.5 Applying and Structuring Oxide Layers 45
- 2.5.1 Thermal Oxidation 46
- 2.5.2 Oxidation by Deposition 47
- 2.5.3 Oxide Structuring by Etching 47
- 2.5.4 Local Oxidation 49
- 2.5.5 Reference to Physical Design 51

2.6 Doping 52
- 2.6.1 Background 52
- 2.6.2 Diffusion 52
- 2.6.3 Ion Implantation 54
- 2.6.4 Reference to Physical Design 56

2.7 Growing and Structuring Silicon Layers 58
- 2.7.1 Homoepitaxy 58
- 2.7.2 Heteroepitaxy and Polysilicon 61
- 2.7.3 Reference to Physical Design 62

2.8 Metallization 62
- 2.8.1 Fundamentals 62
- 2.8.2 Metallization Structures Without Planarization 65
- 2.8.3 Metallization Structures with Planarization 67
- 2.8.4 Reference to Physical Design 71

2.9 CMOS Standard Process 73
- 2.9.1 Fundamentals: The Field-Effect Transistor 73
- 2.9.2 Process Options 76
- 2.9.3 FEOL: Creating Devices 78
- 2.9.4 BEOL: Connecting Devices 82

References 82

3 Bridges to Technology: Interfaces, Design Rules, and Libraries 83

3.1 Circuit Data: Schematics and Netlists 84
- 3.1.1 Structural Description of a Circuit 84
- 3.1.2 Idealizations in Circuit Descriptions 86
- 3.1.3 Circuit Representation: Netlist and Schematic 87

3.2 Layout Data: Layers and Polygons 91
- 3.2.1 Structure of Layout Data 91
- 3.2.2 How to Read a Layout View 95
- 3.2.3 Graphics Operations 97

3.3 Mask Data: Layout Post Processing 102
- 3.3.1 Overview 102
- 3.3.2 Chip Finishing 102
- 3.3.3 Reticle Layout 105
- 3.3.4 Layout-to-Mask Preparation 107

3.4 Geometrical Design Rules . . . 111
3.4.1 Technological Constraints and Geometrical Design Rules . . . 111
3.4.2 Basic Geometrical Design Rules . . . 111
3.4.3 Programmed Geometrical Design Rules . . . 115
3.4.4 Rules for Die Assembly . . . 116
3.5 Libraries . . . 119
3.5.1 Process Design Kits and Primitive Device Libraries . . . 119
3.5.2 Cell Libraries . . . 121
3.5.3 Libraries for Printed Circuit Board Design . . . 123
References . . . 125

4 Methodologies for Physical Design: Models, Styles, Tasks, and Flows . . . 127
4.1 Design Flow . . . 127
4.2 Design Models . . . 133
4.2.1 Three-Dimensional Design Space . . . 134
4.2.2 The Gajski-Kuhn Y-Chart . . . 136
4.3 Design Styles . . . 139
4.3.1 Full-Custom and Semi-Custom Design . . . 139
4.3.2 Top-Down, Bottom-Up and Meet-in-the-Middle Design . . . 144
4.4 Design Tasks and Tools . . . 146
4.4.1 Creating: Synthesis . . . 146
4.4.2 Checking: Analysis . . . 147
4.4.3 Eliminating Deficiencies: Optimization . . . 148
4.5 Physical Design Optimization and Constraints . . . 148
4.5.1 Optimization Goals . . . 148
4.5.2 Constraint Categories . . . 149
4.5.3 Physical Design Optimization . . . 150
4.6 Analog and Digital Design Flows . . . 151
4.6.1 The Different Worlds of Analog and Digital Design . . . 151
4.6.2 Analog Design Flow . . . 155
4.6.3 Digital Design Flow . . . 157
4.6.4 Mixed-Signal Design Flow . . . 158
4.7 Visions for Analog Design Automation . . . 160
4.7.1 A "Continuous" Layout Design Flow . . . 160
4.7.2 A "Bottom-Up Meets Top-Down" Layout Design Flow . . . 162
References . . . 164

5 Steps in Physical Design: From Netlist Generation to Layout Post Processing . . . 165
5.1 Generating a Netlist Using Hardware Description Languages 165
5.1.1 Overview and History . . . 165
5.1.2 Elements and Example . . . 167
5.1.3 Flow . . . 168
5.2 Generating a Netlist Using Symbolic Design Entry . . . 169
5.2.1 Overview . . . 169
5.2.2 Elements and Examples . . . 170
5.2.3 Netlist Generation . . . 173
5.3 Primary Steps in Physical Design . . . 174
5.3.1 Partitioning and Floorplanning . . . 175
5.3.2 Placement . . . 180
5.3.3 Routing . . . 183
5.3.4 Physical Design Using Symbolic Compaction . . . 187
5.3.5 Physical Design Using Standard Cells . . . 187
5.3.6 Physical Design of Printed Circuit Boards . . . 189
5.4 Verification . . . 194
5.4.1 Fundamentals . . . 195
5.4.2 Formal Verification . . . 198
5.4.3 Functional Verification: Simulation . . . 199
5.4.4 Timing Verification . . . 200
5.4.5 Geometric Verification: DRC, ERC . . . 202
5.4.6 Extraction and LVS . . . 205
5.5 Layout Post Processing . . . 208
References . . . 210

6 Special Layout Techniques for Analog IC Design . . . 213
6.1 Sheet Resistance: Calculating with Squares . . . 213
6.2 Wells . . . 216
6.2.1 Implementation . . . 216
6.2.2 Breakdown Voltage . . . 218
6.2.3 Voltage-Dependent Spacing Rules . . . 219
6.3 Devices: Layout, Connection, and Sizing . . . 220
6.3.1 Field-Effect Transistors (MOS-FETs) . . . 220
6.3.2 Resistors . . . 224
6.3.3 Capacitors . . . 227
6.3.4 Bipolar Transistors . . . 230
6.4 Cell Generator: From Parameters to Layout . . . 233
6.4.1 Overview . . . 233
6.4.2 Example . . . 234

6.5 The Importance of Symmetry 236
6.5.1 Absolute and Relative Accuracy: The Big Difference 236
6.5.2 Obtaining Symmetry by Matching Devices 237
6.6 Layout Matching Concepts 239
6.6.1 Matching Concepts for Internal Device Fringe Effects 240
6.6.2 Matching Concepts for Unknown Gradients 247
6.6.3 Matching Concepts for External Device Fringe Effects 249
6.6.4 Matching Concepts for Known Gradients 250
6.6.5 Matching Concepts for Orientation-Dependent Effects 252
6.6.6 Summary of Matching Concepts 254
References 255

7 Addressing Reliability in Physical Design 257
7.1 Parasitic Effects in Silicon 257
7.1.1 Substrate Debiasing 258
7.1.2 Injection of Minority Carriers 261
7.1.3 Latchup 264
7.1.4 Breakdown Voltage, aka Blocking Capability, of p–n Junctions 266
7.2 Surface Effects 267
7.2.1 Parasitic Channel Effects 267
7.2.2 Hot Carrier Injection 270
7.3 Interconnect Parasitics 272
7.3.1 Line Losses 272
7.3.2 Signal Distortions 273
7.3.3 Crosstalk 275
7.4 Overvoltage Protection 275
7.4.1 Electrostatic Discharge (ESD) 276
7.4.2 Antenna Effect 286
7.5 Migration Effects in Metal 290
7.5.1 Electromigration 291
7.5.2 Thermal Migration 292
7.5.3 Stress Migration 293
7.5.4 Mitigating Electromigration 296
7.5.5 Mitigating Thermal and Stress Migration 299
References 301

Index 303

Chapter 1
Introduction

Layout design—or *physical design*, as it is also known in the industry and referred to throughout the book—is the final step in the design process for an electronic circuit. It aims to produce all information necessary for the fabrication process to follow. In order to achieve this, all components of the logical design, such as cells and their connections, must be generated in a geometric format (typically, as collections of rectangles), which is used to create the microscopic devices and connections during fabrication.

This chapter gives a sound introduction to the technologies, tasks and methodologies used to design the layout of an electronic circuit. With this basic design knowledge as a foundation, the subsequent chapters then delve deeper into specific constraints and aspects of physical design, such as semiconductor technologies (Chap. 2), interfaces, design rules and libraries (Chap. 3), design flows and models (Chap. 4), design steps (Chap. 5), analog design specifics (Chap. 6), and finally reliability measures (Chap. 7).

In Sect. 1.1, we introduce several of the most common fabrication technologies for electronic systems. The central topic of this book is the physical design of integrated circuits (aka *chips*, ICs) but hybrid technologies and printed circuit boards (PCBs) are also considered. In Sect. 1.2 of our introduction, we examine in more detail the significance and peculiarities of this related branch of modern electronics—also known as *microelectronics*. In Sect. 1.3, we then consider the physical design of both integrated circuits and printed circuits boards with a specific emphasis on their primary design steps. After these opening sections, we close the introductory chapter in Sect. 1.4 by presenting our motivation for this book and describing the organization of the chapters that follow.

J. Lienig and J. Scheible, *Fundamentals of Layout Design for Electronic Circuits*,
https://doi.org/10.1007/978-3-030-39284-0_1

1.1 Electronics Technologies

All electronic circuits comprise *electronic devices* (transistors, resistors, capacitors, etc.) and the metallic interconnects that electrically connect them. There is, however, a wide range of different fabrication technologies available to physically realize such electronic devices; these technologies can be classified into three main groups:

- *Printed circuit board technology*, which can be subdivided in
  - Through-hole technology (THT),
  - Surface-mount technology (SMT),
- *Hybrid technology*, often subdivided in
  - Thick-film technology,
  - Thin-film technology,
- *Semiconductor technology*, subdivided in
  - Discrete semiconductor components,
  - Integrated circuits.

To each of these technologies are added myriad extra features and custom designs for different use cases. Take, for example, automobile electronics, where a very high degree of robustness is required, or cellphones, where extreme compactness is a key requirement. We next examine the most important of these technologies in further detail.

1.1.1 Printed Circuit Board Technology

The *printed circuit board* (*PCB*) is the most widely used technology for electronic packaging. It mechanically supports and electrically connects electronic devices that are typically soldered onto the PCB.

Circuit Board
The basic element is an electrically isolated circuit board, known as the substrate core, and typically made of glass-fiber-reinforced epoxy resin. Everybody has seen this green board at some point. Papers stabilized with phenolic resins are also an option. (This approach was particularly widespread in the early years of electronics.) Paper-based circuit boards are only suitable for very low-spec applications, are rarely used anymore, and are not covered further in this book.

The circuit board has two primary functions: (i) to provide the foundation upon which the electronic devices are physically mounted, and (ii) to provide a surface upon which the interconnects for electrically connecting the devices can be constructed.

The interconnects are etched out of a metallic layer that has been applied to the substrate surface. The metal layer is made of copper and can be applied to one side of the circuit board, or to both sides. Copper is the material of choice here, as it has several very beneficial properties: (i) it is an excellent electrical conductor; (ii) it lends itself well to etching; and (iii) it can be soldered easily. (Devices are soldered in place on the board, and simultaneously the chip pins are electrically connected to the interconnects on the board with solder.)

Fabricating Interconnects

How interconnects are fabricated is visualized in Fig. 1.1 and explained below with reference to steps (a) to (i) in the figure. The substrate core that is the foundation of the board is sometimes also referred to as the *carrier substrate*, as it "carries" (i.e., "holds" in place) the electronic devices and interconnects.

A substrate core is first coated with copper, and then a layer of photoresist is applied (a to c). The photoresist has a special property, namely that after exposure to light, it can be dissolved with a fluid called the *developer*. In the next step we use a *mask*, which is a (transparent) glass plate onto which the image of the desired interconnect has been applied as an opaque layer to the bottom surface of the glass plate (shown in black in Fig. 1.1d).

This mask is then positioned over the PCB (d) and exposed with light (illuminated sectors in yellow and shaded sectors in gray, Fig. 1.1e). The shaded area produces an image of the desired interconnect on the PCB. The resist at the exposed areas thus becomes dissolvable (the areas in pale blue in Fig. 1.1f); this exposed resist can then be dissolved and washed away with the developer. The remaining unexposed

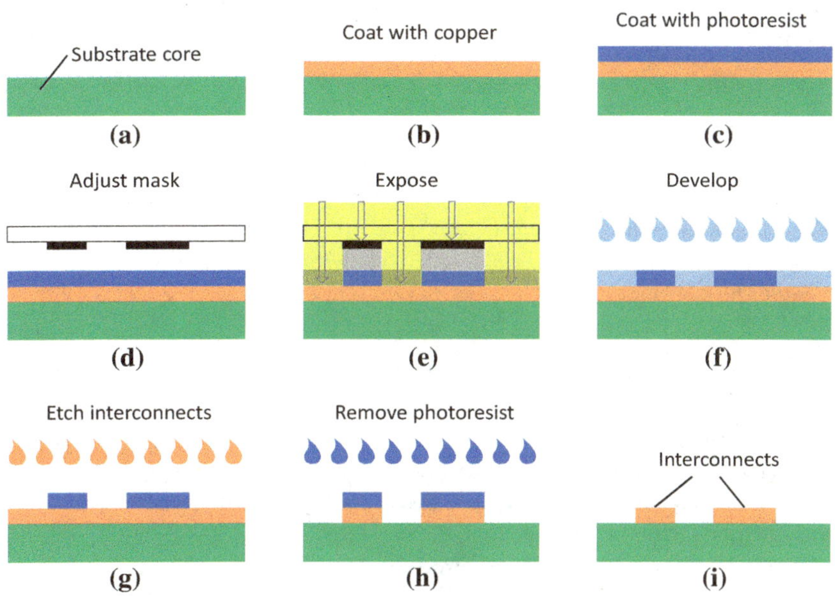

Fig. 1.1 Sectional view of the creation of interconnects on a printed circuit board (PCB) by photolithography and subsequent etching

areas retain their photoresist, which protects the copper layer underneath against etching in the next step (g), such that the etching agent only removes the copper on the unprotected areas. We say the resist "masks" the etching. After etching, only the copper remains at the prior unexposed areas; the remaining photoresist is washed away with a suitable fluid (h). Through this process, we have created in the copper layer the interconnect structure that was patterned on the mask (i).

When only a few PCBs are required, such as for prototypes, the interconnects are sometimes not formed by etching, but by mechanically milling the metal layer.

Multi-Layer Printed Circuit Boards

Boards can be constructed of several stacked substrate cores, and in this case are called *multilayer PCBs*. Figure 1.2 shows an example of a multilayer board, consisting of three cores and six routing layers (the top and bottom of each of the three cores). The cores are glued together with a bonding agent, which also acts as an electrical isolator between the opposing copper layers of neighboring substrate cores, to prevent short circuits.

Plated-through contacts, known as *vias*, are then used to electrically connect different routing layers. Vias are created at the beginning of the manufacturing process by drilling holes through the core layers. The via walls are later coated with copper to make them electrically conductive. These vias are labeled as *buried*, *blind* and *through-hole vias* depending on where the holes are located (see Fig. 1.2).

Mounting Technologies

Two different technologies are primarily used for mounting components on PCBs:

- Through-hole technology (THT), and
- Surface-mount technology (SMT).

THT makes use of devices that have leads to make the electrical contacts. These leads are inserted in holes, which are through-hole vias, and soldered in place (see Fig. 1.2, left). With SMT, the devices instead have metal pads for connecting to the surface of the board (shown in black in Fig. 1.2). Associated with these mounting technologies—THT and SMT—are respectively *through-hole devices* (*THDs*) and *surface-mount(ed) devices* (*SMDs*).

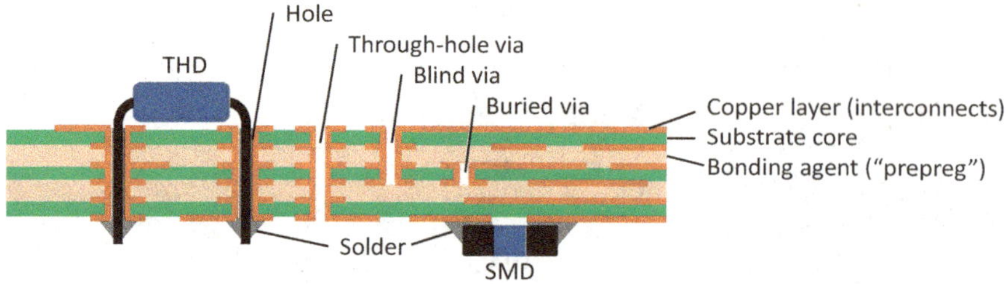

Fig. 1.2 Cross-section of a multilayer board with six routing layers

The two mounting technologies can also be mixed. Component placement systems can handle SMDs much more easily than THDs. In addition, much higher packing densities can be achieved with SMDs, as they are smaller and can be mounted on both sides of a PCB. These advantages make surface-mount technology (SMT) the more widely used approach today.

Besides discrete devices, integrated circuits (ICs) can also be mounted on PCBs. In general, however, they must be "packaged" in an enclosure. Unpackaged ICs (aka *bare dies)* are sometimes mounted directly on PCBs; in this case, however, the stability of the connection is critical due to the different thermal expansion rates of semiconductors and boards.

1.1.2 Hybrid Technology

Hybrid technology is an approach in which some of the electrical components are physically separate devices that are then mounted on the carrier substrate, while other components are created directly on the carrier substrate during fabrication. Thus, we get the name "hybrid technology".

A variety of carrier substrates are used in hybrid technology to construct the substrate core; ceramic, glass and quartz are commonly used. SMDs can be mounted on these carrier substrates; however, THDs cannot be mounted, as through holes for mounting purposes are not drilled in these substrates.

Interconnect traces are produced differently on hybrid-technology substrates than on PCBs. The two technologies typically used are *thick-film technology* and *thin-film technology*. In thick-film technology, conductive pastes are applied in a silk-screen printing process and then burnt in. In thin-film technology, the conductive material is vaporized or sputtered[1] onto the substrate. Interconnects are subsequently formed using a photolithographic process much like the process described above for PCBs.

The electrical conductivity of the deposited layers can be adjusted over a large range, which allows electrical resistors to be produced along with interconnects using these technologies. Resistivity values can be finely tuned with a laser; resistance can be increased by trimming the outer regions of the deposited interconnect back perpendicular to the current flow until the desired resistance value is reached.

Interconnect crossings and capacitors can also be realized, such as by alternately stacking conductive and insulating layers. Capacitors can also be created by intertwining comb-like interconnects within a metal layer. An example of this printed capacitor is given in Fig. 1.3.

Example: LTCC Technology

A widely used variant of the thick-film hybrids is *LTCC* technology, which stands for *low temperature co-fired ceramics*. The LTCC fabrication process is representative of

---

[1] Sputtering is a physical process whereby microscopic particles of a solid material are ejected from its surface by bombarding the solid with high-energy ions.

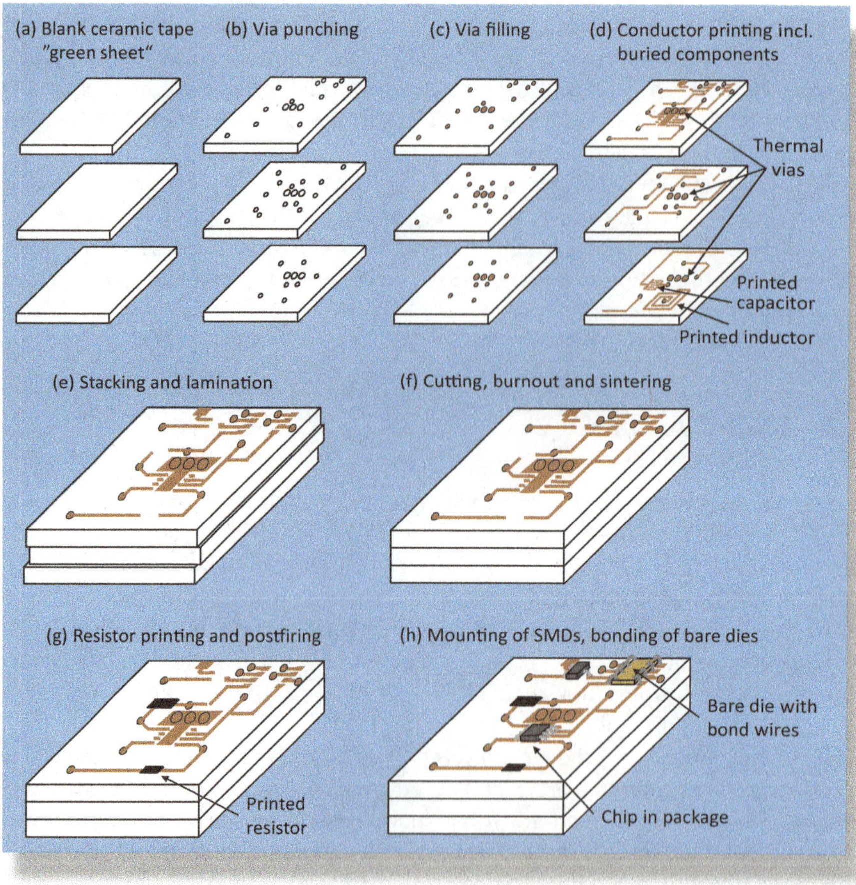

Fig. 1.3 Fabrication of an LTCC hybrid circuit with printed capacitor, inductor and resistor (LTTC: low temperature co-fired ceramics)

many other technology variants and we will discuss it below. The LTCC fabrication process is depicted in Fig. 1.3a–h.

LTCC technology does not use a prepared ceramic substrate. Instead, fabrication starts with films, in which the ceramic mass in powder form is bonded with other substances. These films are called *green sheets* (a). As we will see below, these films will be made rigid during subsequent processing steps, to produce a board that will become a component in the larger system. First, holes for vias are punched in the sheets (b). These are then filled with conductive paste (c). The interconnect geometries are formed with conductive paste in a silk-screen printing process (d). The green sheets are then stacked on top of one another and laminated by heating them up slightly. They are thus bonded together (e). The stack is then cut to size, pressed together and fired and sintered in an oven (f). Some of the additives escape from the sheet stack during this process and the hybrid shrinks and is sintered to a ceramic plate, which can contain several interconnect layers in its core. Pressure continues to be applied to the laminated stack during burnout and sintering, so that the shrinking effect is almost completely in the z-axis and the lateral dimensions are not affected. Resistors and conductive surfaces for contacting SMDs and ICs are then printed onto the surface and burnt in (g).

Finally, the SMDs and ICs are mounted (h). The SMDs are fixed in place with a conductive glue or by reflow soldering. The ICs can be mounted unpackaged as bare dies because of the similar thermal expansion rates of carrier substrate (ceramic) and the dies' silicon. They are electrically connected using bond wires running from contact surfaces on the IC, so-called *pads*, to contact surfaces on the carrier.

Among the benefits of hybrid technology over printed circuit technology are: (i) greater mechanical stability (when exposed to extreme shocks and vibrations in automobiles, for example); (ii) higher packing density (due to the use of bare dies); and (iii) better dissipation of thermal losses.

Thermal dissipation is mainly achieved with LTCCs by mounting the hybrid on a heat sink to maximize the thermal coupling over the entire hybrid surface. Good heat dissipation from the top of the hybrid to its bottom can be further improved by deploying so-called *thermal vias*—these are plated-through contacts designed specifically to transmit heat.

While hybrid technology has advantages as noted above (e.g., greater mechanical stability, packing density, and thermal dissipation), it has the disadvantage of higher manufacturing costs, as compared to PCBs.

1.1.3 Semiconductor Technology

With the technologies we have discussed thus far, the electronic devices must be wholly or partially added from an external source. With semiconductor technology, on the other hand, an entire electronic circuit can be built as a single unit. Here, all electronic devices and all electrical connections are created in the fabrication process itself. This type of circuit is fully integrated—this is where the name *integrated circuit* (*IC*) originates—in a monolithic semiconductor die. These single, small flat pieces, comprised mostly of silicon, are also called "chips".

We can also use semiconductor technology to construct purely discrete (i.e., single) electronic devices. Typical examples are diodes or active devices, such as transistors and thyristors, for driving very large currents in power electronics. If we examine these devices more closely, we find that they are composed of many similar devices connected in parallel on the chip. Protective circuitry that remains hidden from view, but which supports the device's performance characteristics, is also often integrated in the chip.

What are Semiconductors?—Physical Aspects of Semiconductor Materials
While semiconducting materials can conduct electrical current, their electrical resistivity at room temperature is quite high. However, their conductivity increases exponentially with rising temperature. This thermal characteristic is very different from standard interconnects (metals) and is a key property of semiconductors. We will now delve deeper into the underlying physics.

Free moving charge carriers are required for a current flow. In solids, these charge carriers are electrons. Hence, the question is: "How do we get enough 'free' electrons in semiconductors?" Electrons orbit the atomic nucleus, and their energy levels increase the further they are away from the nucleus. It is also a fact that they can only have certain energy states called *shells* (electron shells) that expand into so-called *bands* in constellations of many atoms. The most exterior band in a substance is also called the *valence band*. If electrons in the valence band (so-called *valence electrons*) can be sufficiently energized to enable them to jump to the next higher band, they can move freely there and thus increase the electrical conductivity of the material. Therefore, this band is also labeled as *conduction band*.

Valence and conduction bands are particularly close together in conductive materials, such as metals; they can even overlap (see the orange area in Fig. 1.4). In this case, many electrons can jump to the conduction band. Hence, metals are excellent conductors. In the case of insulators, on the other hand, the energy gap ΔE between the valence and the conduction band (the so-called *band gap* or *band distance*) is so large that it is an almost insurmountable threshold, and there are practically no electrons in the conduction band (blue area in Fig. 1.4).

The main characteristic of semiconductors is that their band gap lies between these two extremes (central green area in Fig. 1.4). On the one hand, this gap is so large that only very few electrons in the valence band have enough energy at

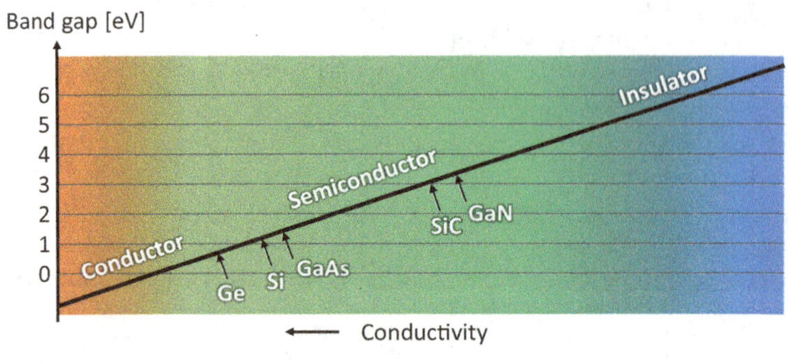

Fig. 1.4 Materials in the "conductor", "semiconductor" and "insulator" categories as a function of the band gap between valence and conduction band. The values are given for typical semiconductor materials at 300 K. (SiC can take values between 2.4 eV and 3.3 eV, depending on the formed crystal lattice. The value for crystal lattice "6H" is shown.)

room temperature to reach the conduction band. On the other hand, the conduction band is close enough that a temperature rise on the order of a few hundred degrees kelvin above room temperature provides enough energy to increase the number of free electrons, and thus the conductivity, by many orders of magnitude.

The conductivity increases not only because of the free electrons in the conduction band, but also due to the creation of electron holes—known as *defect electrons* or *holes*—in the valence band. These holes can be easily filled with valence electrons from neighboring atoms; the holes then disappear, generating new ones in the "electron-delivering" atoms. A current flow—known as *hole conduction*—can then take place as a result of this chain movement of valence electrons. However, generating free-charge carriers (electrons and holes) by thermal means is not our main concern here, rather we want to highlight the underlying physics for a better understanding of our primary intention.

Let us now take silicon as a typical example (Fig. 1.5). For silicon, the band gap between the lower edge of the conduction band E_C and the top edge of the valence band E_V is given by $\Delta E = E_C - E_V = 1.1$ eV. Silicon is *4-valent*—in other words, the valence band of a silicon atom contains four electrons. If we replace a silicon atom with an atom of a 5-valent element, such as phosphor, arsenic or antimony, the extra electron will not "fit" into the valence band of the surrounding silicon crystal. It has an energy level E_D, which is just slightly below the silicon conduction band; it is so close to this band that at room temperature it has enough energy to allow it to enter this conduction band (Fig. 1.5, left). Hence, introducing 5-valence impurity atoms into the silicon allows us to increase its conductivity.

Instead of implanting 5-valence impurity atoms, we can also implant 3-valence impurity atoms, such as boron, indium or aluminum, to increase the conductivity. In this case, the impurity atom is at energy level E_A, which is just slightly above the valence band edge E_V of silicon; hence, this impurity atom can very easily accept a fourth electron from a neighboring silicon atom (Fig. 1.5, right). The number of holes in the silicon valence band is thus increased. These holes can be viewed as positive charge carriers that are available for current flow, as they can move freely.

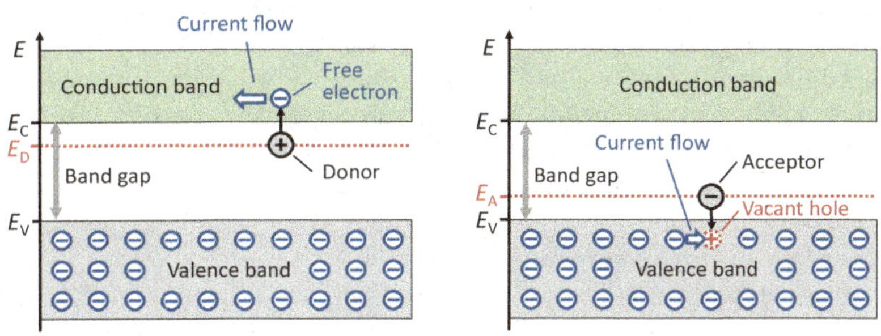

Fig. 1.5 Generating free-charge carriers and thus current flow in a semiconductor (silicon). The creation of free charge carriers by doping with donors is shown on the left and with acceptors on the right, resulting in conduction by electrons (left) and holes (right)

Implanting impurities into a substance is called *doping*. Five-valence impurity atoms are called *donors*, as they release an electron (into the conduction band). They are listed in the column to the right of silicon in the periodic table. Three-valence impurity atoms are known as *acceptors* because of their ability to accept valence electrons from neighboring atoms. They are found in the column to the left of silicon in the periodic table.

A semiconductor containing donors is called *n-doped* and one containing acceptors is called *p-doped*. In areas of n- and p-doping, the additional electrons are trapped by the additional holes. They are said to be *recombined*. Donors and acceptors effectively neutralize each other here.

Any residual donors or acceptors are key to conductivity. The semiconductor is said to be *n-conductive* when donors are in surplus, as the current flow is mostly due to electrons, i.e., negative charge carriers. The semiconductor is said to be *p-conductive* when there are surplus acceptors, as here it is mainly the holes—that can be viewed as positive charge carriers—that cause the current flow. The more abundant charge carriers are called *majority carriers*, while the correspondingly less abundant charge carriers are called *minority carriers*.

The Use of Semiconductors

Extremely pure semiconductor material in monocrystalline form is required for producing integrated circuits. All atoms must be physically arranged in a continuous regular structure. This type of structure must be manufactured as it does not occur naturally. It is produced in the form of crystal ingots that are then machined into very thin slices, or *wafers*, which are used for chip fabrication. One wafer can hold a huge number of chips—many hundreds to tens of thousands, depending on the chip and wafer sizes; the chips are all fabricated together on the wafer. At the end of the fabrication process, individual rectangular chips, or *dies*, are cut from the wafer.

Figure 1.6 shows a finished wafer under a microscope. The dies have been separated from one another and are held in place by an adhesive foil (so-called "blue tape").

The most widely used material in the semiconductor industry is silicon. Other semiconducting materials are utilized for special purposes in the industry. Typical examples include: gallium arsenide (GaAs) and silicon germanium (SiGe) in RF circuits; and gallium nitride (GaN) and silicon carbide (SiC) in power electronics. Two 4-valent elements are combined both in SiC and SiGe; a 3-valent element is combined with a 5-valent element in GaAs and GaN. The resulting crystalline structure again behaves like a 4-valent element.

Integrated Devices and Interconnect Traces

Integrated devices are produced by successively doping a wafer in different ways. The doping operations can be altered in the following ways: (i) different impurities (there are generally several different types of donors and acceptors); (ii) different concentrations (the number of impurity atoms per unit volume); (iii) the penetration depth (up to some μm), and (iv) the doping location.

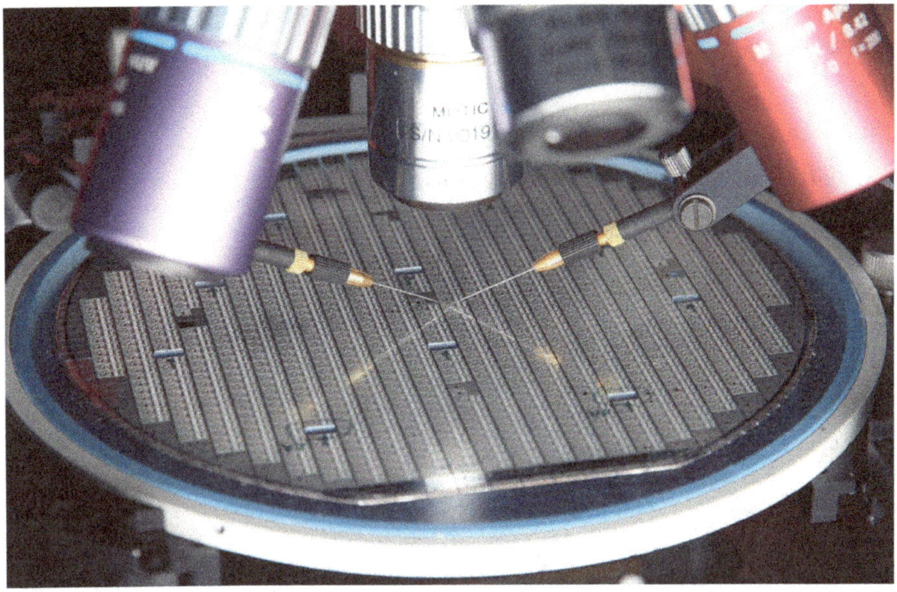

Fig. 1.6 Wafer under microscope; one chip is contacted by two probes for testing

Integrated devices can be produced using simple processes having less than 10 doping operations. In more complex processes there can be more than 20 doping procedures. For example, an additional layer of the base material is often deposited on the wafer surface between doping with impurity atoms. (This process is called *epitaxy*.)

The doping process is conceptually similar to the photolithographic process for PCBs presented in Sect. 1.1.1, in that masks are again used to selectively alter the composition of the surface of an area. For PCBs we masked areas to create interconnect traces (using subsequent etching, etc.). Here we also use masks, but ones that are far more detailed, with feature sizes that are on the order of nanometers, to selectively deposit atoms on the surface of the wafer, effectively implementing different doped areas.

Different types of electronic devices, such as transistors, diodes and resistors, can be created by implementing differently doped areas. However, care needs to be taken here. When we talk about "devices" on chips, we should be aware that we are only talking about different sections of a monolithic semiconductor die. These sections are designed in such a way that the electrical interaction of the doped areas contained in each section produces the behavior of the desired electronic device. In contrast to the devices on a printed circuit board (PCB), the devices on a chip are never isolated from one another. This can always cause the devices on a chip to interact. These interactions must be considered in the design flow as will be discussed in detail in Chap. 7.

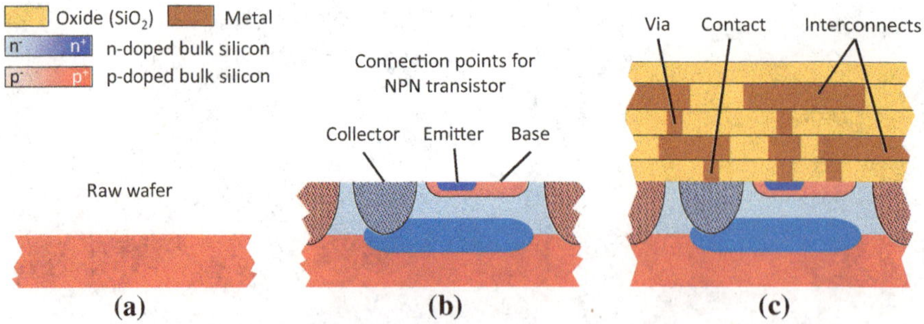

Fig. 1.7 Sectional view of an IC chip of a typical NPN transistor **a** at the beginning of the process, **b** after "front-end-of-line" (FEOL) and **c** after "back-end-of-line" (BEOL) in the semiconductor process. n-doped regions are drawn in blue, p-doped regions in red. Metallic layers are brown and insulating layers ocher

Figure 1.7 shows a sectional view of a chip. A wafer is slightly less than 1 mm thick. The electrically active parts, however, are located in a very thin layer on one of the two chip surfaces. This area of approximately 1 to 2% of the wafer thickness is depicted in Fig. 1.7; the figure also shows the fabrication stages, which are explained below.

First, the electronic devices are constructed on the wafer surface using doping operations. Semiconductor manufacture begins with a *raw wafer* (see Fig. 1.7a). All doping is carried out in this so-called "front-end-of-line" (*FEOL*) part of IC fabrication[2] and an epitaxial layer is applied as well, if required. The result is exemplified with an NPN bipolar transistor[3] in Fig. 1.7b. The regions doped with impurities are shown in color. We always show n-conductive regions in blue and p-conductive regions in red in this book. The colors indicate that the raw wafer (a) was originally p-doped and the epitaxial layer (b, on top of the raw wafer) n-doped.

Second, the electrical devices are interconnected by constructing metallic and insulating layers. This process is commonly labeled as "back-end-of-line" (*BEOL*). Here too a photolithographic process involving masks is used to create interconnects. This process is also conceptually similar to the way interconnects were created on PCBs (as presented in Sect. 1.1.1), but here the masks are again far more detailed with nanometer-sized features. Insulating layers (shown in ocher) and metallic layers (brown) are alternately stacked on top of one another and structured in the back-end-of-line (BEOL) part of fabrication. *Interconnects* and *through contacts* are formed in this step.

---

[2]While the term "front-end-of-line" (FEOL) refers to the first portion of any IC fabrication where the individual devices are patterned, "back-end-of-line" (BEOL) comprises the subsequent deposition of metal interconnect layers. Both are discussed in Chap. 2.

[3]Bipolar transistors are devices whose operation depends on the two types of charge carriers (electrons and holes). We will cover these devices and their operation more fully in Chap. 6.

The result of the BEOL is shown in Fig. 1.7c for two interconnect layers. The devices are electrically connected on the silicon surface through *contact holes* (*contacts*) in the bottom insulating layer. The through contacts between neighboring metal layers are known as *vias* in IC chips as well as in PCBs. (Please note that we use the term *through contact* to label any vertical connection between layers. Referring to ICs, we further differentiate between *contacts* that connect the devices to the (first) metal layer and *vias* that connect (two) metal layers.)

As illustrated in Fig. 1.7, photolithography is used to create all structures on the chip, regardless of whether it is delineating doping areas in the FEOL (front-end-of-line), or the formation of vias and interconnects in the BEOL (back-end-of-line). Photolithography plays a key role in fabrication, as we also saw in the production of PCBs in Sect. 1.1.1. We shall cover the process steps in semiconductor fabrication in detail in Chap. 2.

1.2 Integrated Circuits

1.2.1 Importance and Characteristics

Since the first integrated circuits (ICs) appeared in the 1960s, microelectronics has developed at a breathtaking pace. It has long become a key technology across all our technological advances. It has a huge effect on all our lives and will continue to do so. What are the drivers behind this massive relentless creative power? We will now dig deeper into the important properties of ICs that continue to propel these developments, and try to answer this question.

The idea of integrating electronic circuits on a single piece of semiconductor material was first expressed towards the end of the 1950s by Jack Kilby [3] and Robert Noyce [5] independently of one another. The first commercial semiconductor chip was produced in 1961: it was a logical memory element (called a *flipflop*) with four transistors and five resistors [1].

This was the birth of microelectronics and the beginning of the modern computing era. From that time forth, semiconductor technology went from strength to strength, accompanied by unabating IC miniaturization. This miniaturization is the driver for a series of effects, that are mutually supportive, and whose cumulate effect, upon closer scrutiny, continues to amaze.

The on-going reduction in the footprint of individual devices means the chips use less power, operate faster, and can accommodate an increasing number of functions as the individual devices can be packed more densely. These are all logical developments that are easy to understand. Why the extra functions should be cheaper is not so easy to fathom though. We can explain it as follows. It is a fact that semiconductor technologies and chip space become more expensive with increasing miniaturization. Nonetheless, the extra costs are more than recouped as individual functions require less chip space due to the miniaturization. You get more bang for your buck—effectively "more functionality for the same price"—with each new chip generation.

There is another effect that is less obvious but is nonetheless a very important aspect of the chip success story. Ever increasing integration density in chips has a very positive effect on the reliability of electronic systems: every incremental reduction in the number of solder points and every discrete device you can do without reduces the probability of a system failure. An entire chip is only a single device as far as the downtime risk goes. (Recall that integrated devices are essentially only sections of a semiconductor chip.) The semiconductor chip thus represents a single device, and a single point of failure; as a result, systems built using such semiconductor chips have fewer possible points of failure, which leads to higher reliability.

Let us try to imagine the implications of these effects: If you split the electronics from a modern mobile phone into discrete (i.e., individual) electronic components, and place them on PCBs, for example, you would need a huge industrial building to hold them. This monstrous "device" would not only be unwieldy and therefore useless, it would also be prohibitively expensive and highly unreliable. If a single device or connection in this hypothetical system failed, the system would fail as well. It would likely be permanently down, which would mean that (on the bright side) at least you wouldn't need the entire power plant required to run it!

Let us list these six effects again: continued improvements in microelectronics make electronic systems smaller, faster, more economical, more intelligent, cheaper and more reliable.[4] Normally, these performance characteristics cannot all be improved together—as evidenced in other technology sectors, such as the automotive industry, for example. They typically hamper each other, and engineers must find the best compromise for each use case. In contrast, all these performance characteristics have improved in unison in microelectronics, which explains its persistent and sustainable success.

1.2.2 Analog, Digital and Mixed-Signal Circuits

Today's chips are highly complex. The first thing to be aware of is that most of them contain both digital and analog circuits. Not only do these two types of circuits operate fundamentally differently from one another, they differ greatly in their suitability for existing design flows and semiconductor technologies.

Let us look at digital circuits first. They are much more technically amenable than their analog counterparts because they handle exclusively discrete signal values. These typically are binary signals, with only two differentiable values, which are

[4] We should mention here that downscaling to lower technology nodes in semiconductor fabrication has reached a point where aging effects are becoming increasingly critical. One of the more acute concerns is interconnect degradation by migration effects, in which the electrical current flowing through the IC can slowly erode the miniature physical structures. Preventive measures for these effects are needed especially in the physical design flow. We shall deal with this topic fully in Chap. 7.

normally interpreted as the dual binary digits “1” and “0”, or the logical values “true” and “false”.

Digital logic can be implemented electrically with two (arbitrary) voltage levels. These given voltage levels must only be reached within a given tolerance. A range is defined between the logic states so that they can be clearly demarcated. Flawless operation can be achieved by waiting for a sufficiently long period to ensure that all logic states have settled at a defined value. (This is accomplished by setting the clock rate accordingly.) Hence, the requirements for these devices (typically unipolar CMOS[5] transistors) to switch logic states can be low.

Modern ICs in digital electronics integrate numerous cores and the necessary peripherals directly on the chip itself. These chips can contain more than ten billion transistors today (2020). Figure 1.8 shows the Intel® “i7 Haswell-E™” from 2014 as an (historic) example. The chip, which was fabricated in a 22 nm node, has a surface area of 355 mm^2. It contains eight cores and consists of a total of 2.6 billion transistors [6]. These chips often look like images of cities taken from satellites orbiting the earth; it is amazing to realize that the level of complexity similar to that of a huge city (which would stretch over an entire continent!) has been precisely manufactured on a piece of silicon the size of your fingernail.

Aside from digital circuitry, electronic systems require analog circuit devices, too. They mediate between the abstract digital data processing and our real world—with its myriad physical parameters whose values and periodicities change *continuously*

Fig. 1.8 Intel microprocessor in 22 nm technology node with eight core processors

[5]CMOS is an acronym for “complementary metal oxide silicon”. CMOS technology incorporates two complementary n-type and p-type unipolar transistors, which are manufactured as metal-oxide-silicon layers. We will cover them in detail in Chap. 2.

in contrast to *discrete* digital states. This "task sharing" is analogous to biological organisms. Here, in addition to a brain for information processing, every organism needs (i) sensory organs to scan the environment, (ii) internal energy supplies, and (iii) appendages to physically act on the environment. Similarly, in any mechatronic or electronic system there is a need for analog circuitry, to support its inherent digital information processing. These systems (i) scan analog sensor inputs and convert them to digital data, (ii) supply the system with current and voltage, and (iii) act upon the computed data to control external displays, speakers, valves, electric motors, and the like. All these tasks are performed by circuits with one thing in common: they handle and produce analog signals.

It should be noted that the CMOS technology used in digital circuits can be used as well for many analog tasks. In addition, there are many applications where devices with special performance characteristics are required. Among these custom-designed devices are bipolar transistors—characterized by high cutoff voltages, robustness, usage of temperature dependence—, and special power transistors with very low resistance in the "on" state and the ability to conduct very large currents. These custom circuits were implemented using different, separate chips in the early semiconductor years. They were manufactured using semiconductor processes tailored for the respective devices.

The situation has changed in recent years. Semiconductor technologies, in which *all types* of devices needed for an integrated system can be manufactured on a single chip, have been available since the 1990s. Examples of these "mixed technologies" are *BICMOS* (bipolar transistors and CMOS) and *BCD* (bipolar transistors, CMOS and DMOS[6]).

Due to today's high degree of integration, the combination of digital and analog parts of a circuit chip is standard practice. Now most chips are *mixed-signal* chips; they are fabricated in CMOS or BICMOS depending on the specification. If they include power transistors as well, then they are called *smart power* ICs which are fabricated in the above-mentioned BCD technology.

Figure 1.9 shows a smart power chip from 2018 for an automotive control unit from Robert Bosch GmbH®. All system functions are integrated in this single chip: analog circuits for sensor scanning ("Sense"); internal power supply ("Supply"); power stages for actuator control ("Act"); digital information processing ("Think") implemented with *standard cells*.[7] A chip containing all these different types of electronic modules is called an *SOC* (*system on chip*) [2]. The chip illustrated in Fig. 1.9 is manufactured in a BCD technology in the 130 nm node, with a surface area of 34 mm^2 and containing 164,000 devices in the analog circuit parts and some

[6]DMOS stands for "double diffused metal oxide silicon". This is a fabrication technology for unipolar transistors that switch large currents in power electronics. We can implement exceptionally low on-state resistances on the order of mΩ with DMOS transistors.

[7]Designing with standard cells is a very efficient and hence popular design flow for ICs. We will introduce and discuss standard cells and design flows in Chap. 4.

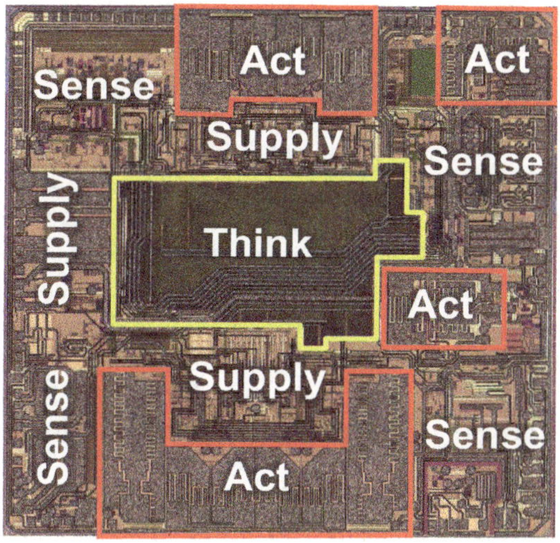

Fig. 1.9 Bosch smart power chip in 130 nm BCD technology (BCD: bipolar transistors, CMOS and DMOS) for automotive electronics

3 million transistors in the digital part (yellow box). The external power supply is 14 V, and the chip breakdown voltage[8] is 60 V.

1.2.3 Moore's Law and Design Gaps

We have seen how semiconductor downscaling, the relentless push to smaller chip structures, is central to the microelectronics evolution. So-called *technology nodes* (also referred to as *process nodes*) are categorized according to the smallest structure size that can be implemented reliably and reproducibly on a wafer. There is however no universally accepted definition of the scale in question here. Determining and specifying this feature size differs from manufacturer to manufacturer. For example, via holes, or the smallest permissible wire width, or the smallest permissible active length of a unipolar transistor (defined as the distance between the transistor's source and drain), are some of the smallest structures found on chips. Nevertheless, we can at least estimate the approximate size of these dimensions from the smallest feature size specification.

Fabrication processes for semiconductors are very complex, and every effort is made to avoid tinkering with established production processes. Reducing the feature size is a huge undertaking. Hence, miniaturization is not a continuous process; rather, it takes place in clearly defined steps. Experience has shown that transitioning to a smaller structure size is economically viable if the number of manufacturable devices

[8]As we will further explain in Chaps. 6 and 7, the breakdown voltage of an insulating material defines the maximum electric field that the material can withstand without breaking down, i.e., without conducting some amount of electricity.

per unit surface area can approximately be doubled. This means that the devices' surface areas must be halved without loss of functionality.

As described above, CMOS transistors used in digital circuits have the lowest functional requirements—which is also the reason why this type of transistor lends itself so well to miniaturization. Since the late 1970s (when CMOS technology matured), the surface areas of CMOS transistors have been successfully halved every two to three years, while their internal structures have remained unchanged.[9]

Hence, economical miniaturization steps have regularly been achieved with structural downscaling by a factor of $(1/\sqrt{2})$. This process is also called *shrinking*. Technology nodes have scaled approximately in this way since "1 μm processes" (that is, processes that allow a 1 μm structure size) first appeared. As mentioned earlier, these technological milestones are referred to as "process nodes" or "technology nodes".

Figure 1.10 shows the technology node evolution for different semiconductor processes on a logarithmical scale since 1970. CMOS technology (the brown/red plot) has clearly been the main driver in these advances. The plot is based on when the first microcontroller chips became available in the respective feature sizes. The semiconductor processes for other applications (shown in blue) follow this "leading edge technology" at different time intervals. All curves plot the averaged long-term trend based on real data. These plots are not meant to give precise information on dates and feature sizes, but serve as a guide to the trends.

As we have seen, the miniaturization depicted in Fig. 1.10 has paved the way for integrating ever increasing numbers of devices on a single chip and, hence, more and more functions as well. This progression is also shown in Fig. 1.11, also beginning

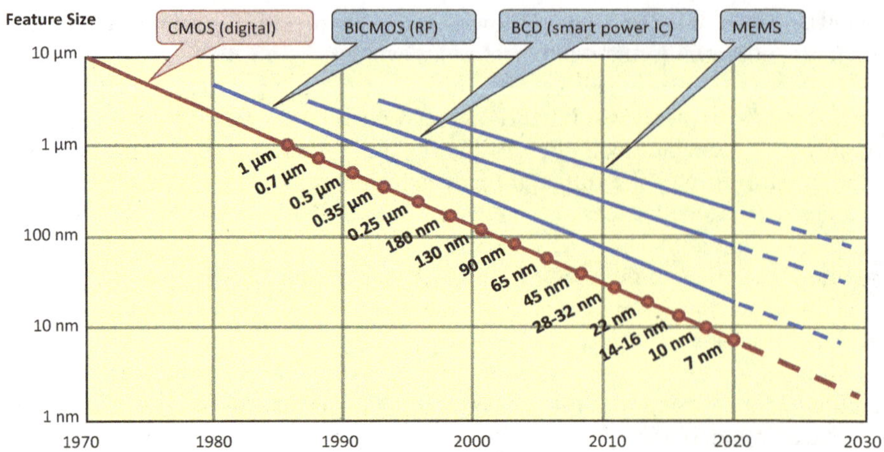

Fig. 1.10 Plots of the smallest manufacturable feature sizes over time for different technology nodes. Marked points indicate typical "process nodes"

[9]This statement applies to structure sizes greater than approximately 20 nm. For smaller sizes, a different internal structure is required for unipolar transistors. FinFETs—which are beyond the scope of this book—are used in this case.

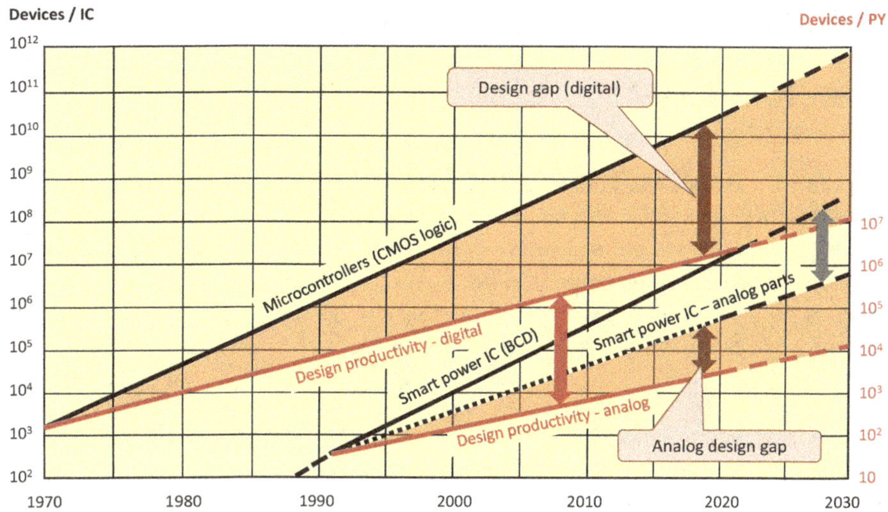

Fig. 1.11 Approximate growth rates for devices per IC (black, left scale) and design productivity in devices per person-year (red, right scale) for digital chips (above) and smart power ICs which contain both analog and digital parts (bottom). The digital design gap, the analog design gap and the gap between analog- and digital-design productivity (red vertical arrow) are also shown

in the 1970s. The chart shows the exponential increase in the number of devices integrated on a chip (black curves, left scale).[10]

The director of research and development at Fairchild Semiconductor Inc., Gordon Moore stated right at the beginning of the chip-miniaturization narrative that the number of devices on a chip doubled every year and he plotted his observation in a similar chart back in 1965 [4]. He predicted in his publication as well that this trend would continue into the foreseeable future. This exponential trend was then given the name "*Moore's Law*" in the early 1970s—the first microcontrollers emerged at that time—when it became clear that the trend had by then become established. (In 1975, Moore revised the forecast to doubling every two years, a compound annual growth rate of 41.4%.)

In Sect. 1.2.1, we discussed the amazing effects of miniaturization that have caused Moore's Law to still apply today. So far, we have discussed all this in the context of the end user and the chip fabricators. But there is one important aspect we haven't touched on yet, and this brings us to the topic of this book: there is much more to be done before these amazing little chips can be manufactured and ultimately used—they need to be *designed* first!

Designing ICs, i.e., laying out their specific physical dimensions, is a huge challenge. The first digital chips used for switching logic gates were drawn in circuit diagrams and the photomasks for them were designed manually or with simple

[10]We use "devices/IC" and not, like most other authors, "transistors/IC" as a unit of scale, as many other types of devices besides transistors are used in mixed processes. The data for "transistors/IC" and "devices/IC" is almost the same for logic chips, though.

drafting software. This type of manual design was quickly superseded in the 1980s because it was too inefficient. Accompanying the exponential growth in complexity in microelectronics, significant work was done in academia and industry to provide IC designers with powerful software tools and innovative design techniques. This field is known as *electronic design automation*, or *EDA* for short.

The effectiveness of digital IC designers could be increased considerably with EDA. While the design process for integrated logic circuits is currently highly automated (i.e., many steps are routinely performed by software programs), the work involved in designing logic chips is nevertheless increasing all the time. This conundrum can be quantified and visualized by considering the number of devices on a chip and the total effort required for its development measured in person-years, and then calculating the quotients. This metric is called *design productivity*. It is plotted in red in Fig. 1.11 and refers to the scale on the right. Although the increase in design productivity is exponential, as well, the rate of increase falls far short of Moore's law. In other words, the mean IC complexity and the design productivity continuously drift apart. This phenomenon is known as *design gap*.

The design gap in digital logic design has been widely written about. It is visualized in Fig. 1.11 by the upper shaded area and the brown vertical arrow, and is one of the toughest and most urgent problems in microelectronics. A major effect of the design gap, along with the rise in IC development costs, is that the number of designers working in chip design must be constantly increased, as the design lead time cannot be extended due to market pressures. Upwards of 1,000 engineers—and they are often spread around the globe—are required today in a typical project team to launch a new computer chip on the market.

We want to address another similar problem now—the *analog design gap*. This gap started emerging around the turn of the century. It is now a critical issue that affects all chips with both digital and analog circuitry. These are mixed-signal and smart power chips that—as already mentioned—make up most of all current chip designs. These ICs also comply with Moore's Law, and although the growth in absolute device numbers lags that of digital (logic) chips, the growth rates are similar.

The increase in device numbers is primarily due to the growing digital circuitry in these mixed-signal designs. The situation for smart power ICs is shown in the lower solid black line in Fig. 1.11. More than 90% of the devices in a modern smart power IC are digital. The highly automated EDA processes tailored for digital design (indicated by the upper red line) are ideally suited for designing these digital sections.

In contrast, the situation with the design of analog circuit devices is very different: their numbers are growing too, but at a slower pace (dotted line). As described in Sect. 1.2.2, analog signals are signals of continuous amplitude and time that must be undistorted as much as possible. IC designers must consider the effects of multiple noise sources on their design: noise can disturb analog signals and cause malfunctions. Designers use analog design methods and take steps in the physical layout design to suppress noise as much as possible. As there are a multitude of different sources of interference, they must consider the physical interactions in their entirety and apply all available design options. This design challenge is so complex that it is very difficult to model mathematically. Hence, the search for an automated solution has proved unsuccessful thus far. To this day, designing analog ICs is mainly based on designer experience and has remained a primarily manual task.

About 90% of the entire development costs of mixed signal and smart power ICs are due to the analog parts, although, based on the number of devices they contain, these parts are only a very small percentage (typically <10%, indicated by the gray vertical arrow in Fig. 1.11) of the chip. This means that analog design productivity is two to three orders of magnitude lower than digital design productivity. This is clearly shown in Fig. 1.11 by the distance between the red lines (red vertical arrow).

Analog circuit design for mixed signal and smart power chips has become a bottleneck, and urgent action is needed to improve the analog design flow in order to prevent further widening of the analog design gap (shown by the lower shaded area and the brown vertical arrow). We shall propose some measures to tackle this issue in physical design at the end of Chap. 4.

1.3 Physical Design

1.3.1 Main Design Steps

A greatly simplified schematic of the design flow for electronic systems is given in Fig. 1.12. It starts with a *specification*, where the desired system functions and performance characteristics for the intended task are set out. Besides the standard signal representations in time and frequency domains, other characterizations, provided as text, figures, tables, and the like, are included in order to specify the design goal as accurately and comprehensively as possible. The specification describes *what* the system is intended to do, with a focus on inputs and outputs; for example, part of a

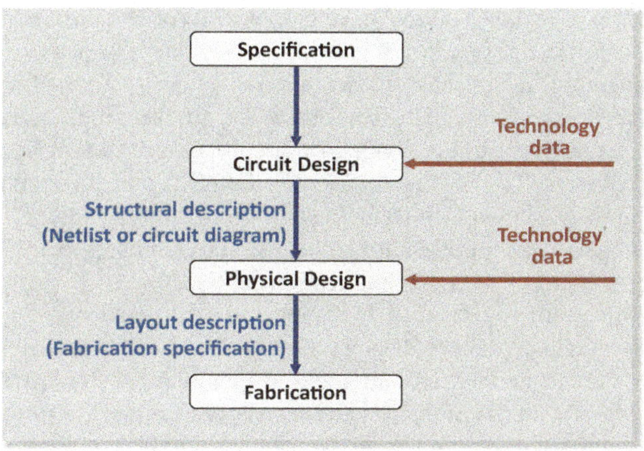

Fig. 1.12 A simplified schematic of the main design steps for an electronic circuit

specification might be: “The system shall receive two digital inputs, using pins 0–7 and 8–15, at a frequency of 1.2 GHz, and produce their multiplicative product as a single 16-bit digital output using pins 16–31, in no more than 5 clock cycles.” The specification describes the task to be performed, but does not define *how* it is to be done. (How the system provides or performs the required functionality is determined as part of the design process in subsequent steps.)

In general, the electronic system is designed to the specification in two major steps, which are outlined next.

Circuit Design

During the circuit design, the functional specification is used to create an electrical circuit that correctly implements the required functionality. For digital circuits this design process will likely include multiple levels of top-down decomposition, where higher-level functionality is broken down iteratively into simpler and simpler lower-level functions, each of which can finally be implemented by a functional unit (for example, an AND adder, a comparator, or a register). The result of this design process is captured in a structural description of the electronic system.

Thus, a *structural description* of the electronic system, such as an IC, is produced as an output of *circuit design*. All the required electrical functional units of the circuit and the electrical connections between them (*nets*) are listed in the structural description. Each net represents an electrical connection—in essence, a wire—that connects an output of a first functional unit to an input of one or more secondary functional units. Due to the complexity of modern electronic systems, this so-called “netlist” is typically organized in a hierarchical tree structure. As such, it contains functional units in the form of standard electronic devices as well as so-called *circuit blocks* or *function blocks* specifying sections of the circuit as subsets of the integrated system.

This structural description, generated during circuit design, serves as the input data to the next major step, the physical design of the circuit. The structural description can be in the form of text as a *netlist*, or in graphical form as a *circuit diagram* (also known as a *schematic diagram* or a *schematic*).

A circuit diagram is a pictorial representation of a netlist, in which the functional units are depicted as symbols and the nets as lines. A simple circuit diagram is drawn as an example on the right in Fig. 1.13. The circuit comprises four basic electronic devices (two resistors R1, R2, two capacitors C1, C2) and a function block (an operational amplifier, shown as a triangle symbol). A circuit diagram and a netlist are effectively equivalent representations of a structural description. Which of the two formats is used depends on the application at hand.

The circuit diagram is produced by graphics-based data entry in a schematic editor in predominantly manual design styles—often used for analog ICs or for PCBs. Symbols for resistors, transistors, etc. stored in *symbol libraries* are loaded and placed on the circuit diagram; the devices are then connected (Fig. 1.13, right). The circuit structure is stored as a netlist in the design tool.

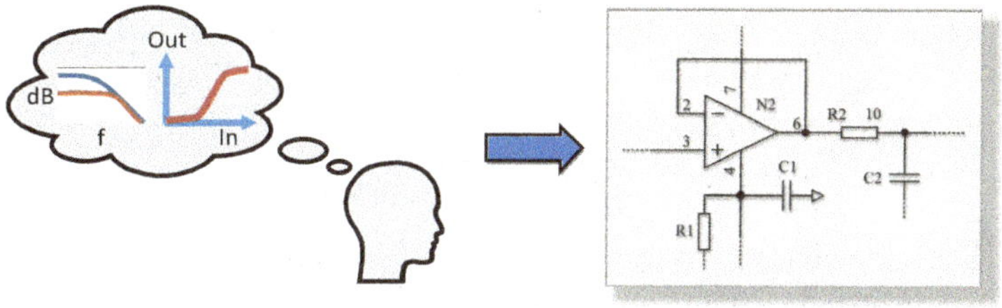

Fig. 1.13 Circuit design: from the specification (left) to the circuit diagram (right)

In highly automated design styles, such as digital IC design, the circuit design data is generated by synthesis procedures and is output as a netlist. A circuit diagram is often not required in these cases.

Physical Design

The purpose of the physical design (aka layout design) is to produce a *fabrication specification* from the structural description of a circuit. The circuit is then manufactured (that is, it is physically realized) based on this specification. *Optimization goals* are pursued and given boundary conditions must be met by this transformation from a structural description to a fabrication specification ("layout"). We discuss these next.

These boundary conditions can be classified into process- and project-specific ones. *Process*-specific boundary conditions describe the options and boundaries of the deployed fabrication technology. They are included in the *technology data* in Fig. 1.12 and must be considered in all designs realized in the given fabrication technology. *Project*-specific boundary conditions, on the other hand, only apply to the product being developed. They are produced during circuit design as another result supplementing the structural description and are generally known as *constraints*. They are instructions for the IC designer concerning the proper functioning of a circuit or the required reliability. In Chap. 4, we shall examine optimization goals and constraints in the context of different design models and design styles.

Along with the fabrication specifications, the result of physical design also defines the exterior appearance—in particular, the dimensions—of the designed electronic system and its component parts. Therefore, the design result is also known as a *layout*, as it is a (physical) interpretation or layout of the (abstract) structural description. The physical design is therefore also called *layout design*.

1.3.2 Physical Design of Integrated Circuits

As we have seen, in semiconductor technology all structures are constructed on a wafer using photolithography. Central to the photolithographic process are masks, which are used to iteratively create microscopic structures on the wafer—the individual electronic devices are created using FEOL processing, and the multi-layered interconnects are created using BEOL processing, as described previously in reference to Fig. 1.7. The masks contain the production templates for the structures that are to be constructed on the wafer. The purpose of IC physical design is to generate these geometrical structures, which are subsequently used to create masks.

The result of the physical design process is an *IC layout*—a complete image of the chip to be fabricated, which is composed of a stacking or layering of these mask images. This IC layout defines the physical implementation of all circuit elements on an IC chip. Among these elements are (i) the internal structure of the devices, (ii) the devices' placement or arrangement on the chip, (iii) the interconnects, (iv) contacts and vias, and (v) the bond pads (located at the chip margin) for the electrical connection with the environment.

Figure 1.14 on the right side shows a small section of a chip layout, as it would appear in a graphics editor used in layout design. Each graphics element is assigned to a so-called *layer*, which generally corresponds to a mask. The designer uses layer-specific colors, line thicknesses and fill patterns to represent the graphics elements in the graphics editor to help distinguish which elements belong to which layer. For example, the yellow "Contact" layer in Fig. 1.14 shows the mask features used to make holes in the bottom insulating layer that lies directly above the silicon surface. These holes are filled with metal to generate an electrical connection between the silicon surface and the bottom metal layer (see also Fig. 1.7c). These connections are positioned where the electronic devices in silicon must be electrically contacted; this is also the reason why these holes are called *contacts* or *contact holes*.

The physical design of digital and analog IC circuits differs greatly, as we describe below.

Considering the physical design of digital circuits, standard elements with an immutable internal structure are stored as *cells* in a library. These cells, such as logic gates and memories, provide standard functions and their layouts have already been produced. More complex logic blocks are often stored in the library as *macro cells*, essentially larger cells that provide higher-level functions, such as an adder or multiplier, or even higher-level application-specific components, such as ones that implement e.g. communication protocols. Instances of these cells are placed on

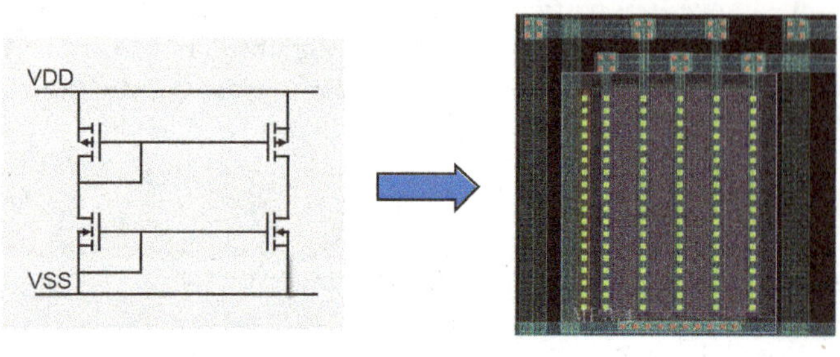

Fig. 1.14 Visualization of the physical design (layout design) of an integrated circuit: converting the structural description of a circuit (a circuit diagram in this case, left) to geometric data, which forms the basis for producing the masks (layout, right)

unoccupied surface areas in the first step in physical design—the *placement*. The interconnects, that make the electrical connections between the cells, are designed in the next *routing* step. Both steps are almost fully automated, which is why they are referred to as *layout synthesis*. A netlist that is suitable as input data for the software programs is used here. We shall examine the processes in (digital) layout synthesis and their application more closely in Chap. 4.

The situation for the physical design of analog circuits is totally different: here, as has been explained in Sect. 1.2.2, most of the design work is still performed manually to this day. Specifically, a circuit's functionality depends very much on the individual device designs and how they are physically placed relative to one another.[11] The layout engineer must understand the circuit and how it works in order to make the right decisions regarding the physical design and placement of the devices. A circuit diagram of the structural description should therefore be drafted when designing analog ICs. This circuit structure allows designers to more easily understand the electrical relationships (the *circuit topology*) and thus the functionality implemented by the circuit.

The first step in designing physical analog circuits is to implement the devices. In this step, each device is sized according to the electrical parameters in the circuit diagram and adapted to other structural criteria for the specific use case. Generators—these are scripts that can automatically produce different layout variants using various parameters—are available for this task. Some of these parameters are defined by the circuit diagram. Other parameters allow the layout engineer to invoke further automated adaptations.

In the placement step, these (analog) devices are arranged on the chip to facilitate subsequent routing. Placement is performed almost entirely manually with the graphics editor due to the complexity of the task and the requirements to be considered. Automated routing software is not widely used in practice either for the subsequent routing step, as the expertise of the layout designer is often needed here for some critical parts of the layout. We shall go into the physical design of analog circuits in more depth in Chap. 6.

For both digital and analog physical design, the finished layout is stored as a graphics file and used to generate the masks, typically a separate mask for each layer. More information will be given on this topic in Chap. 3.

Once the physical design has been completed, it is essential to verify its correctness before committing the design to silicon in the manufacturing process that follows. Many automated verification algorithms are available to check the physical design result, i.e., the layout. Two of these algorithms are obligatory as they verify key quality requirements in chip development. Indeed, without them it would be impossible to flawlessly fabricate modern highly complex ICs. First, the layout is checked for compliance with technological constraints—specified as design rules—by the *design rule check* (*DRC*). The manufacturability of a chip layout is confirmed with this verification technique. Second, the *electrical verification* (also called *layout versus*

[11] *Matching*—a technique used in analog chip layout—plays a key role here. We shall discuss this very important topic fully in Chap. 6.

schematic check, LVS) verifies that the specifications contained in the structural description have been correctly realized in the layout. Specifically, the LVS checks (i) that the devices are properly connected electrically; (ii) that the correct types of devices are in the layout; and (iii) that the devices are correctly parameterized. We shall examine the operation and use of these and other verification tools in Chaps. 3 and 5.

1.3.3 Physical Design of Printed Circuit Boards

In Sect. 1.3.2 we focused on the physical design of chips; these chips are then mounted on printed circuit boards (PCBs) and interconnected to realize the final system. Thus, the physical design of PCBs focuses on the design of a wiring substrate for mounting and electrically connecting electrical and electronic devices. In contrast to IC chip design, the devices are not an integral part of the technological fabrication process for the circuit, rather they are independently provided from an external source.

As explained in Sect. 1.1.1 and illustrated in Fig. 1.2, PCBs are made up of many conductive and non-conductive layers connected by vias. These layers are mapped into corresponding data structures in the design tool—in a similar manner to IC physical design. In addition to these layers used for connecting the devices, further layers are required for fabrication, such as solder resist and solder paste masks or placement imprint. The design of the graphics structures in all these layers is known as "PCB physical design" or "PCB layout design".

The input data for the layout design of a PCB are specified by a circuit diagram. The following steps are required to produce a PCB layout:

(1) Specifying the placement of the devices,
(2) Defining the size of the board and number of the routing layers,
(3) Specifying the location of the interconnects (wires) on the wiring substrate (the routing step) and arranging the via positions.

All the necessary structures for physically attaching and electrically connecting a device on the PCB are known collectively as a *footprint*, sometimes also called a *land pattern*. The footprint describes *pads* (contact surfaces) and the placement imprint as polygons; it also contains the necessary holes (for THDs) and through contacts (vias). Footprints are stored in *footprint libraries*.

Creating the *layout file* for the PCB is the first step in the layout process. Here, the devices are extracted from the circuit diagram and the matching footprints are loaded from the footprint library. The footprints are then placed—in other words, their locations on the PCB are defined (Fig. 1.15).

The physical routes for the electrical connections (interconnects) between the devices have not yet been determined. They are shown symbolically as "rubber bands" (Fig. 1.15, right). Next, the shape and location of the interconnect tracks, and the layer to which they belong, are defined in the routing step (Fig. 1.16).

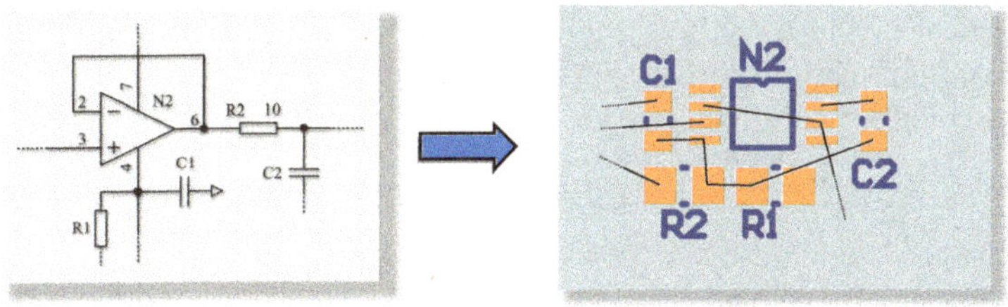

Fig. 1.15 First step in the PCB layout process: Specifying the placement of the devices on a PCB, originating from a circuit diagram (left)

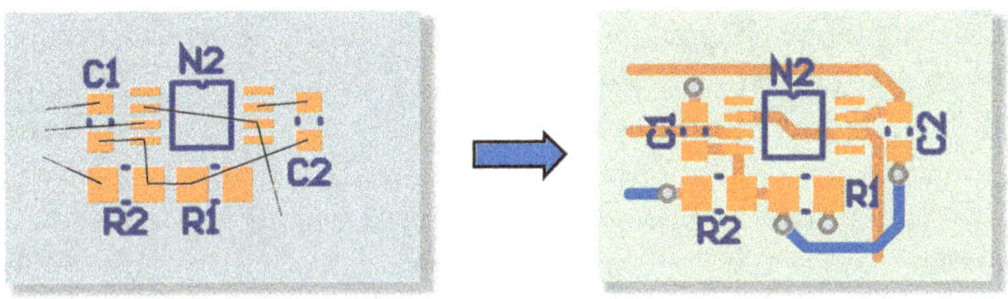

Fig. 1.16 Second step in the PCB layout process: Specifying the routing of the interconnects which includes arranging the via positions and the layer allocation

After placement of the devices and the routing step, the layout design is verified. Specifically, the layout is checked for short circuits (electrical rule check, ERC) and for compliance with constraints necessary for proper functioning and manufacture (e.g., design rule check, DRC).

The required fabrication data is then exported. In contrast to an IC layout, where the entire finished layout is stored in one graphics file, different files and formats are needed for PCB manufacture. The fabrication data consists of a set of *Gerber files* that describe the conductor tracks for the individual layers, the solder resist and solder paste mask, and the placement imprint consisting of polygons. A *drill file* containing the diameters and coordinates of all drill holes in the PCB is required, as well. Finally, a *pick-and-place file* containing the location and alignment of the devices is generated for automated placement of the devices in the assembly process.

1.4 Motivation and Structure of This Book

As described throughout this chapter, design components are instantiated as geometric representations during physical (layout) design. In other words, all electronic devices, cells, gates, transistors, etc., are realized with fixed shapes and sizes, assigned locations (placement), and have appropriate wiring connections (routing) completed in metal layers. The result of layout design is a set of manufacturing specifications that must subsequently be verified prior to the actual manufacturing process.

The physical layout directly affects circuit performance, area, reliability, power consumption, and manufacturing yield. As such, the quality of layout design significantly influences the quality of the resulting electronic circuit, irrespective of whether it is an IC or a PCB.

The continuing miniaturization leads to increasing design problems that must be overcome by the layout designer, and include worsening parasitic disturbances, and more and more technological constraints. As a result, the demand for experienced layout designers continues to grow. At the same time, the need for new methodologies and tools in layout design is also increasing.

This book addresses all these challenges. It presents the fundamental knowledge of layout design from the ground up, from technological constraints to reliability requirements. Chapter by chapter, the book provides the awareness a layout designer must possess to convert a structural description produced during circuit design into the physical layout of transistors, cells, devices and wires on the surface of the chip or the board.

While all relevant aspects of layout design are covered (digital and analog, IC and PCB layout), the reader will notice in some parts a focus on analog layout design. This is due to the stronger need for manual work in analog design: here, the expert knowledge of the actual layout designer is more crucial than in the often fully-automated digital design flows. Nevertheless, this book is intended to provide the fundamentals of physical design irrespective of its specific application, as the basic knowledge is the same for all abstraction layers.

This chapter provided a basic and introductory grounding in the technologies, the tasks and the methodologies needed for designing the layout of an electronic circuit.

The following Chap. 2 introduces in detail the engineering know-how to transform silicon into devices and thus, integrated circuits. This knowledge is crucial for any IC layout designer, as the boundary conditions that must be considered during layout design result directly from the specific semiconductor technology subsequently applied for manufacturing the electronic circuit. This chapter should give the reader the requisite understanding of the technology for which the layout is targeted.

Chapter 3 describes the data interfaces of layout design. A layout designer must be aware of the implications of these "bridges to technology". In this chapter, we first introduce circuit and layout data structures, such as netlists, layers and polygons. We also investigate the specific links that exist between layout design and targeted technology, such as mask data, design rules and libraries. A special emphasis is paid to the layout post-processing flow. Here we explain all the steps needed to transpose the layout data of an IC into the mask data (i.e., the specification for IC fabrication), including chip finishing, reticle layout, and graphic manipulation processes to achieve compliance of the graphic data with manufacturing requirements.

After covering technologies and the way the design process interfaces to them, physical design is dealt with in Chap. 4. Here, the flow, constraints and strategies of today's state-of-the-art physical design are introduced. We investigate the various types of constraints, the design models and styles, and discuss the analog-digital design gap. A detailed look into the specifics of analog design, including its outlook, are also presented. In summary, this chapter provides the basic knowledge any engineer must possess about physical design methodologies.

Due to its high complexity, physical design is split into several primary steps. These steps, which transform a netlist into optimized layout data, are dealt with one by one in Chap. 5. We first provide an overview on how to generate a netlist, either by using hardware description languages in digital design, or by deriving it from a schematic as is common in analog design. Then the physical design steps, such as partitioning, floorplanning, placement and routing, are presented in detail.

After physical design is completed, the layout must be fully verified to ensure correct electrical and logical functionality. Some problems found during physical verification can be tolerated if their impact on chip yield is negligible. In other cases, the layout must be modified, but these changes must be minimal and should not introduce new problems. These options for layout verification are discussed in Chap. 5 as well. We also touch on layout post-processing methodologies, such as resolution enhancement techniques (RET), that might impact physical design.

While the physical-design steps presented thus far are universal, analog circuits require additional layout techniques. Any analog layout designer must be fully aware of these analog design techniques, so we introduce the most common analog devices, cell generators, symmetry and matching principles in Chap. 6.

As reliability of our circuits is becoming a growing concern, the final Chap. 7 summarizes reliability aspects that are of relevance in layout design. We start by presenting reliability issues that can lead to *temporary* circuit malfunctions. We discuss in this context parasitic effects in the bulk of silicon, at its surface, and in the interconnect layers. Afterwards, we deal with the growing challenges of preventing ICs from *irreversible* damage. This requires the investigation of overvoltage events and migration processes, such as electromigration, thermal and stress migration. The goal of this chapter is to summarize the state of the art in reliability-driven design and related mitigating measures. This knowledge can be applied by a circuit designer to increase the reliability of the generated layout.

References

1. L. Berlin, *The Man Behind the Microchip: Robert Noyce and the Invention of Silicon Valley* (Oxford University Press, 2005), ISBN 978-019516343-8. https://doi.org/10.1093/acprof:oso/9780195163438.001.0001
2. R. Fischbach, J. Lienig, T. Meister, From 3D circuit technologies and data structures to interconnect prediction, in *Proceedings of 2009 International Workshop on System Level Interconnect Prediction (SLIP)* (2009), pp. 77–84. https://doi.org/10.1145/1572471.1572485
3. J. Kilby, Patent No. US3138743: Miniaturized electronic circuits. Patent filed Feb. 6, 1959, published June 23, 1964
4. G.E. Moore, Cramming more components onto integrated circuits. *Electronics* **38**(8), 114–117 (1965). https://doi.org/10.1109/N-SSC.2006.4785860
5. R.N. Noyce, Patent No. US2981877: Semiconductor device and lead structure. Patent filed June 30, 1959, published April 25, 1961
6. https://en.wikipedia.org/wiki/Transistor_count. Accessed 1 Jan 2020

Chapter 2
Technology Know-How: From Silicon to Devices

As we have seen in Chap. 1, the purpose of physical design is to produce all data necessary to fabricate an electronic circuit. If we compare this task, for example, with the design of a mechanical product, it is equivalent to the work of a mechanical design engineer. Clearly, a mechanical engineer needs to be very familiar with the options (or lack thereof) for the manufacturing technologies that can be used, to produce good outcomes.

For this reason, we discuss the fabrication technologies for IC chips in this chapter. Our objective is not to comprehensively describe the very complex, state-of-the-art semiconductor technologies; the reader will find ample information on this topic in the technical literature referenced in the text. Rather, we will focus on the main process steps and especially on those aspects that are of particular importance for understanding how they affect, and in some cases drive, the layout of ICs. All our analyses in this chapter will be for silicon as the base material; the principles and understanding gained can be applied to other substrates as well.

Following a brief introduction to the fundamentals of IC fabrication (Sect. 2.1) and the base material used in it, namely silicon (Sect. 2.2), we discuss the photolithography process deployed for all structuring work in Sect. 2.3. We will then present in Sect. 2.4 some theoretical opening remarks on typical phenomena encountered in IC fabrication. Knowledge of these phenomena is very useful for understanding the process steps we cover in Sects. 2.5–2.8. We examine a simple exemplar process in Sect. 2.9 and observe how a field-effect transistor—the most important device in modern integrated circuits—is created. To drive the key points home, we provide a review of each topic at the end of every section from the point of view of layout design by discussing relevant physical design aspects.

J. Lienig and J. Scheible, *Fundamentals of Layout Design for Electronic Circuits*,
https://doi.org/10.1007/978-3-030-39284-0_2

2.1 Fundamentals of IC Fabrication

The semiconductor material used for the fabrication of integrated circuits (ICs) is prepared in the form of thin, monocrystalline slices, so-called *wafers* (Sect. 2.2). Many ICs arranged in rows and columns on the surface are produced simultaneously per wafer. At the end of the process, individual ICs are "singulated", which is performed by cutting the wafer with vertical cuts perpendicular to one another. The resulting ICs are small rectangular plates, which is why they are also called *chips*. One wafer can produce many hundreds to many tens of thousands of chips depending on the wafer size and required circuits (see Fig. 1.6 in Chap. 1).

Processing a wafer requires many (typically several hundred) individual manufacturing steps, which are executed in series. Fabrication in a wafer fabrication facility (or *fab* for short) can take months. These steps can be divided into the three categories: (i) *doping*, (ii) *deposition* and (iii) *removal* of materials. The processes involved are typically repeated many times, which results in the hundreds of steps. These processes can be restricted to certain wafer regions by *structuring*. We discuss these processes next.

Doping. In this process, *dopants* (*acceptors* or *donors*) are implanted in the wafer. The extent of p- or n-conductivity is determined here. Dopants are generally introduced "selectively". In other words, the procedure is performed so that it only impacts certain parts of the wafer surface. We will get to know these processes in Sect. 2.6. The results of these processes are the required (laterally structured) doped areas and the vertical doping profiles.

Deposition. As part of processing, additional layers (e.g., silicon-dioxide, metal) are deposited on the surface of the wafer. The wafer surface is said to "grow" with deposition. The entire wafer surface is affected in most cases, with some exceptions such as *local oxidation* (Sect. 2.5.4).

Removal. Material is normally removed by etching, that is, by chemical means. A layer is structured in many cases by etching. If material is to be removed globally, this can be done solely by etching or by etching along with a mechanically-activated removal process. The former is called *bright etching*. In the latter, the wafer surface is leveled by so-called *CMP* (*chemical-mechanical polishing*, Sect. 2.8.3).

Structuring. Structuring is required when the above manipulations must be selectively applied to produce lateral structures. The effects of the manipulations can be restricted to the desired regions with this process. These manipulations are then called *masked* processes. This means that certain regions are "protected" against the effects of a manipulation. The aforesaid applies to all of the above three processes, i.e., for targeted doping, selective deposition and the selective removal of material. The structuring for a *masking* is produced by photolithography (Sect. 2.3).

2.2 Base Material Silicon

Silicon (Si) is used as the base material for the vast majority of chips. As mentioned at the beginning of this chapter, we will limit our treatise to this material. The reason why silicon is so widely used is that it has many useful properties:

- Silicon has an ideal band gap of 1.1 eV[1] for most circuit applications. This value is so high that intrinsic conduction is not triggered in silicon until over 200 °C (392 °F). It is therefore widely used, as typical temperatures encountered in most technical applications are below this. On the other hand, this value is low enough that field-effect transistors with very low threshold voltages can be easily manufactured using silicon.
- Silicon dioxide (SiO_2) is a very stable and intrinsic oxide with good insulating properties, which can easily be manufactured. SiO_2 is used as an insulator between interconnects and as a dielectric for capacitors and field-effect transistors.
- Silicon is a good heat conductor. This is a vital prerequisite for miniaturization, where power losses from minuscule structure sizes must be quickly dissipated to prevent overheating (and thus intrinsic conduction or catastrophic damage).
- Silicon can be grown easily in large monocrystals, which can then be cut into wafers. The atoms in the monocrystals are arranged in all directions with absolute regularity and without breaks. This is a key prerequisite for the use of silicon as a base material in IC fabrication, as otherwise lattice irregularities (grain boundaries and lattice defects) could cause unwanted current paths.

Silicon is a natural resource available only in an oxidized and impure state, primarily as common sand. As such, it must first be freed from bonded oxygen and cleansed of impurities. It is then "grown" as a monocrystal. This is achieved by melting the silicon, while maintaining the temperature only just above its molten temperature of 1414 °C (2577 °F). A small silicon monocrystal is then put on the surface of this molten mass. As soon as the seed crystal makes contact with the molten mass, silicon atoms from the molten mass become attached to it. These silicon atoms then cool slightly and form the same structure as the seed crystal. By moving the seed crystal very slowly away from the molten mass, this monocrystal grows further. The silicon can be doped as well during this operation by adding doping agents to the molten mass.

The two most common technologies for crystal growth, the *Czochralski*[2] *process* and *zone melting*, are shown schematically in Fig. 2.1. In the Czochralski process, the silicon monocrystal is drawn from a silicon melt, whereby it is rotated about the axis along which the pulling force acts. The silicon monocrystal is, in effect, "grown" from the bottom as it is drawn vertically. In the zone melting method, in contrast, a very thin strip of a polycrystalline bar is melted by an annular heater. The atoms are aligned in this thin strip during this operation. The heater and thus the melting zone

---

[1]One electron absorbs a kinetic energy of 1 eV along a potential gradient of 1 V.

[2]Named after Polish scientist Jan Czochralski, who invented the method in 1915 while investigating the crystallization rates of metals.

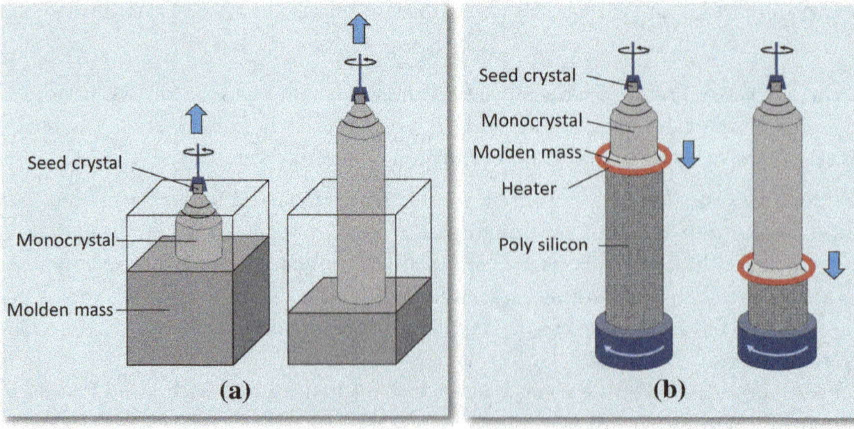

Fig. 2.1 Methods for fabricating highly purified, monocrystalline silicon: **a** Czochralski method, **b** crucible-free zone melting

are slowly moved along the bar. The two rigid parts of the bar are rotated in opposite directions about their axes. Hence, the resulting monocrystal is in effect produced in-place.

The methods shown in Fig. 2.1 must be carried out under vacuum or in an inert-gas atmosphere. The monocrystal is formed in these two processes as a round bar. Wafers approximately 1 mm in thickness are cut from the bar by internal diameter sawing (IDS). The wafers are sent as base material to a wafer fab for IC manufacture. Typical wafer diameters are between 200 and 450 mm.

2.3 Photolithography

2.3.1 Fundamentals

As mentioned at the outset, photolithography is used in all structuring process steps. Its purpose is to transfer a two-dimensional image of the required structures onto the wafer surface, so that subsequent processing (e.g., implantation, etching) can be applied to a restricted area.

The wafer is first coated with a thin radiation sensitive film, called a *photoresist* or *resist*. A photomask is then exposed to light to project a black and white image of the desired structure on the photoresist. The solubility of the photoresist changes w.r.t. a specific fluid, called a *developer*, at the regions exposed to light. In the *developing process,* the soluble areas of the photoresist are removed from the wafer with this fluid. The insoluble parts of the photoresist remain.

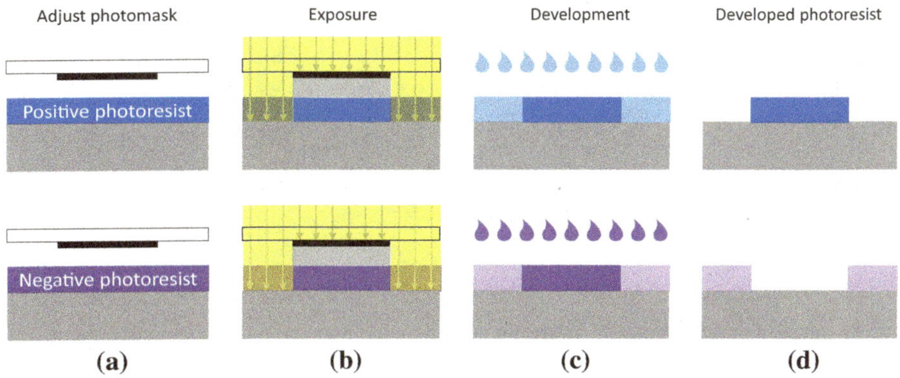

Fig. 2.2 Schematic representation of the photolithography with positive resist (above), where the exposed regions are removed, and negative resist (bottom), where the exposed regions are kept

Figure 2.2 shows this process, which is explained in detail in Sect. 2.3.2. The desired structure is now present in the photoresist. This structure then serves as a mask for the real wafer process step.

Masking can be performed in two ways:

- In some process steps, the (remaining) photoresist itself acts as the mask. In these cases, the mask is called a *resist mask*.
- In other process steps, where the (remaining) photoresist cannot directly assume the masking task, an intermediate step is required. Here, the structure in the developed photoresist is transferred by etching to a layer underneath. This layer then acts as a mask for the actual process step.

2.3.2 *Photoresist*

Photoresist is deposited on the wafer by means of spin coating. Liquid coating material is applied to the center of the wafer, which is rapidly rotated. The centrifugal force causes the fluid to be distributed on the wafer. The solvent for setting the viscosity in the fluid is vaporized to produce a layer of constant thickness. Spraying the resist onto the wafer is another method that is sometimes used.

Photoresists are radiation-reactive polymers. Positive and negative photoresists are available. The solubility is greatly increased due to the reconstruction and deconstruction of molecular chains caused by exposure to light in the case of positive photoresists. The effect is the opposite with negative photoresists where exposure causes a greater cross linking of the molecular chains which in turn greatly reduces the solubility. Photoresists can be produced to change their solubility only in a specific, tightly bounded wavelength range.

The characteristics of the exposure (wavelength, intensity, duration), of the photoresist (layer thickness, sensitivity to light), and of the developer must be finely tuned. In the case of the positive photoresist, the exposed regions are removed, and the unexposed regions remain when the resist is developed (see Fig. 2.2, top row). In the case of the negative photoresist, the exposed regions remain, and the unexposed regions are removed (see Fig. 2.2, bottom row). In each case, the remaining photoresist serves as a mask for the next process step.

2.3.3 Photomasks and Exposure

A *photomask* is used to expose a wafer. A photomask is a sheet of glass, on which a black-and-white image of the structures to be processed is applied to an opaque layer made of chromium. When this photomask is exposed, a shadow is cast on the exposed wafer to produce the desired image. There are two types of exposure: *direct exposure* and *projection exposure*.

Direct Exposure

With direct exposure, the photomask is placed above and in close proximity to the wafer, either in direct contact with the photoresist (*contact exposure*) or close to it (*proximity exposure*). The principle is similar to that shown in Fig. 2.2b. The result is in both cases a 1:1 (full scale) image produced by simple shadowing. Hence, the structure sizes on the photomask must correspond with the structures on the wafer (so-called "1X photomasks").

Contact exposure is unsuitable for volume manufacture as it can cause damage to and soiling on the resist and the photomask. Proximity exposure differs in that there is no contact between photomask and resist. However, the distance between mask and resist in proximity exposure, which cannot be reduced indefinitely but must lie between 10 and 40 μm, can cause resolution degrading.

If the dimensions of the structures being imaged are similar in size to the exposure wavelength, significant diffraction phenomena may occur. Direct exposure therefore approaches its limit for structure sizes to be imaged at this order of magnitude. The limit is even bigger (approximately 3 μm) for proximity exposure. Hence, the structure sizes required in state-of-the-art semiconductor technologies that are smaller than this value cannot be imaged with direct exposure; this process is therefore not deployed any more today.

Projection Exposure

Projection exposure was developed to overcome the above-mentioned direct-exposure issues. Here, the pattern on the photomask is projected through lenses onto a wafer coated with photoresist, as illustrated in Fig. 2.3. This system brings with it two fundamental benefits: (i) photomask and wafer are physically separated and (ii) the image can be optically reduced in size during the projection—thus improving

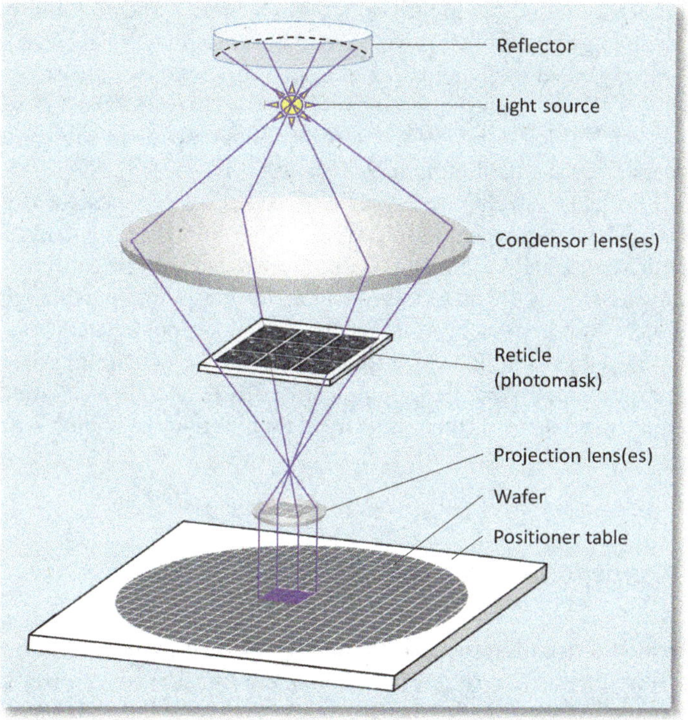

Fig. 2.3 Exposing a wafer with step-and-repeat technology. Using one reticle, multiple exposures are made to cover the full patterned area

imaging accuracy. Among the most commonly used photomasks are 4X, 5X and 10X masks, in other words, photomasks with imaging ratios of 4, 5 and 10 to 1.

A small part of a wafer can only be exposed in one step with this method, as the sizes of the photomasks and lenses are limited.[3]

Photomasks therefore have structures for one or a few chips. These types of photomasks are called *reticles*. (A reticle is a special type of photomask where the data for only part of the final exposed area is present. However, the term "reticle" has become a synonym for "photomask" as state-of-the-art photomasks are seldom designed for an entire wafer.)

A wafer is exposed in many single steps in the so-called "step-and-repeat technique". The wafer is transported on a positioner table under the projection optics. This type of exposure equipment is called a *wafer stepper* or simply a *stepper*. Figure 2.3 shows the stepper principle of operation.

---

[3]To expose a 200 mm wafer with a 5X-photomask in only one step, the mask diameter would have to be 1 m. The same applies to the optics required. Photomasks and optical devices of these sizes are non-viable from both a technical and commercial perspective.

Ultraviolet light is used for higher optical resolution. Exposure and photomask technology are being continuously developed to improve imaging accuracy, and such innovations have pushed the boundaries of optical lithography so that feature sizes of approximately 10 nm can be patterned with the widely used exposure wavelength of 193 nm (argon-fluoride laser). When we talk about chips with such small feature sizes, we are entering the field of *nanoelectronics*.

In addition to these developments, there are other approaches for downscaling structure sizes. The optical resolution can be further improved by using ultraviolet light of shorter wavelengths. Mirror systems are needed for projections using light of shorter wavelengths as the materials available for lenses are increasingly opaque.

In addition to this approach, systems for directly exposing wafers with electron beams are used as well. However, this is a very time-consuming process as all structures are individually "written" on the wafer. It is nonetheless an economical technique for small lot sizes, as there is no need for photomasks, which are expensive to produce.

2.3.4 Alignment and Alignment Marks

Integrated devices are built progressively in a series of interacting structuring steps. This means that the position of a reticle (i.e., a photomask) for a given layer must always be accurately aligned with the structures already on the wafer. For example, when the devices have been implemented, the contact holes must be accurately placed at the locations where the device's connection areas are situated.

The location and attitude of the wafer with respect to the exposure reticle must be adjusted before every exposure in the step-and-repeat process. Alignment marks on the reticle are automatically detected by optical means for this purpose (Fig. 2.4). These marks are geometrical figures, such as crosses which, like all other geometrical elements on the reticle, produce structures on the wafer according to the actual process step. If these structures caused by alignment marks can be optically detected,

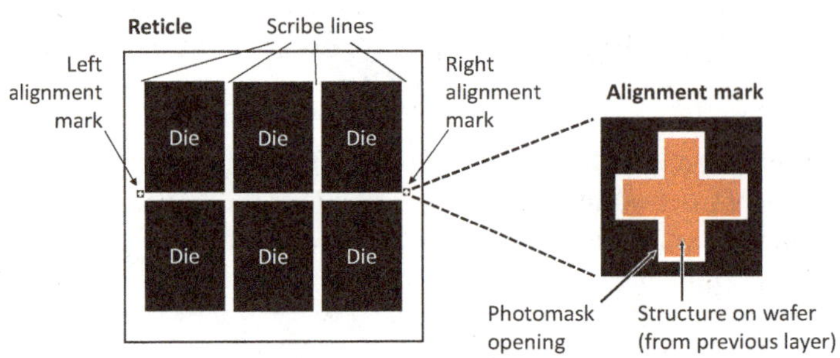

Fig. 2.4 Alignment marks on a reticle (photomask) with six chip structures

the wafer position can be aligned with the help of alignment marks on subsequent photomask(s).

The alignment mark on the right in Fig. 2.4 is an opening in the photomask in the shape of a cross. The wafer underneath is adjusted such that the structure (which is also in the shape of a cross and is shown here in orange) on the wafer is located in the middle of this opening.

There are however some structuring steps, such as ion implantation doping procedures (Sect. 2.6), for example, that leave no reliable optical traces. This is not an issue, though. Since there are always deviations in every adjustment arising from mechanical tolerances, adjustments are made in as many process steps as possible using the *same* alignment mark on the wafer, so that the deviations do not accumulate. An alignment mark from a subsequent process step is used when a structure cannot be detected any more. The next steps then refer to this "newer" alignment mark.

Besides the correct locations of the alignment marks, the correct angular orientation of the wafer must be checked as well during the alignment procedure. Hence, the alignment marks are located at two points on the reticle that are far apart, to ensure the wafer attitude is correct.

As the alignment marks are not functional chip structures, they are placed outside the chips. To facilitate (subsequent) chip dicing, the chips have a clearance of 50–100 μm from one another. This clearance is called "sawing trench", "saw street", or "scribe line". The alignment marks are placed in this clearance space (Fig. 2.4, left).

2.3.5 Reference to Physical Design

The price for the degree of miniaturization currently reached with structure sizes of the order of nanometers is the extremely high effort required to create the photomasks derived from the layout design to perform exposure. Photomasks for modern semiconductor processes can be very expensive.

It is worth remembering that a single design flaw can render photomasks and wafers worthless. This financial hit in the event of a fault is accompanied by a significant delay in development time. Troubleshooting and a new production run can easily take six months or more. The financial losses are even more serious when one also considers the delays in bringing the product to market.

The critical conclusion here is: *the physical design of an IC chip must be absolutely flawless!* Every effort must be made to detect design risks, and suitable and effective measures must be put in place to prevent them. In this regard, automated verification algorithms, which we will cover in Chaps. 3 and 5, are a crucial part of physical design. The reader should refer to Chaps. 6 and 7 for numerous other options for assuring layout quality.

2.4 Imaging Errors

As we have seen in Chap. 1 (Sect. 1.3.2), layout data determines the structures on the photomasks. So far in Chap. 2, we have talked about how these mask structures are converted to wafer structures via photolithographic images and subsequent targeted process steps. If we compare the structures created "on" and "in" the wafer with the original layout structures, different types of deviations become evident. We next discuss these unavoidable deviations, and examine how we can deal with them in the design of the layout.

There are three categories of imaging errors: (1) *overlay errors,* (2) *diffraction effects* and (3) *edge shifts*. Overlay errors and diffraction effects occur during exposure, while edge shifts occur in subsequent structuring process steps. Although we cover these process steps in detail in the following Sects. 2.5–2.8, we first introduce these imaging errors here.

2.4.1 Overlay Errors

Due to mechanical tolerances and measurement inaccuracies that can arise during alignment, the wafer and the photomask (reticle) cannot be positioned with absolute accuracy relative to the exposure device. The result is that structures on the photomask are not exactly mapped to the wafer during exposure, in accordance with the layout template.

Possible exposure faults are depicted in exaggerated form in Fig. 2.5, where we illustrate (a) displacements and (b) rotations that can occur w.r.t. the required positions. Wafer and photomask are moved along the optical axis to adjust the depth of focus. If the distances between masks, lenses and wafer are changed during this focusing step, the layers will be scaled relative to one another (c). If the photomask is tilted w.r.t. the optical axis, the perspective will be distorted (d).

Another cause of overlay errors is the fact that wafers and photomasks—like all materials—expand when heated. If wafer exposures take place at different temperatures, spacings between structures on the wafer will change. The result are displacements (see Fig. 2.5a), the extent of which depends on where they occur on the

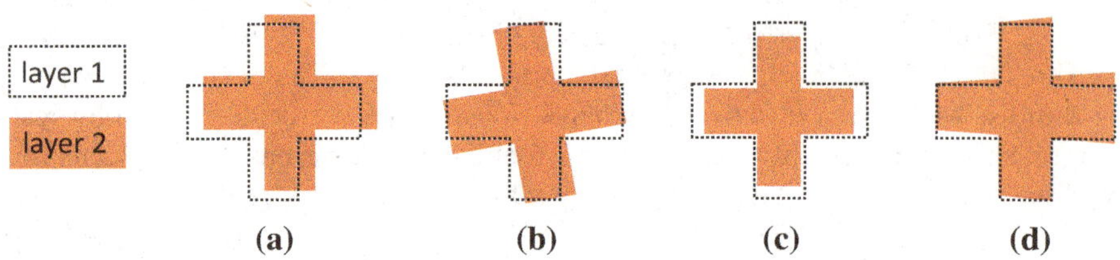

Fig. 2.5 Possible overlay errors between two layers: **a** displacement, **b** rotation, **c** scaling, **d** perspective distortion

exposed wafer. To minimize this effect, the exposure temperatures should be kept as constant as possible throughout the manufacturing run.

As it is not always possible to maintain these temperatures at a constant, this cause of overlay errors cannot be completely eliminated. Sometimes the wafer is irreversibly deformed in high temperature steps in the process. The degree to which these deformations occur differs from wafer to wafer. This effect can cause displacements and scaling (see Fig. 2.5a, c).

The type and extent of these overlay defects cannot be predicted. Their effects are cumulative and can only be limited to certain scopes by efforts that focus on specific devices and the fabrication process. Clearly, the maximum permissible overlay error in a semiconductor process should not exceed the minimum feature size. It should typically be much less than this value.

Let us look at a typical overlay fault to illustrate how a layout can be affected. In chip fabrication, contacts must always be fully covered with metal to ensure proper electrical connection. As contacts and metallic interconnect layouts are structured with different photomasks, this "design rule" for full coverage must also be satisfied when overlay errors occur.

How overlay errors are handled in layout design is shown in Fig. 2.6. An *enclosure* design rule is specified for creating the layout that stipulates that all structures in the "Contact" layer are covered by structures in the "Metal1" layer and that they overlap on all sides by a minimum value. This minimum value is the same as the maximum permissible overlay error, that is, the deviation that can occur in the worst case when all overlay errors are superimposed. This *design rule* is satisfied for the layout structure in the middle of Fig. 2.6. The right-hand side of the figure shows a possible situation encountered in fabrication. Here, the assumed overlay error is exaggerated for clarity sake.

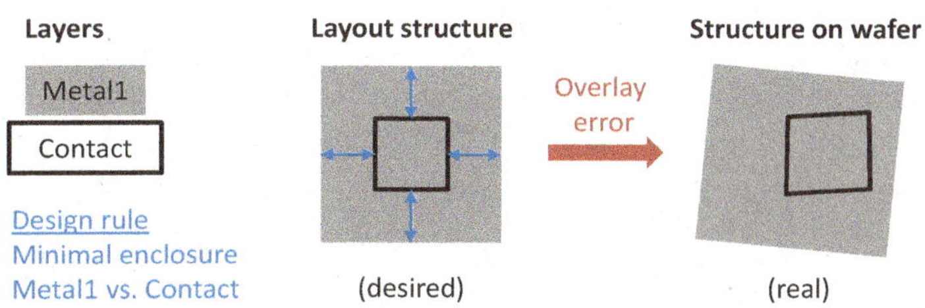

Fig. 2.6 Handling of unavoidable overlay errors in physical design. Contact holes must be fully covered by metal under all circumstances; this requires a design rule for a sufficiently large "minimal enclosure" between metal and contact layer that takes possible shifting, rotation, scaling, and perspective distortion into account

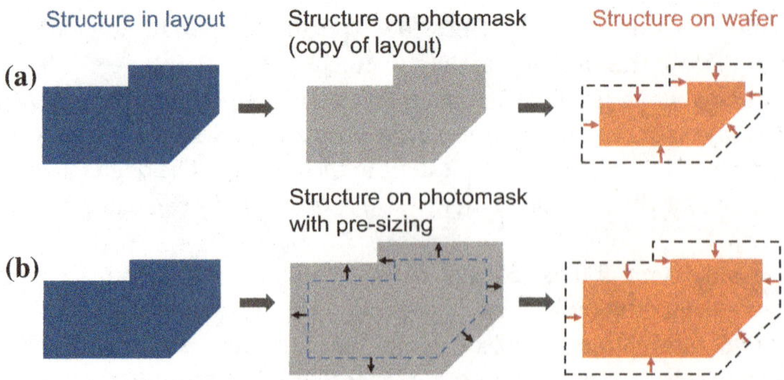

Fig. 2.7 Edge shift inwards in the wafer process; **a** without pre-sizing, **b** with pre-sizing

2.4.2 *Edge Shifts*

Effects occur in some technology layers that can cause the graphics elements on the processed wafer to be enlarged or shrunk w.r.t. the associated graphics elements in the layout. (Please note that we are not talking about scaling, where the dimensions of the elements are changed by a certain *factor* as with a "Zoom".) The changes with these enlarging/shrinking effects are *additive*: the structure's boundary lines are shifted outwards by a specific value (positive shift) or inwards (negative shift). As the individual structures in the layout are typically modeled as polygons (i.e., as geometrical elements bounded by edge strips), we call this category of imaging errors *edge shifts*.

Figure 2.7 shows a simple example of a process step, where a negative edge shift occurs. The structure on the wafer shrinks w.r.t. the element on the photomask (shown with red arrows in the figure on the right).

The sizes of these edge shifts are layer-specific, and they are defined for every semiconductor process. The effect can therefore be compensated for by *pre-sizing* when creating the photomask geometry. Specifically, the prepared layout data are modified automatically in a *layout to mask preparation process* (Chap. 3, Sect. 3.3.4) which is a part of the *layout post process* (Chap. 3, Sect. 3.3).

If, for example, an edge shift of value k occurs in the process, the edges of the layout geometries are shifted by a value k before these new data are transferred to the photomask. This operation is shown in the bottom row (b) in Fig. 2.7.

2.4.3 *Diffraction Effects*

Due to the wave properties of light, diffraction phenomena occur at the structural edges of the chromium layer on the photomasks, which limit the optical resolution of the photolithography. The smaller the feature size (for a constant exposure wavelength), the more the diffraction effects impact the accuracy of the imaging. We demonstrate these effects with an L-shaped layout structure in Fig. 2.8.

In the top row, the layout element is transferred unchanged to the photomask (in gray). Clearly, the shape of the area exposed in the photoresist (blue) deviates to a greater degree from the shape of the photomask opening and thus from the desired layout structure, as the feature size shrinks (moving from left-to-right in the figure).

As long as the wavelength of light is greater than the feature size, the small fillets at the corners are negligible (see Fig. 2.8, top left).[4] However, as soon as the ratio of the feature size to the wavelength gets below 1 (so-called *sub-wavelength lithography*), significant *line-end shortening* occurs. There is a marked increase in *corner rounding* as well (see Fig. 2.8, top center and right).

These diffraction effects can be corrected by slightly enlarging the photomask opening at underexposed places and slightly shrinking it at overexposed places. This procedure is another example of preemptive measure and is known as *optical proximity correction* (*OPC*).

These corrective measures can be defined with simple rules based on structural shapes, as long as the diffraction effects are of the same orders of magnitude as those shown in the middle of the figure. So-called "hammer heads" are attached to the ends of "thin" lines if the line width is approximately the same as the feature size. Square elements, so-called "serifs", are added (this means more light) to the outer corners, or

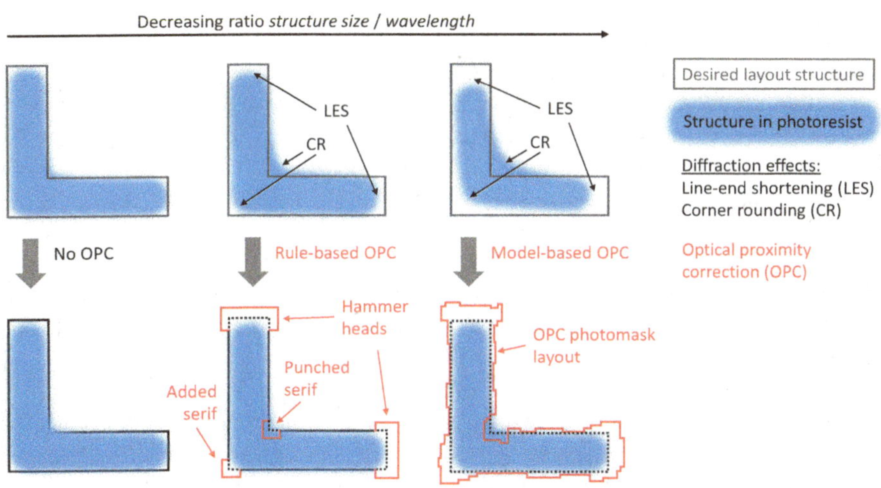

Fig. 2.8 Diffraction effects in photolithography (top row) and possible corrective measures using optical proximity correction (OPC, bottom row). The imaging errors increase as the ratio of feature size to optical wavelength decreases (from left to right)

[4]These fillets have a positive electrical effect because the local field strength increases at the outer corners are capped. The same applies to current density increases at inside radiuses of bent interconnects.

"punched" (means less light, hence also called "jogs") from the inner corners. These measures, known as *rule-based OPC*, are illustrated in Fig. 2.8, bottom center. In addition, the change in line widths, caused by interference from many parallel lines, can be corrected by rule-based OPC (not shown in Fig. 2.8).

If the feature sizes w.r.t. the wavelength of light shrink further, the extent of imaging errors increases (see Fig. 2.8, top right). Rule-based OPC is not a solution in this case, as the exposure result is now increasingly impacted by surrounding features. Corrections to be made on the photomasks must then be calculated for all structures individually. The algorithms deployed here are based on models that describe the wave-optical effects. A result of such a *model-based OPC* is shown in Fig. 2.8, bottom right.

2.4.4 Reference to Physical Design

There are two types of imaging errors: deterministic and stochastic. While the former can be predicted in advance, the latter cannot. Hence, they are handled in different ways and impact the layout design process differently, as well.

Deterministic Imaging Errors

Edge shifts and diffraction effects are examples of deterministic imaging errors, as they are known in type and scope in advance. We can therefore take *preventive corrective action* in these cases, as described above. In particular, the layout data are altered graphically in an automated layout post processing step when the layout design is complete and before the photomasks are produced. The modifications in question are: (i) edge shifts that compensate the edge shifts that occur in the technology, and (ii) OPC measures, both performed in the *layout to mask preparation process* (Chap. 3, Sect. 3.3.4).

The purpose of these corrective measures is that the graphics elements in the layout are updated so the resulting structures appear on the wafer as they should. This strategy avoids unnecessary complications in layout design. For one thing, it saves labor as the layout designer doesn't have to perform these adjustment tasks. For another, the layout is easier to "read", which also saves time and money, and what's more, helps enhance the quality of the layout results.

A word of caution here, though. Not all edge shifts occurring in the semiconductor process are handled and neutralized by pre-emptive operations. Some layout features will look different than features created on the wafer. We will flag these issues when we meet them later in the book.

Calculations for model-based OPC are very CPU intensive due to the highly complex nature of the layout patterns. The required computational overhead can be so large that it is advisable to allocate the necessary computer time for this task in the project schedule. Fortunately, the calculations do not depend upon one another and can run in parallel, speeding up the process.

As we have seen, the smallest technologically achievable feature size is a consequence of the wave-optical properties of the radiation used for the photolithography. Even if this boundary can be further reduced by many technology measures, there will always be a boundary of accuracy for a given process. This technological constraint is realized in the layout design by specifying design rules that prescribe specific minimum dimensions for features *within a layer*. These minimum dimensions apply (i) to the width of the geometrical elements to ensure they can be exposed (i.e., the features on the wafer do not "disappear") and (ii) to the spacing between two neighboring geometrical elements so that they can be safely separated (i.e., the features on the wafer do not "merge"). These design rules are known as *minimum width* and *minimum spacing* rules (Chap. 3, Sect. 3.4 and Chap. 5, Sect. 5.4.5), respectively.

Stochastic Imaging Errors

Overlay errors are stochastic defects, in other words, they cannot be accurately predicted. One can only ensure, that, as described above, by complying with tolerances for devices and process regulation, the sum of all deviations does not exceed a specified limit. Further key design rules that should be considered in physical design are derived from this maximum permissible overlay error.

Rules for overlay errors describe specific minimum dimensions that refer to features on *different layers*. The rules prescribe minimum values

(1) for two overlapping geometrical elements (to ensure the features also overlap on the wafer),
(2) for the case where one geometrical element is enclosed by another (so that one structure covers another on the wafer, as shown in Fig. 2.6), or
(3) for spacing between two geometrical elements (to ensure there is a clearance between features on the wafer, or at least no contact between them).

These respective design rules are known as (1) *extension* and *intrusion* rules, (2) *enclosure* rules, and (3) *spacing* rules (Chap. 3, Sect. 3.4.2 and Chap. 5, Sect. 5.4.5).

2.5 Applying and Structuring Oxide Layers

One of the big advantages of silicon over other semiconductors is that it forms a very stable intrinsic oxide: silicon dioxide (SiO_2). Silicon dioxide, which for the sake of simplicity we will generally refer to as "oxide" in the following, has many beneficial properties.

Oxide is an excellent electrical insulator and it acts as a dielectric for capacitive applications. It is mechanically stable and hence suitable for a robust layer structure. From a process perspective, it is easy to make, and it serves as a masking layer for many process steps, as we shall see. It is also transparent. This is a useful factor in fabrication, as alignment features can be detected beneath the oxide. This also enables many applications, such as LEDs (silicon can emit light), solar cells and photodiodes (light can penetrate silicon from outside).

Different processes are available for producing and structuring oxide layers. We will examine these now.

2.5.1 Thermal Oxidation

Thermal oxidation is used to form oxide with the silicon on the wafer surface. Once an oxide layer has formed, it can only continue to grow if oxygen atoms diffuse through it until they reach the silicon underneath. Consequently, as the thickness of the oxide layer increases, the pace of oxide growth decreases. Analysis has shown that oxide grows from the original silicon surface by approximately 44% into the silicon (this is the same as the amount of silicon used) and by about 56% outwards (Fig. 2.9).

There are two different thermal oxidation processes: *dry oxidation* and *wet oxidation.*

Dry oxidation
Wafers are heated in an oxidation furnace and exposed to pure oxygen (O_2) at 1000–1200 °C (approx. 2000 °F). The oxide grows very slowly according to the formula $Si + O_2 \rightarrow SiO_2$ and produces good quality oxide with few vacancy defects. This process is used for the very thin *gate oxides* (*GOX*) in field-effect transistors and for dielectrics in capacitors (Chap. 6).

Wet oxidation
In this process, the oxygen first flows through boiling water. The wafer is thus exposed to steam, as well. The reaction takes place as per the formula $Si + 2H_2O \rightarrow SiO_2 + 2H_2$ at 950–1000 °C (approx. 1800 °F) and is much faster than dry oxidation. It is more difficult to control however and produces a lower quality oxide. Wet oxidation is therefore more widely used to produce the *field oxide* (*FOX*). This is the first thick oxide layer that is created directly on the silicon. It is used to laterally insulate regions from each other. In old processes, the field oxide was only produced where there were no devices in the silicon. These "non-active" regions were also called "field" regions, and is where the name derives from.

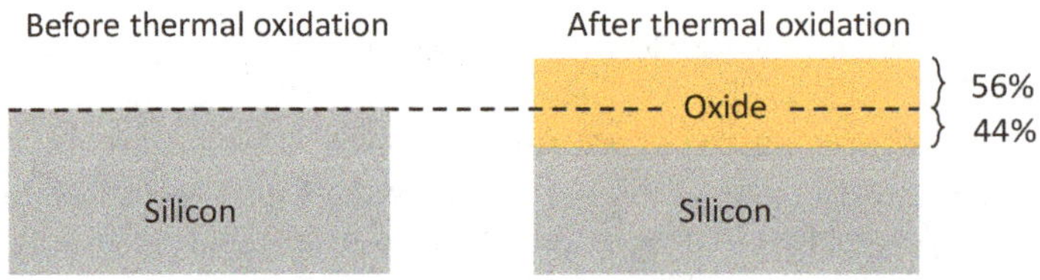

Fig. 2.9 Thermal oxidation of silicon (Si) to silicon dioxide (SiO_2)

2.5.2 *Oxidation by Deposition*

In the just discussed thermal oxidation, the silicon is obtained from the wafer surface and "consumed". However, if the silicon surface is covered by other layers, the oxide for additional oxide layers must be deposited. Silicon must be added from outside as well as the oxygen. A number of different deposition methods are available, discussion of which is beyond the scope of this book. This type of oxidation is deployed to electrically isolate metallization layers from one another.

2.5.3 *Oxide Structuring by Etching*

Etching is a process where material is removed chemically, and is often repeated for different materials in chip fabrication. It is important that etchants that remove material be used *selectively* w.r.t. the substance, i.e., so that they do not etch away other substances (or at least only very slightly).

Figure 2.10 illustrates how an existing oxide layer is structured by etching. A photoresist that is impervious to etching is exposed and developed through a photomask. Two different types of etching can be used: *wet etching* and *dry etching*.

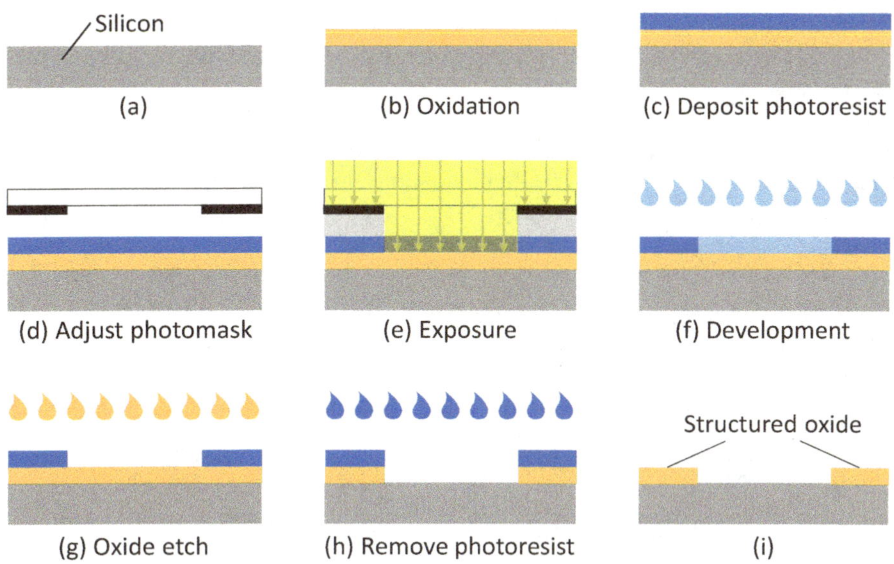

Fig. 2.10 Structuring an oxide layer

Wet Etching

In wet etching, oxide is dissolved and removed by a fluid chemical etching agent. This is a simple and commonly used method, and the fast etching rate can be fine-tuned.[5] The disadvantage of wet etching is that the etching is isotropic, i.e., it acts in all directions. This results in unwanted lateral etching underneath the photoresist.

These so-called *undercuts* mean that the oxide openings are always bigger than the openings in the photoresist. This leads to an edge shift (Fig. 2.11, left). The undercut has a slightly lower etching rate than the vertical etching, as the etching agent cannot circulate as easily under the photoresist and is therefore more highly saturated. A typical value for the lateral undercut is 80% of the etching depth.

Wet etching is not appropriate anymore for imaging typical feature sizes in advanced processes, due to this undercutting effect. Hence, wet etching is only used in these processes for dissolving and removing entire layers.

Dry Etching

Reactive ion etching (*RIE*) is an important dry-etching technique. In outline, the etching agent is ionized and applied as a gas plasma. The ions are set in oscillatory motion via an electrically alternating field. The field is aligned normal to the wafer surface. The chemically active ions oscillate in this direction and etch away material vertically only. There is no edge shift in this process, which is the main advantage of RIE (Fig. 2.11, right).

The etching effect in this process is a combination of a physical effect—where the material to be etched is bombarded with particles in a specific direction—and a chemical effect, i.e., etching. Very fine structures can be created with the RIE process. Furthermore, the resulting trenches can be far deeper than they are wide.

Oxide Steps

If the residual oxide is not removed after oxidation and targeted etching, an *oxide step* (or simply, a "step") is created, which does not disappear in subsequent processing. The term "step" refers to the small step-like elevation produced on the surface of the chip, and should not be confused with a "step" (i.e., a stage) that is performed as part of a process. The formation of steps is illustrated in Fig. 2.12.

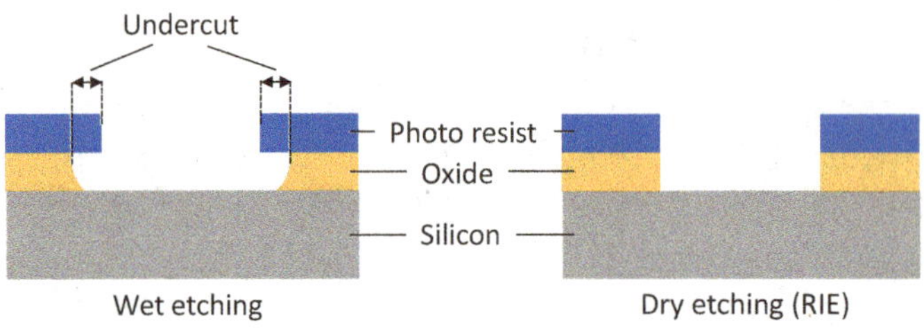

Fig. 2.11 A comparison of wet and dry etching (RIE, reactive ion etching)

---

[5]The etching rate R is the thickness T of the material being etched per unit time t, i.e., $R = T/t$.

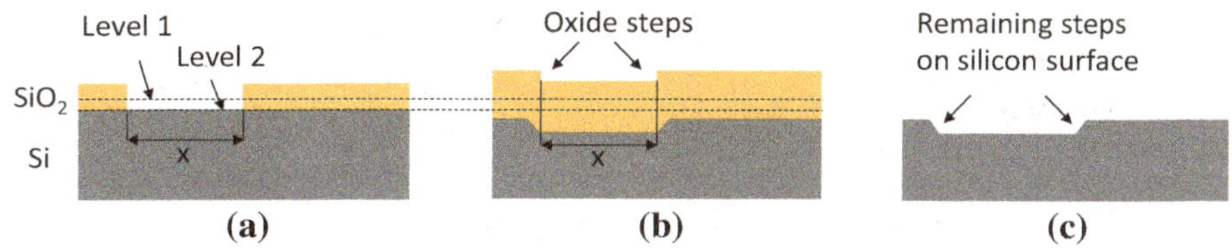

Fig. 2.12 The formation of oxide steps

Figure 2.12a shows the status following the structured etching of the field oxide (the first thick oxide layer). The height of the silicon surface before this oxide layer was thermally produced is at level 1. If another thermal oxidation is performed, the oxide in the opening starting at level 2 grows faster and thus reduces the height of the steps. However, as the surrounding oxide layer has a "head start", a step will remain at the surface (Fig. 2.12b).

A step forms below as well, whose height grows through time, as the oxide growth starts in the opening at the lower level 2. Hence, a step ultimately remains on the wafer surface, if the wafer is bright etched, whereby the entire oxide is removed (Fig. 2.12c).

If thermal oxidation followed by structuring, as described above, is repeated many times, further oxide steps will be created on the wafer surface. The wafer surface will thus become more uneven. This irregular surface makes accurate focusing more difficult during exposure in photolithography, which in turn degrades the imaging of photomask structures on the wafer. This effect prevents progress in the process technology to smaller feature sizes, and ultimately means multiple thermal oxidation followed by structuring cannot be used in cutting-edge technology nodes.

2.5.4 *Local Oxidation*

To ameliorate the oxide-steps problem described above, the *local oxidation* method (*LOCOS*—local oxidation of silicon) was developed to produce the field oxide.

In local oxidation, the field oxide is not structured by masked etching, rather the oxide layer is only allowed to grow in regions where the field oxide is required. Essentially, a material deposition procedure is masked instead of a material removal procedure. Specifically, the regions where the openings in the field oxide should be are protected from oxidation. Silicon nitride (Si_3Ni_4) acts as the protective layer.

Figure 2.13 shows the procedure. Since silicon nitride (abbreviation: "nitride") adheres badly to silicon, a thin oxide layer, the *pad oxide*, is first produced thermally as a bonding agent. A nitride layer is then applied to this pad oxide layer (Fig. 2.13a).

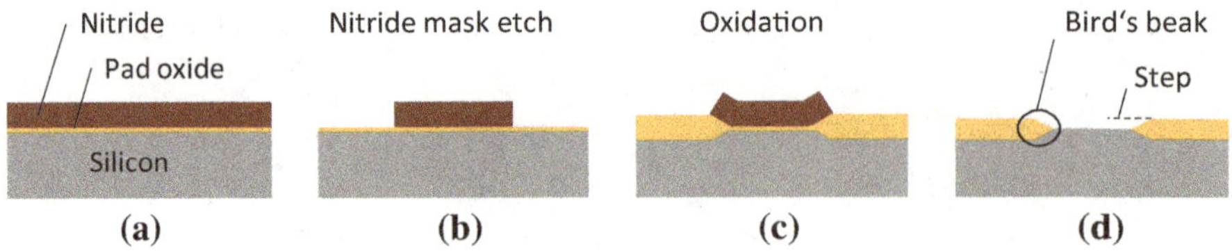

Fig. 2.13 Local oxidation of silicon (LOCOS)

The nitride mask is then etched with a photolithographically structured photoresist (Fig. 2.13b shows the result of this etching). The regions where the field oxide is to grow are now bare. Given that the thermal oxidation is isotropic, the field oxide grows at the nitride boundary in the horizontal direction as well. It pushes its way under the nitride, raising it at the edges (Fig. 2.13c). As the oxide spreads sideways under the nitride, the cross-section of the oxide layer is tapered. Finally, the nitride layer is chemically dissolved. The residual field oxide is peaked at its boundary—which is also known as a "bird's beak" (Fig. 2.13d). As a result, the remaining oxide step (d), is only about half as high as the step when structured by etching (cf. Fig. 2.12a).

Different extensions to the LOCOS technique have been developed to avoid this (small) oxide step as well. Figure 2.14 shows an idealized representation of the principle behind such an extension. The basic idea here is to first reduce the height of the silicon surface to be oxidized so that the original height is reached again after thermal oxidation (Fig. 2.14c). The necessary drop here equals 56% of the desired oxide thickness (Fig. 2.14b). The height could be reduced in two ways: (i) the silicon is masked by the nitride layer and etched or (ii) oxide steps are built as shown in Fig. 2.12 by locally oxidizing twice. In the latter method, the first oxide layer leaves a step at the desired height, after it—the layer—has been fully removed by bright etching. The desired oxide layer is then obtained with a second oxidation.

The local oxidation described here offers many benefits over thermal oxidation structured by etching (Sect. 2.5.3):

- No silicon is consumed inside the oxide openings.
- The oxide step produced by the LOCOS method is only about half as high as the one produced with thermal oxidation (that is structured by etching).

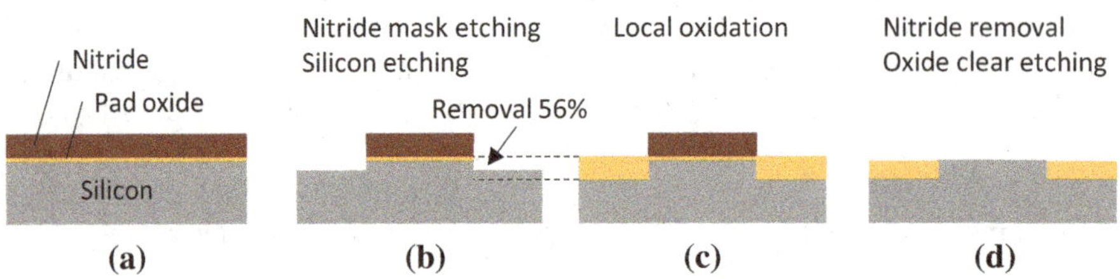

Fig. 2.14 Schematic representation of the extensions to the LOCOS process (idealized)

- The oxide step is inclined and not steep. This has the advantage that layers protruding over the edge of the field oxide (polysilicon, metal) cover the edge better.
- The height of the oxide can be reduced by LOCOS extensions to produce an almost flat surface.

2.5.5 *Reference to Physical Design*

The preceding sections have described several factors that affect oxide structures constructed on the surface of a chip. We next explore how they are handled during the physical design process.

Edge Shifts
If the field oxide is structured by wet etching, undercutting occurs, causing the oxide openings to be *greater* than the masking structure (see Fig. 2.11, left).

If the field oxide is produced by local oxidation, the bird's beak effect appears (see Fig. 2.13d), which causes the oxide openings to be *smaller* than the masking structure.

Both effects are deterministic, i.e., the extent of the undercut and the length of the bird's beak can be determined. The resulting edge shifts can therefore be taken into account by suitable pre-emptive corrections in the post-process when the masks are generated. Given that the practices to deal with these issues differ between various semiconductor processes, the field-oxide openings in the layout may or may not correspond to the real topography on the chip. We recommend that the reader refer to the documentation for the semiconductor process (known as the *process design kit*, or *PDK*) for guidance with this issue.

Through contacts (i.e., contacts and vias) are exclusively dry etched in processes nowadays, i.e., there are no edge shifts. Edge shifts produced when through contacts are wet etched in legacy processes are typically compensated by pre-sizing in the layout post process. Through contacts therefore always show the real dimensions in all semiconductor processes in the layout.

Oxide Steps with Through Contacts
Irrespective of the etching process, oxide steps are always produced when etching through contacts. This causes significant problems in the fabrication of metallization layers in modern processes for which there are nonetheless effective counter measures. We shall look at these matters in more detail in Sect. 2.8, when we deal with metal coating.

2.6 Doping

2.6.1 Background

Variously doped semiconductor regions are the basis for all semiconductor devices. With doping, a (i) 5-valent foreign substance ("donor") or (ii) a 3-valent foreign substance ("acceptor") is introduced into the semiconductor crystal to increase the conductivity of the crystal by (i) a surplus of electrons or (ii) a surplus of defect electrons (so-called *holes*). Silicon is primarily doped with phosphor, arsenic and antimony as donors for n-type doping, and with boron as an acceptor for p-type doping. We have already dealt with the physical fundamentals in Chap. 1 (Sect. 1.1.3); Fig. 2.15 visualizes this once more.

Doping can occur (i) by implanting external atoms during wafer fabrication, (ii) by implanting external atoms when growing silicon on the wafer (epitaxy), and (iii) by implanting external atoms from an outside source in the wafer surface. We shall examine the last case (iii) here, doping through the wafer surface, which can be achieved through three techniques: *alloying, diffusion*, and *ion implantation*. We will only focus on the two latter techniques.

2.6.2 Diffusion

The general term "diffusion" describes the process whereby a substance spreads out of its own accord due to a concentration gradient. The silicon crystal is a fixed lattice of atoms. For this crystal to be doped by diffusion, the silicon atoms and the atoms of the foreign substance must be so thermally excited that they can move within the lattice. To achieve this, the wafers are heated up to approximately 1200 °C (2190 °F) in a diffusion furnace. The dopant, which is supplied via a carrier gas, diffuses into

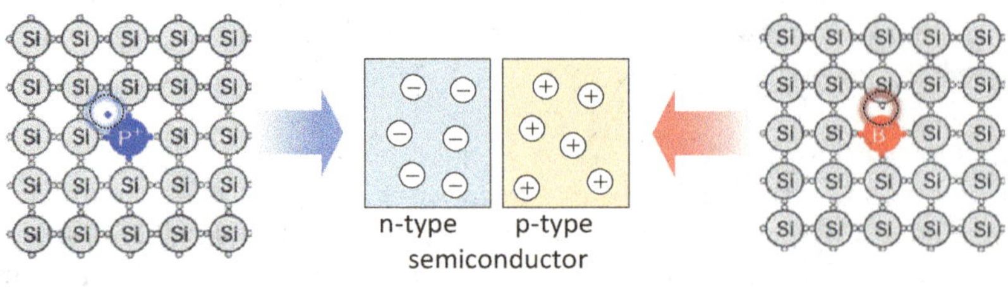

Fig. 2.15 Producing variously doped semiconductor regions by introducing a 5-valent donor (phosphorus, left) to generate free electrons (left) or a 3-valent acceptor (boron, right) to generate holes (right) in the silicon. Please note that—despite naming them n-type or p-type semiconductors—the overall number of positive and negative charges remains balanced in both areas. Increased doping leads to increased conductivity of the semiconductor (Si) due to the higher concentration of carriers

the wafer surface, as the dopant concentration in the wafer is much lower than in the gas.

A dopant atom is electrically effective as a donor or acceptor whenever it reaches a regular lattice site. This can occur in two ways: either it occupies an empty lattice site, or it replaces a silicon atom on its lattice site.

A structured oxide layer is used as a mask for specific diffusion. Dopants cannot pass through the oxide, which is also resistant to the high temperatures. The dopants only penetrate wafer regions that are accessible through oxide openings (Fig. 2.16, left). There is a concentration gradient for the dopant in the silicon in the form of the characteristic "diffusion with an inexhaustible source" shown in Fig. 2.16 (right). The concentration is $C(z = 0)$ at the silicon surface due to the carrier gas and is maintained at a constant level over time in the diffusion furnace. The concentration rises in all areas as time elapses, while it decreases with growing depth z.

As the dopants typically diffuse in all directions (this is called *isotropic diffusion*), the impurity atoms also move sideways underneath the masking oxide (see Fig. 2.16, left). This effect is called *outdiffusion*. The result is an edge shift, and the lateral expansion of a doped area is always greater than the corresponding oxide opening.

As we have seen, thermal oxidation "consumes" the silicon at the surface. In regions where oxidation is preceded by doping steps, some of the doped areas are lost again. This loss always occurs at the surface of the silicon, which typically has the maximum dopant concentration (see Fig. 2.16, right).

The above loss becomes critical if one wishes to perform many doping procedures in series with different masks (each by diffusion), as silicon is then repeatedly consumed by the repeated oxidation required for the mask. Increased step formation occurs as well at the surface (Sect. 2.5.3, "Oxide Steps"). Ion implantation, which is covered next, was developed to overcome these doping issues arising from diffusion.

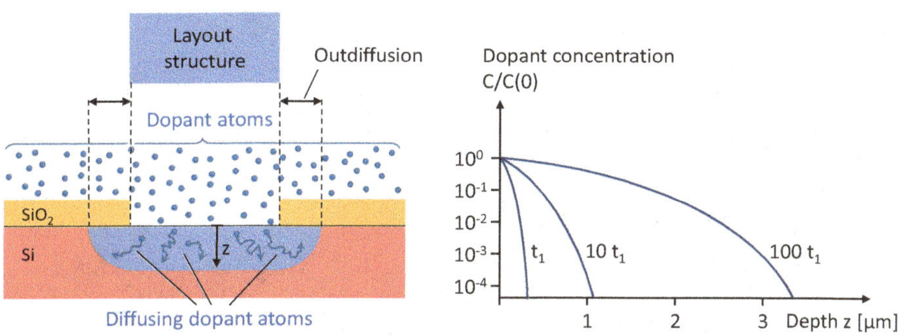

Fig. 2.16 Silicon doping through diffusion. Layout structure (top left), cross-sectional view of the diffusion zone (bottom left), doping profiles for "diffusion with an inexhaustible source" at three different times (right)

2.6.3 Ion Implantation

In this method, ionized dopant particles are accelerated to high speed in an electrical field and hurled at the wafer. The principle behind this technique is shown in Fig. 2.17.

The required ion type must first be selected from the particles emitted by the ion source. The particles are selected by a magnetic field that is normal to the ion beam. The (positively charged) ions are forced by the Lorentz force into a circular trajectory whose radius is a function of the particle mass. All unrequired particles are blocked by a simple aperture diaphragm and only the required types of ions are allowed through. This resulting beam is controlled by an electrical field and "written" onto the wafer.

The kinetic energy of the ions causes them to penetrate the wafer surface like projectiles (Fig. 2.18a). Collisions with silicon atoms cause them to decelerate. The penetration depth can be very accurately adjusted by the accelerating voltage. The entire operation is carried out at room temperature. The developed photoresist can therefore be used as a direct mask for selective doping. The masks in this case are called *resist masks*.

Channels are formed in the crystal lattice by the regular arrangement of the silicon atoms. If the dopant ions flow exactly between the atomic nuclei in the direction of these channels, they are only slightly slowed down and can penetrate deeper than intended into the substrate along these channels.

Two options are available to prevent this so-called *channeling effect*:

- The wafers are inclined by a suitable angle w.r.t. the ion beam so that the ions cannot enter the lattice channels. A disadvantage of this approach is that there is a low asymmetric doping concentration underneath the edge of the photoresist.

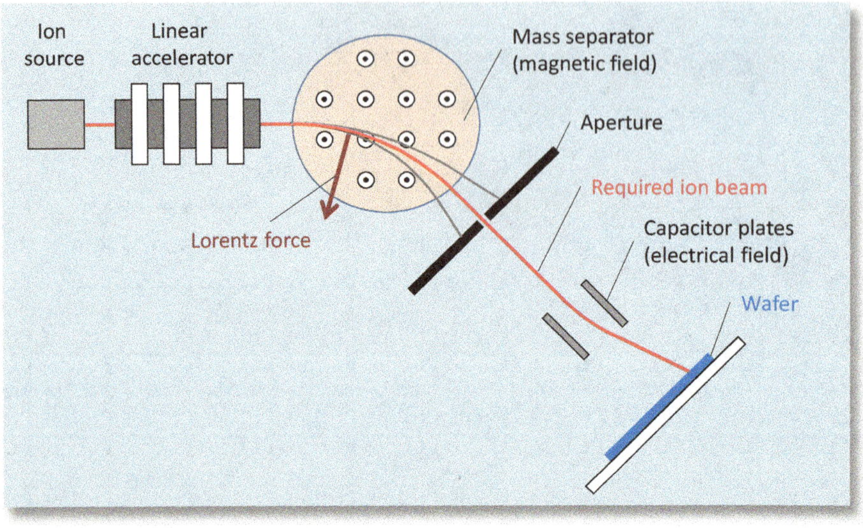

Fig. 2.17 Schematic representation of a system for ion implantation

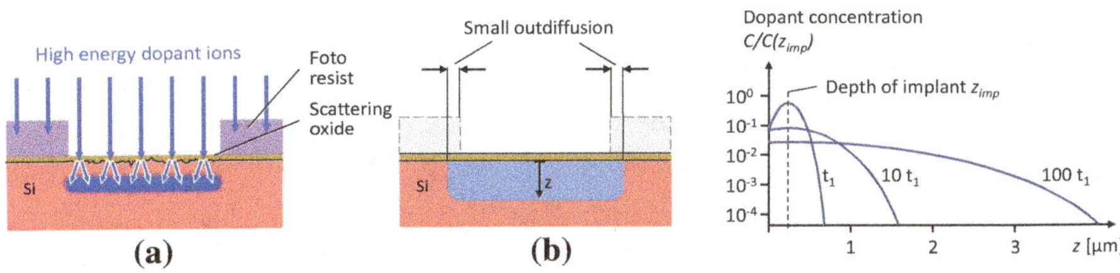

Fig. 2.18 Doping by ion implantation (prior photolithography is not shown); **a** implantation of the ionized dopant masked by the developed photoresist; **b** after removing the photoresist and crystal healing. Doping profiles for "diffusion with an exhaustible source" at three different points in time are shown at the right

- A thin oxide is applied to the wafer surface, which deflects the ions, and therefore prevents them from arriving in parallel streams. This scattering effect is shown in Fig. 2.18a.

The lattice structure on the surface is slightly damaged by the particle bombardment. Consequently, following the removal of the photoresist, the silicon crystal must be annealed in a temperature step at 800–1000 °C (approx. 1650 °F, Fig. 2.18b). Silicon atoms that were knocked from their lattice sites and the impurity atoms are embedded again in the crystal lattice by this so-called *tempering process*. It is only at this point that the dopants become activated electrically. A diffusion occurs during this process, causing an outdiffusion, as well. As this is a fast step, only a very slight edge shift occurs. This is illustrated in Fig. 2.18b, where the location of the removed photoresist is drawn with dashed lines.

If we want to extend the doping to greater depths, this must be done by diffusion. However, this will again cause outdiffusion and thus an edge shift. The diffusion source is in this case limited to the amount of implanted dopant. Only these atoms can diffuse; no new atoms are added. The doping profile is derived from the so-called "diffusion with an exhaustible source" characteristic (see Fig. 2.18, right). While the dopant concentration decreases near the surface during the diffusion process (this is caused by atoms diffusing away from the surface), it increases in deeper regions.

In contrast to doping by diffusion (Sect. 2.6.2), where many wafers are treated simultaneously in the diffusion furnace, each wafer must be treated individually in the implantation method we have described here. Additionally, a much larger fabrication facility than a diffusion furnace is required for implantation; and the latter is also much more expensive than the former. It is, nonetheless, only through the technological benefits of implantation that feature sizes could, and can still be, reduced at all. Wafers are therefore almost exclusively doped by implantation in cutting-edge semiconductor processes. The additional costs of implantation are recouped many times over by gains in surface area. We summarize the technological benefits of doping by ion implantation briefly here again:

- Masking is performed with resist masks. The two issues with masking with structured oxide are thus avoided. That is, there are no oxide steps, and the region near the silicon surface, which is the most heavily doped region, is not lost.

- Doping by ion implantation is more accurate (approximately ±5%).
- Only a slight outdiffusion occurs. The only exception is doping that is performed by subsequent diffusion to reach greater depths.
- Post diffusion—caused by multiple diffusion-based doping procedures—of existing impurities can be avoided. More degrees of freedom are thus available for configuring doping profiles.

2.6.4 Reference to Physical Design

Edge Shifts

The sideways expansions of the doped areas caused by the outdiffusions are always greater than the mask (oxide opening or resist opening). Hence, we obtain an edge shift. The size of the edge shift depends on the temperature and duration of the diffusion process and on the tendency of the dopant to diffuse. A characteristic value, typically between 70–80% of the diffusion depth, can be assigned for this purpose to every diffusion procedure in a semiconductor process. This enables the edge shift to be compensated by an appropriate pre-sizing operation.

This pre-sizing is generally *not* deployed for layers that define doped areas. These so-called *doping layers*[6] do not therefore show the lateral expansion of the resulting doped regions in the layout, instead they show the doping mask. Therefore, an outdiffusion—if present—must always be "added mentally" when creating and "reading" a layout!

As we have seen, progress to increasingly small feature sizes has been achieved in semiconductor technology by replacing isotropic processes, such as wet etching and diffusion, with anisotropic processes, such as dry etching and implantation, to reduce edge shifts or to avoid them altogether.

Spacing Rules

Let us return to the phenomenon of outdiffusion, which is not directly visible in the layout. Outdiffusion can also be the reason for spacings that are defined (in design rules) greater than one would expect from the feature size (in case of one layer) or from the maximum overlay value of a process (in case of two layers). This is due to design rules that need to take outdiffusions into account.

Furthermore, doped regions must also have extended spacing requirements for electrical reasons, such as avoiding short-circuits. Smart power processes are a case in point. They have a much higher electric strength than CMOS processes for pure logic circuits. Hence, spacing values between the doping areas specified by the design

[6]Layers that define the doped areas have always been known historically as "diffusion layers" among academics and in industry, even though the substrates are doped by implantation. We wish to dispense with this deceptive term in this book by calling these layers "doping layers".

rules are often determined by electrical requirements and not fabrication-related considerations. We shall discuss this topic in more detail in Chap. 6 (Sect. 6.2).

Vertical p–n Transitions Produced by Diffusion and Ion Implantation

As we have seen, diffusion and implantation always produce inhomogeneous doping. Here, the dopant concentration (greatly) decreases with increasing depth, as the concentration curves in Figs. 2.16 and 2.18 illustrate (note the logarithmical scales). This has a few important consequences, which we wish to investigate further now.

The different types of doping are indicated in cross-sectional views by colors, shading or hatching. These different doping regions are always uniform in the figures and they give the impression of homogeneous doping. This is not however the case due to the concentration gradient referred to previously. Shown in the cross-sectional views are the regions where the indicated doping type is predominant; these regions often also contain other types of doping. The regional boundaries are the places where these majorities change. In the case of two neighboring p- and n-doped areas, the drawn area boundaries indicate the p–n transition. Concentrations of acceptors and donors are balanced at these boundaries. The transitions are not abrupt, but seamless.

Figure 2.19 shows an example with four different doped areas, where we represent n-type conductivity in blue and p-type conductivity in red. (We follow this color convention throughout the book.) There are three p–n transitions in this example. We indicate areas of low dopant concentration with light colors, and areas of high dopant concentrations with dark colors. The superscripts ("+") and ("−") on the conductivity types "n" and "p" signify that the dopant concentration is particularly strong ("+") or weak ("−")—this is also the convention followed in the technical literature. The concentration curves in Fig. 2.19 (right) illustrate the continuous transitions between n-type and p-type conductivities. The p^--type concentration curve is constant because the wafer was homogenously doped beforehand.

It should be noted that the concentration curves in Fig. 2.19 tell us something else as well. More and more dopant accumulates irreversibly every time the wafer is doped. This is an irreversible process. Vertical p–n transitions can only be produced by targeted redoping (i.e., by diffusion or implantation) of doped areas. Here, the present concentration must be *overcompensated* by new doping such that the complementary dopant becomes the majority.

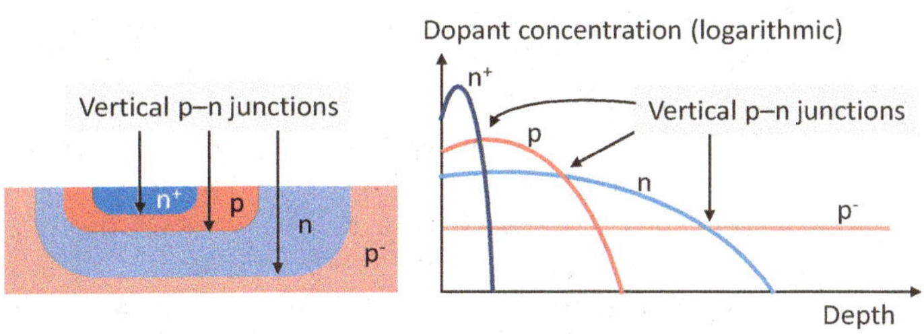

Fig. 2.19 Vertical p–n transitions created by multiple selective doping steps, whose majority concentrations do not change abruptly, but continuously

"Reliable" redoping—to change a n-type region into an p-type region, for example—requires a concentration of acceptors that is noticeable greater than the concentration of donors. Hence, regions that have undergone multiple redoping are always heavily doped.

2.7 Growing and Structuring Silicon Layers

Silicon layers can be grown on the wafer surface by different deposition methods. This process is called *epitaxy*. The atomic crystal structure of the newly grown silicon layer is determined by the composition of the wafer surface. There are two types of epitaxy: *homoepitaxy* and *heteroepitaxy*. While homoepitaxy is a kind of epitaxy in which a crystalline film is grown on a substrate of the same material and thus continues the substrate's lattice structure (Sect. 2.7.1.), heteroepitaxy is performed with materials that are different from the substrate material (Sect. 2.7.2).

2.7.1 Homoepitaxy

Layer Growth

If the new silicon atoms are deposited on the monocrystalline silicon surface, they adopt the atomic structure of the silicon in the substrate. The existing monocrystalline structure is thus propagated in the new layer. This process is called *homoepitaxy*.

This procedure is typically called *epitaxy* for short in semiconductor jargon. When someone talks about "epitaxy", or "epi", they nearly always mean homoepitaxy, i.e., growing monocrystalline layers.

Epitaxially grown layers can be variously doped by implanting impurity atoms, whereby their doping may differ from the doping of the base material. Clearly defined abrupt vertical p–n transitions that extend across the entire wafer can thus be created. Buried doped areas, known as *buried layers,* can be created too by selectively doping the base material on the surface before the epitaxial step.

Figure 2.20 shows this procedure with a p-doped wafer, on which a lightly n-doped epitaxial layer is grown. The selective doping before silicon deposition (n-type conductivity in this example) can be carried out by ion implantation (a_1) or diffusion (a_2). As the epitaxy process needs very high temperatures, the doping diffuses further during epitaxial growth and spreads into the epi layer as well. Steps are taken to minimize this unwanted outdiffusion especially upwards in the substrate by using arsenic or antimony as donors, as these elements tend to diffuse less than phosphor because of their larger size.

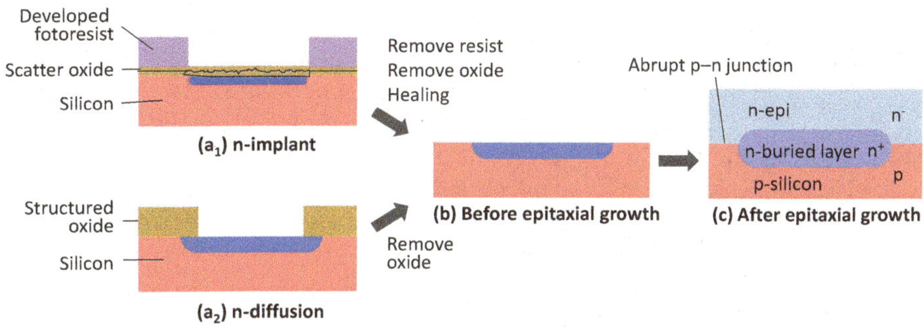

Fig. 2.20 Creating an epitaxial layer with a buried doped layer. (The prior photolithography and the oxide etching for (a_2) are not shown.)

Structuring

As we have seen in Chap. 1 (Sect. 1.1), the IC devices are situated in a thin layer that is only a few μm thick on the monocrystalline wafer. This layer is only approximately 1–2% of the wafer thickness. It can be formed by one or many epitaxial layers, as discussed above. There are some simple semiconductor processes without any epitaxy.

The epitaxial layer was normally not structured in the first decades of IC technology. Devices in close proximity to one another were electrically isolated on the surface by the field oxide and under the surface by p–n junctions[7] in reverse bias. As device miniaturization progressed, the relative space requirements of these passive field areas increased and became a barrier to further increases in integration density.

To deal with these issues, so-called *trench isolation* methods were developed with which the silicon could be structured in regions, which were insulated, laterally at least (to a given depth) from each other. In particular, silicon trenches are etched in the surface and filled with other materials so they can be used as electrically isolating barriers between neighboring devices. These trenches can be kept very narrow, so that devices can be more densely packed. This type of dielectric insulation has the added advantage that it affects the electrical behavior to a much lesser extent compared to insulation techniques based on reverse biased p–n junctions. It is thus much less susceptible to faults.

Depending on the semiconductor process at hand, the depth of the trench isolation ranges from a few hundred nm (e.g., for CMOS logic chips) up to several μm (e.g., for BCD chips[8] in automobile electronics). There are two main types of trench isolation: *shallow trench isolation* (*STI*) and *deep trench isolation* (*DTI*), although

[7] A thorough understanding of blocked p–n junctions is essential for physical design. We shall cover this topic later along with design rules (Chap. 6, Sect. 6.2) and reliability measures (Chap. 7).

[8] The Bipolar-CMOS-DMOS (BCD) process technology is typically used to make products where high power or voltage must be controlled by a digital controller. BCD technology incorporates analog components (Bipolar, CMOS), digital components (CMOS) and high-voltage transistors (DMOS) on the same die (Chap. 1, Sect. 1.2.2).

a boundary between these two designations is not clearly defined. There are many different process options for trench isolation, some of which are very complex.

The principles of fabrication for shallow and deep trenches are shown in Figs. 2.21 and 2.22, respectively. Some details have been omitted as they are not required for the basic understanding of the processes involved.

Trenches are typically produced by reactive ion etching (dry etching), as deep and narrow trenches can be etched in this way (shown in (a) in both figures). As explained in Sect. 2.5.3, the etching is masked by a lithographically structured photoresist. The etching depth is controlled by the etching period as silicon cannot act as an etch stop here since it is the only material in the substrate.

In shallow trench isolation (STI), the trenches are completely filled with oxide, either completely by thermal oxidation or—in order to save the silicon in the wafer—by brief thermal oxidation followed by oxide deposition (Fig. 2.21b). In deep trench isolation (DTI), often only the walls of the trenches are coated with enough oxide for electrical insulation (Fig. 2.22b). The rest of the trench is then filled with polysilicon (Fig. 2.22c), as it has a very similar thermal expansion coefficient to monocrystalline silicon. Any mechanical strain arising from temperature changes are therefore significantly reduced.

Materials that have been deposited on the surface during processing and are not required are subsequently removed. A special technique is employed for this purpose: chemical etching across the entire surface combined with mechanical material removal, the so-called *chemical-mechanical polishing* (*CMP*). This is a technique

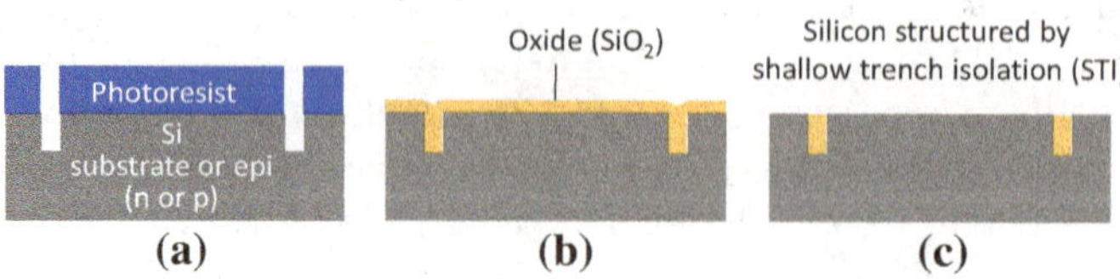

Fig. 2.21 Schematic representation of the fabrication of "shallow" isolation trenches; **a** reactive ion etching (RIE) masked by developed photoresist (prior photolithography is omitted), **b** removing photoresist and oxidation, **c** etching back the oxide and planarizing the surface by chemical mechanical polishing (CMP)

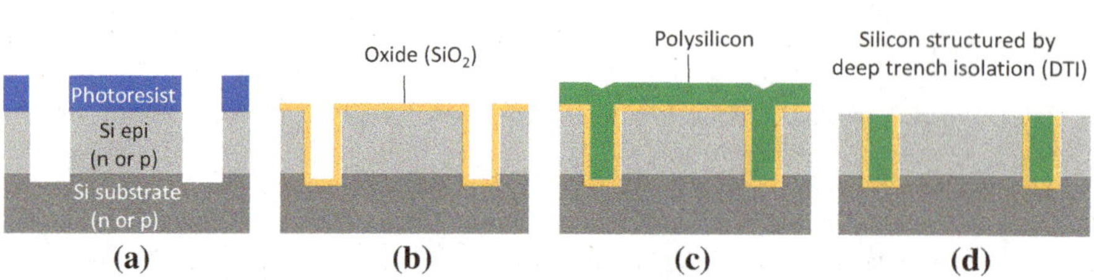

Fig. 2.22 Schematic representation of the fabrication of "deep" isolation trenches; **a** reactive ion etching (RIE) masked by developed photoresist (prior photolithography is omitted), **b** removing photoresist and oxidation, **c** deposition of polysilicon, **d** etching back the polysilicon and the oxide and planarizing the surface by chemical mechanical polishing (CMP)

used in different process steps for planarizing the wafer surface. We shall discuss this technique in the next Sect. 2.8.

Trench-isolation techniques only insulate sideways. To completely insulate a device dielectrically, it must be surrounded on all sides with insulating material. The biggest challenge here is creating a "buried" oxide layer. A number of different techniques are available known as *silicone on insulator* (*SOI*). We shall not discuss these techniques further here; we refer the reader to the literature, for example, [3].

2.7.2 Heteroepitaxy and Polysilicon

Layer Growth

In *heteroepitaxy*, the material structure on the wafer surface differs from the growth crystals, i.e., the materials are different from each other.

In almost all cases, the "different substrate" on which silicon is being deposited is oxide, which is an amorphous, i.e., *non-monocrystalline* material. Hence, there is no preferred direction for the deposited silicon to align with. Many small crystallization cores form first that grow independently in different lattice alignments. These cores grow to a so-called *polycrystalline silicon*, known as *polysilicon* or *poly* for short, as the deposition process progresses. Polysilicon is therefore composed of minute crystallites, also called "grains". Polysilicon deposition is as such a typical case of heteroepitaxy.

The grain size depends on the process parameters and can be as small as a few hundred nm. Different lattice alignments meet at the grain boundaries. Because of the current leaks at these boundaries, polysilicon is not suitable for producing useful p–n junctions, but only for basic current conduction.

Despite this issue, poly has many uses. For example, polysilicon structures are suitable for use as (i) control electrodes for field-effect transistors ("gates"), (ii) ohmic resistors, and (iii) electrodes for capacitors.

Structuring

Polysilicon is structured to produce the devices mentioned above by means of photolithography followed by etching. Reactive ion etching (RIE) is used in state-of-the-art processes to realize more accurate and more detailed structures.

Polysilicon structures are produced in the fabrication process before the metallization layers. They are to be found insulated by oxide between the monocrystalline silicon and the metallization layers. We illustrate the structuring procedure in Fig. 2.23 with a process using STI for the field oxide (Sect. 2.7.1). It is clear from Fig. 2.23a that polysilicon is not only deposited on the thick field oxide, but also on the *gate oxide* (*GOX*), a very thin oxide layer, that serves as a dielectric for capacitors and field-effect transistors. The name "gate oxide" for this thin oxide layer comes from the latter use case. Steps (b)–(f) in Fig. 2.23 show the structuring of the poly layer by photolithographically masked etching in anisotropic reactive ion etching (RIE).

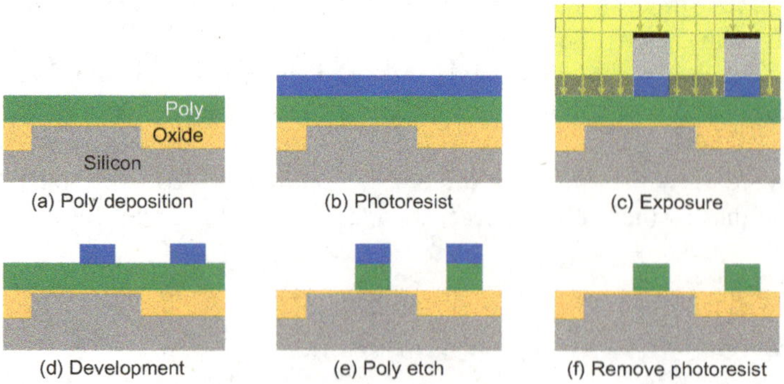

Fig. 2.23 Deposition and structuring of polysilicon for building devices

2.7.3 Reference to Physical Design

Aside from making devices, if the polysilicon is located above the field oxide it can be used as well as an interconnect. That said, great caution should be taken here. Although generally speaking, polysilicon is very heavily doped, the ohmic resistance of a poly interconnect is typically approximately three orders of magnitude higher than that of a metal interconnect with the same dimensions. Using polysilicon as an interconnect should therefore only be considered in exceptional cases, to connect a device to a nearby contact, for example.

An interconnect (between devices) made of polysilicon is usually only recommended for very small currents and short distances. One must keep a very close eye on the voltage drop (also called *IR drop*) caused when routing with polysilicon.

2.8 Metallization

2.8.1 Fundamentals

The process steps covered in Sects. 2.5–2.7 are part of the *front-end-of-line* (*FEOL*) in IC fabrication. FEOL is the first portion of any IC fabrication where the individual devices, i.e., transistors, capacitors and resistors, are patterned.

Following these steps, the devices must be connected together electrically as per the nets to build the electrical circuit. This is done in the *back-end-of-line* (*BEOL*). BEOL is thus the second portion of IC fabrication where the individual devices are connected. As we will see, this is done by alternately stacking oxide layers (for insulation purposes) and metal layers (for the interconnect tracks). The vias between layers and the interconnects on the individual layers are thus formed using a structuring process.

Hence, two layers need to be designed for each metallization layer in the layout design: (i) a first layer to define the through contacts in an oxide layer and (ii) a second layer to define the interconnect layouts located above this oxide layer. This portion of the layout design, i.e., the design of the structures in these layers, is called *routing*. Essentially, this step fabricates the paths of the nets that connect the devices.

Routing Layers

Many metallization layers are needed to implement the nets for the complex circuits in today's chips. While FEOL layer designations differ greatly from manufacturer to manufacturer and from process to process, the BEOL layer designations are widely standardized, as shown schematically in Fig. 2.24. The metal layers are generally numbered in the order they are fabricated—that is, from bottom to top. The same applies to the layers containing through contacts with the peculiarity that the through contacts between neighboring metal layers are called *vias* and the bottom through contacts that make the electrical contacts between the first metal layer and the silicon surface, and thus to the devices, are eponymously called *contacts*.

Hence, for every routing layer there are a pair of two corresponding layers. These layer pairs (Contact + Metal1, and Via[n] + Metal[n+1]) responsible for routing are also shown in Fig. 2.24.

Materials

From a fabrication perspective, the material used for the metallization layers must be easy to deposit and structure, and should also adhere well to oxide. The material should also meet the following requirements with regard to applications:

- High electrical conductivity (for low parasitic IR drops),
- High current-carrying capacity (to support miniaturization),
- Good contact with silicon (for connecting devices electrically),
- Good contact with the environment (for electrically connecting the chip as a whole),

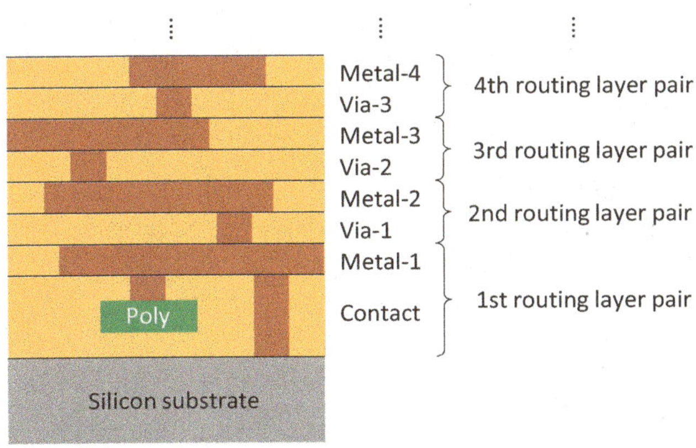

Fig. 2.24 Sectional view of the routing layers with typical layer designations

- Low susceptibility to corrosion, and mechanical stability (for long time to failure),
- Potential for multi-layer routing (to save chip surface area and ease layout design).

Aluminum (Al) is the material that best meets these requirements. By enriching aluminum with other materials by a few percent, some of the above properties can be slightly improved. A widely used option is the enrichment with silicon and copper (known as AlSiCu). As a result, aluminum has been the material of choice for metallization for a long time.

However, for shrinking feature sizes, aluminum falls short of expectations. One of its main drawbacks is that the parasitic resistance increases inordinately due to the smaller interconnect cross-sectional areas. Hence, aluminum has increasingly been replaced in state-of-the-art processes by copper, which has a lower specific resistance. (Its advantage in this regard over pure aluminum is approximately a factor of 2.) In addition, electromigration occurs to a much lesser extent in copper than in aluminum, with the result that copper can accommodate comparatively higher current densities. This means that much smaller interconnects can be designed in copper than in aluminum.

Due to this reduction in cross-section, the surface area formed by the interconnect and thus the parasitic capacitance per unit length is reduced with copper as well. Hence, analog circuits benefit from less crosstalk (Sect. 7.3.3) between interconnects in neighboring metallization layers. This effect has a positive impact on digital circuit technology too. The propagation time of a signal is proportional to the product of the parasitic resistance and the parasitic capacitance of a conductor ($R \cdot C$). Therefore, the use of copper enables much faster response times.

The benefits of copper, however, are equally matched by a considerable array of serious issues and challenges. Copper has the unfortunate negative property that it contaminates almost everything it comes in contact with. It thus diffuses very quickly into the surrounding oxide. It becomes corroded very easily as well, and not only on its surface like aluminum (which protects it against further oxidation), but throughout. Hence, copper must be shielded against diffusion and oxidation with appropriate protective coatings. In addition, new structuring techniques needed to be developed, as it is difficult to dry etch copper. All these issues and challenges greatly drive up the capital outlay and effort required for fabricating metallization structures with copper, as compared to using aluminum.

The switch from the use of aluminum to copper as interconnect material takes place in 350 nm down to 90 nm technology nodes. Both interconnect materials are found between these two nodes. In fact, the top metallization layer in smart power chips is often produced as an extra-thick copper layer to accommodate interconnects carrying very high currents (of the order of amperes), while the thinner metallization layers underneath are fabricated in aluminum.

2.8.2 Metallization Structures Without Planarization

There are two types of metallization: one with planarization between the creation of individual metallizations and the other without planarization between the creation of the layers. Planarization is the process of leveling or "smoothing out" the surface of the chip before constructing the next layer. For historical reasons and full understanding, we first briefly draw the reader's attention to the latter processes (those without planarization), which are now outmoded. State-of-the-art processes with submicron feature sizes require intermediate planarization steps. We shall deal with these critical processes in more detail in Sect. 2.8.3.

Metallization structures without planarization have only been fabricated using aluminum up to the present. Fabrication comprises the following four process steps:

(1) Application of insulating oxide layer,
(2) Fabrication of contacts/vias in this oxide layer by selective etching,
(3) Application of the metal layer (aluminum enriched with other materials),
(4) Fabrication of the interconnects in this metal layer by selective etching.

These are process steps where material is deposited and selectively removed alternately across the entire surface. The aluminum is vapor deposited primarily by depositing the material normal to the surface, from above. The through contacts (contacts/vias) are metalized simultaneously. The photolithographically structured photoresist is used to mask the material removal processes. Figure 2.25 shows how two successive metal layers are patterned, i.e., the above sequence is performed twice. (The photolithography itself is not shown.) The field oxide in this example is realized with LOCOS (local oxidation of silicon), as this was a typical process option for the "aluminum era".

The example shows the contacting and routing of an NMOS field-effect transistor (backgate contact is omitted) and a separate poly structure. As the structuring process progresses, the surface of the chip becomes increasingly uneven; the diagrams in Fig. 2.25 illustrate how steps are formed by etching. Steps in the oxide are particularly challenging, as it is difficult to cover these edges with sufficient metal in the subsequent metal deposition step. Thus, the via edges are weak points in the electrical circuit (Fig. 2.25, step 3). If the vias are located directly above one another, the step heights are added up; and the electrical contact is further weakened and may eventually be broken (Fig. 2.25, steps 6 and 7). It is therefore often forbidden to locate vias directly above one another (so-called "stacked vias") in these types of processes.

The oxide edges can be better covered with metal by forming bevel via edges. Beveling takes up space though, and is therefore a hindrance to miniaturization, which favors small vias and steep edges. By switching the deposition process from vapor deposition to sputtering, metal can be deposited on vertical edges as well. However, it is still not easy to fill oxide holes with ever decreasing via diameters.

This problem can be solved by first filling the contact holes with tungsten in a separate step and then depositing the aluminum for interconnects. This tungsten

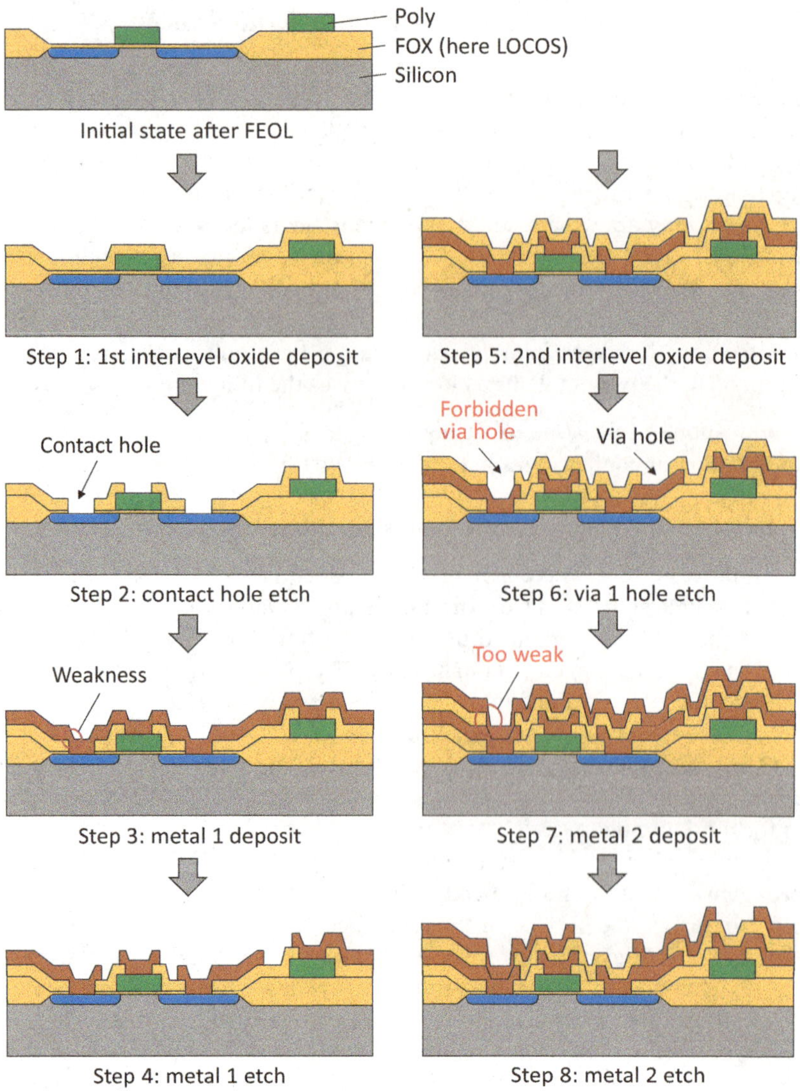

Fig. 2.25 Fabrication of the first two metallization layers without planarization (photolithography is not shown)

is deposited with the *CVD* process (*chemical vapor deposition*), which provides different options [5]. Tungsten is deposited across the surface of the wafer or grown only on the silicon surface (i.e., in the contact hole), depending on which chemical reactions are used. Excess material is then etched back. So-called *tungsten plugs* are thus formed in the contacts. This process for contacting aluminum interconnects is still in use today, as the interconnects can be structured by dry etching to produce clearly defined edges and very fine structures.

2.8.3 Metallization Structures with Planarization

Additive Processes

Aside from these measures to improve edge coverage in metal deposition, there are approaches to compensate the steps resulting from the oxide and metal structuring by adding extra material. One of these approaches is the *reflow* technology, where doped glass material is used. These glass materials liquefy in a high-temperature step and primarily flow into the recesses. The technique can only be used however to smoothen poly structures due to the low melting point of aluminum.

To smoothen aluminum features, the so-called *spin-on glass* (*SOG*)—comprising dissolved silicon-oxide connections—is available. SOG can be applied at room temperature and can be hardened below the molten temperature of aluminum.

All these additive measures only smooth local rough spots due to the relative high viscosity of the glass materials. Hence, they can only relieve step-formation issues, but not fully eliminate them.

The Damascene Process with CMP

The key breakthrough for solving the problems caused by step formation came in the form of a subtractive process. The entire wafer surface can be planarized with this technique. The process is called *chemical-mechanical polishing* (*CMP*). Specifically, the wafers undergo chemical etching combined with mechanical material removal. The surfaces of the wafers are pressed against a rotating polishing table, where etching agent and polish are continuously supplied. Any elevations of the wafer surface are removed until there is a continuous flat surface right across the entire wafer. Specific layers of material can be partially or entirely abraded as well with this technique.

Damascene technique. The CMP process is used in the BEOL for the final planarization of a metallization structure. As such, it is the final step in the manufacturing cycle for producing a layer pair, as depicted in Fig. 2.24. Clearly, this interconnect layout cannot be produced with the process steps described in Sect. 2.8.2 (Fig. 2.25), as the structured interconnects would be removed again (as they protrude above the wafer). In contrast to the process described previously, where the interconnect layout is produced by material deposition across the entire wafer surface followed by selective etching, here a recess is first etched in an oxide layer, which is then filled with metal. The process is thus the same as used for the fabrication of contacts and vias. The excess metal is abraded and removed by CMP leaving only the interconnect layout embedded in the oxide trench. This technique is called the *Damascene*[9] *technique*.

We show the main elements of this technique for structuring copper interconnects in Fig. 2.26. First, the trenches are etched in the available oxide (Fig. 2.26a). Due to the properties of copper described above, a barrier to prevent diffusion and to guard against oxidation is deposited (Fig. 2.26b). Tantalum (Ta), tantalum nitride (TaN) and titanium nitride (TiN) are suitable conductive barriers, also labeled as *metal liner* or *liner layer(s)*.

Copper can be deposited in different ways. If it is deposited electrochemically, a thin copper coat must be applied to the barrier layer, to which more copper atoms can attach themselves (Fig. 2.26c). When the trenches have been filled, the excess copper is removed by CMP and a final protective layer (barrier) is deposited (Fig. 2.26d). The nitride (Si_3N_4) mentioned earlier or amorphous silicon nitride (SiN_x) are suitable compounds for a dielectrical barrier.

This procedure must be performed two times in succession for a metallization layer pair, i.e., once for the contacts/vias, and once for the interconnects. The principle steps in the Damascene process flow are as follows:

(1) Deposit a first insulating layer (to contain the contacts/vias),
(2) Form the holes for contacts/vias by selectively etching this layer,
(3) Deposit a diffusion barrier and a seed layer for copper deposition,
(4) Deposit copper to fill the contact/via holes,
(5) Remove the excess copper and planarize by CMP,
(6) Deposit a protective layer as a barrier to diffusion and oxidation,
(7) Deposit the second insulating oxide layer (to contain the interconnects),
(8) Fabricate the trenches for the interconnects by selectively etching this layer,
(9) Deposit a diffusion barrier and a seed layer for copper deposition,
(10) Deposit copper to fill the trenches,
(11) Remove the excess copper and planarize with CMP,

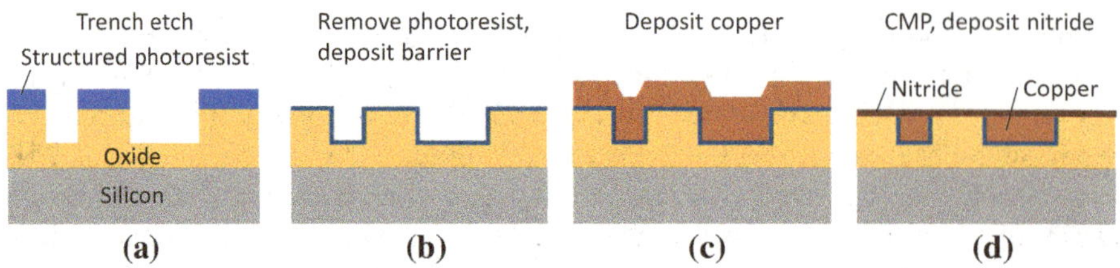

Fig. 2.26 Creating copper interconnects with the Damascene technique

[9]The term "Damascening" refers a historical metalworking artistic practice, in which different metals such as gold or silver are interlayed into a darkly oxidized steel background, to produce intricate designs and patterns.

(12) Deposit a protective layer as a barrier against diffusion and oxidation.

All structuring steps take place in the Damascene process (for both contacts/vias *and* interconnects) by etching insulating layers (steps 2 and 8), in which the metal is then embedded (steps 4 and 10). As such, this process is especially suitable for fabricating metallization structures in copper, which is largely unsuitable for dry etching.

We talk more generally about an "insulating layer" instead of an "oxide layer" in this description. There are two reasons for this:

(1) Other materials are used as well in state-of-the-art processes instead of oxide. These materials have lower dielectric constants than silicon dioxide (so-called "low-k" materials). Parasitic coupling capacitances between interconnects can thus be minimized.
(2) The additional protective coatings must be etched as well when using copper as an interconnect material. We will elaborate on this in the following.

The Damascene process can be deployed for both aluminum and copper. While, in the case of aluminum (as already mentioned), vias are made of tungsten, vias can also be made of copper in the case of copper interconnects. It is only in the first layer, when the copper makes contact with silicon, that both materials must be separated by tungsten.

Dual-Damascene process flow. The Damascene process flow is much more complex than the metallization process presented earlier in Sect. 2.8.2. In the meantime, it has been developed to the *dual-Damascene process*, in which contact/via holes and trenches are filled in a single deposition process. Only one final CMP process is then required. Different variations of the dual-Damascene process are in use. The following is one possible sequence of process steps for copper (Fig. 2.27):

(1) Deposit the first insulating oxide layer,
(2) Deposit a nitride layer,
(3) Structure the nitride layer to a mask for etching the contacts/vias,
(4) Deposit the second insulating oxide layer,
(5) Create the trenches for the interconnects by selectively etching this top oxide and by (nitride-masked) etching the holes for the contacts/vias,
(6) Nitride etching,
(7) Deposit a diffusion barrier and a seed layer for copper deposition,
(8) Deposit copper to fill the contacts/vias and trenches for interconnect structures,
(9) Remove the excess copper and planarize with CMP,
(10) Deposit a nitride layer as a barrier to copper diffusion and oxidation.

The dual-Damascene process as per this sequence of process steps for fabricating a metallization layer with associated vias is shown in Fig. 2.27.

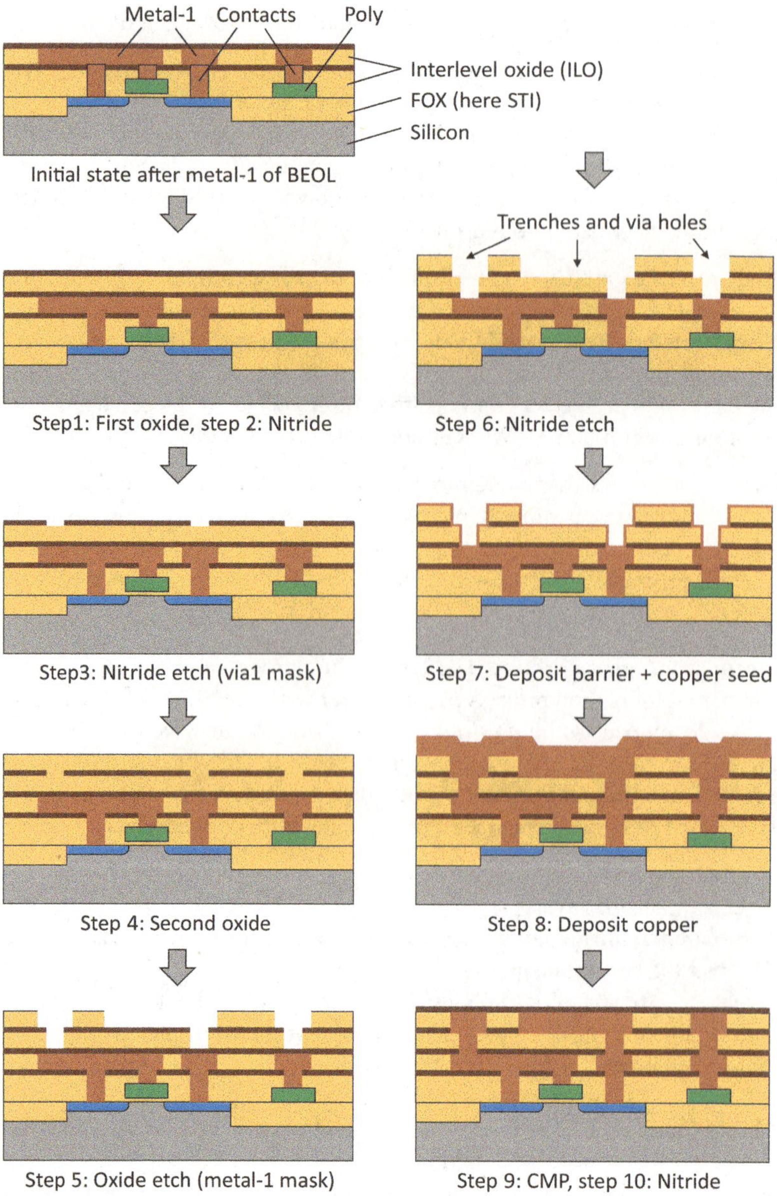

Fig. 2.27 Creating a routing layer (Metal-1), including its vias (Via-1), in copper with the dual-Damascene technique

2.8.4 Reference to Physical Design

The above metallization processes have some major impacts on the physical layout design. We describe key issues below.

Spacing Rules for Interconnect Layouts Without Planarization Technology

We have seen how the metal coverage of the oxide edges of contacts and via holes is a weak point in metallization techniques, requiring extensive planarization steps. It is normally forbidden in these kinds of processes to place vias directly above one another, as the resulting edges would be too steep for proper metal coverage. Hence, so-called "via stacks", featured in layout tools to simplify layouting, are forbidden. In addition to the typical spacing rules that apply between contacts and vias within a layer, further design rules are used in this case. They also prescribe specific spacings between vias in different layers. This adds additional complexity to the routing task.

As more layers are added to the wafer, its surface becomes more uneven in the upwards direction. This degrades the image sharpness in the photolithography, as already explained. The design rules thus prescribe greater minimum widths and minimum spacings for the top metallization layers in these processes.

Density Rules

Many finely structured metallization layers can be produced with the high-quality planarity achieved with the Damascene technique. This is of great value for performing the routing task in the layout design and facilitates the use of automated routing procedures in highly complex digital circuits.

One serious drawback of the Damascene technique w.r.t. the layout design is its use of CMP. With CMP, the ablation depth depends on the properties of the material to be removed, as the material is generally a mixture of different types of materials, such as silicon/oxide or metal/oxide. The problem is that the amount of a particular type of material removed can vary from place to place on the substrate if these different types are inhomogenously distributed throughout the layer. The results are unwanted "indentations" or "hillocks" on the surface. To avoid these surface issues, special design rules must be drawn up that prescribe a mean density that is representative of the materials to be removed and is a function of the surface area. Fittingly, these rules are called *density rules*.

Compliance with the density rules often requires significant additional work in the layout. If the quantity of a given material is too low in a particular region, additional filler structures without any electrical function must be introduced to increase the density. Semi-automated algorithms can sometimes be used here, but this is not always possible. Reducing the density of materials can be even more issue-ridden. For example, slots must be "cut" in very wide interconnects. Spacings between existing structures may have to be increased in some cases. More surface area may then have to be consumed—which is an undesirable outcome. We explain these effects and mitigating measures in more detail in Chap. 3 (Sect. 3.3.2, Fig. 3.16).

In order to reduce the risk of costly and time-critical refinishing, density rules should be applied in early layout phases so that the right design decisions can be made. (Remember, these density rules can still only be *fully* verified at the end of physical design.) The efficient handling of density rules typically requires a lot of experience.

Current Carrying Capacity

Metallization structures must be sized in the layout such that they are permanent current carriers, i.e., they should work reliably during the chip's entire lifetime (see Sect. 7.5.4 in Chap. 7). Contacts and vias are especially challenging in this regard [4]. These concerns apply as much to old processes (due to the poorer metal coverage at oxide edges) as to new ones (vias are generally the smallest structures here).

During routing, simply ensuring that the track width is big enough for a given current load is not enough. If we wish to conduct a current from one metallization layer to another, it is important to provide enough vias in order to increase their reliability [4].

Via Doubling

The probability that, despite outstanding process control, there is a non-functioning via on the chip has increased sharply due to the dramatic increase in the number of vias in state-of-the-art chips; the increased number of vias is a result of shrinking structure sizes and the associated extreme rise in complexity. A single via not making sufficient electrical contact can lead to an entire chip failing with serious consequences for the yield.[10] To address this issue, redundant vias are typically inserted in the design.

In order to effectively mitigate this problem, the number of redundant vias is increased. For example, it is recommended that at least two vias are used for every connection between two metal layers, even if one via would be adequate to meet current carrying capacity criteria. As a result, the failure of one of these vias would have no effect on the yield.

Metal-Semiconductor Contact

At the interface between semiconductor and metal, charge carriers are depleted and thus a depletion zone is formed. This zone is an energy barrier for charge carriers due to the different band structures in the semiconductor and in the metal. The result is a diode-type response, called a *Schottky diode* in this case, much the same as a p–n junction in a semiconductor. By highly doping the semiconductor where it interfaces with the metal, this zone can be made so small that the charge carriers are able to "tunnel through" this barrier (a quantum-physical effect).

There are many other physical effects where the two semiconductor-metal materials systems impact each other. To prevent unwanted effects, additional process steps are required, depending on the individual processes and materials used. Please refer to the literature, e.g., [1] and [6], for further reading on these fabrication steps for designing the materials' interface.

The key aspect w.r.t. the layout is that the contact area in the silicon must be heavily doped. This ensures a linear current-voltage response at the silicon-metal

[10]The "yield" is the ratio of the number of functioning chips to the total number of fabricated chips.

interface. The desired "ohmic" contact response is thus achieved. Typical resistance values for single contacts are in the single to double-digit ohm range.

The Number of Metal Layers as an Optimization Goal
The number of metallization layers can be chosen in state-of-the-art semi-conductor processes. This decision is often taken during physical design. The golden rule of economical technology applies here: "as many as needed, but as few as possible". From an engineering point of view, a specific number of metal layers are needed to solve the routing problem—all interconnections must be realized in accordance with the design rules. In other words, "as many as needed" is mandatory. From an economic viewpoint, on the other hand, "as few as possible" metallization layers should be used to minimize the manufacturing costs. Finding the sweet spot is not easy, especially when it may be possible technically to waive one routing layer, but requires, as a consequence, both additional layout design time and chip surface area.

Semiconductor processes for mixed-signal applications are normally composed of three to five metallization layers. In smart power processes, the top (final) metallization layer is often a thick layer for conducting high currents. Non-standard design rules that define larger minimum widths and spacings for the interconnect structures then apply to this layer.

Modern CMOS processes for purely digital chips, such as microprocessors, typically offer more metallization layers. Leveraging the smallest feature size, the number of interconnects per unit surface area can be maximized in these applications due to the high circuit complexity. Generally speaking, the number of metallization layers on real chips depends primarily on the circuit complexity (see Fig. 1.11 in Chap. 1).

2.9 CMOS Standard Process

We want to apply what we have learned in this chapter by observing the process steps in a semiconductor process. We shall discuss a CMOS standard process for this purpose. We first wish to point out that there is "no such thing" as a CMOS *standard* process, as the process technology is changing all the time with advancing IC downscaling. In addition, there are differences between manufacturers within a technology node. By "standard", we shall restrict ourselves to the steps in the flow needed to fabricate simple, but typical NMOS- and PMOS-field-effect transistors. Real industrial processes will differ from our example in their details, but the basic concepts remain the same.

2.9.1 Fundamentals: The Field-Effect Transistor

First, we wish to explain the principle of operation of the field-effect transistor. The reasons behind some of the CMOS process steps will then become clearer.

Figure 2.28 depicts the cross-section of a simple field-effect transistor—the most common NMOS type, also known as *NMOS-FET* for short. The "N" signifies that the current flow is due to electrons (n-type semiconductor). Defect electrons (so-called "holes") can be neglected in the current flow in an NMOS-FET device. The acronym "MOS" describes the "metal-oxide-silicon" stratification from top to bottom. The top layer, from which the FET control electrode is assembled, has long been made of polysilicon instead of metal. The designation "MOS" has nevertheless continued to remain in use to this day.

The complement of the NMOS-FET, where only holes are responsible for the current flow, is the PMOS-FET.

The "C" in CMOS stands for "complementary" and means that the technology provides both basic types, an NMOS and a PMOS transistor, which are complementary w.r.t. their current carrier types and thus to their usage. These field-effect transistors are *unipolar* transistors as there is always only one type of carrier for the current flow.

The standard field-effect transistor has two similar doped regions, *source* and *drain* (colored blue in Fig. 2.28), spaced apart and embedded in an environment known as a *bulk* or *backgate*. A conductive layer, acting as a control electrode and hence known as a *gate*, is situated above the backgate. This layer is separated from the backgate by a thin oxide layer, the gate oxide (GOX). Because the backgate has complementary conductivity to the source and drain, as is shown in Fig. 2.28, there are two p–n transitions between source and drain that prevent current flow, as at least one transition is always polarized in the reverse direction. A device in this configuration is called an *enhancement type* device, i.e., it is "normally off", as no current flows in this state.

We shall only discuss this type of device below, as it is by far the most widely used type. Furthermore, we shall refer to the NMOS-FET shown in Fig. 2.28 when explaining the electrical control logic. (All statements apply as well to the PMOS-FET, where the conductivity types and the signs for voltages, currents and fields need only be swapped.)

We select 0 V as reference potential and apply it to the source and bulk pads. If we now apply a voltage $V_{GB} > 0$ between gate (G) and backgate (B), the configuration acts like a parallel-plate capacitor with gate and backgate as electrodes. Remember the backgate is p-conductive, i.e., its majority charge carriers are holes. The electrical

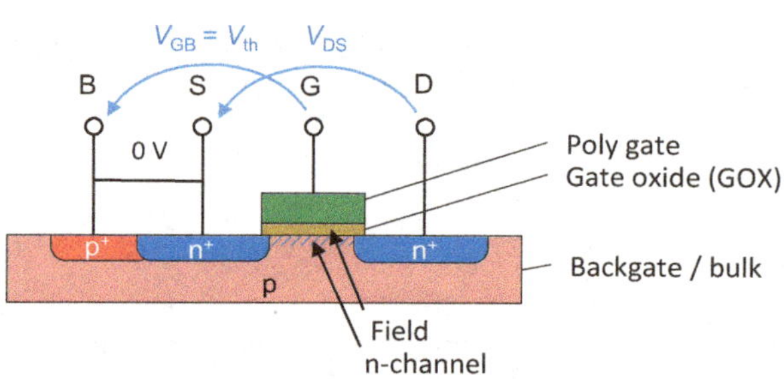

Fig. 2.28 Basic NMOS-field-effect transistor (the BEOL layers have been left out)

field now displaces these holes at the boundary between the backgate and the gate oxide, that is the bottom electrode of the capacitor. The electrons accumulate there to the same extent and form the negative countercharge to the positive charge of the gate electrode. These changes in the numbers of the two types of charge carriers are interdependent, as the relationship

$$n \cdot p = n_i^2 \tag{2.1}$$

applies to the electron density n and hole density p. Equation 2.1 is known as the *law of mass action*.

Here, n_i is the intrinsic (i.e., applicable to undoped semiconductors) charge carrier density, with an approximate value 10^{10} cm^{-3} for silicon at room temperature. This formula states that n and p are always inversely proportional to one another.[11] Figuratively, this is like a scale, with electrons on one side and holes on the other (Fig. 2.29). This scale is evenly balanced for an undoped semiconductor or in case of well-balanced n- and p-dopings. The greater the surplus of one doping type, the more the scale will tilt in one direction.

Now if we go back to our example with the NMOS-FET and increase the voltage V_{GB}, a point is reached when the field is so strong that electrons outnumber holes at the boundary layer. This voltage is called the *threshold voltage* V_{th}. The minority carriers are now the majority carriers, and vice versa. The boundary layer as such

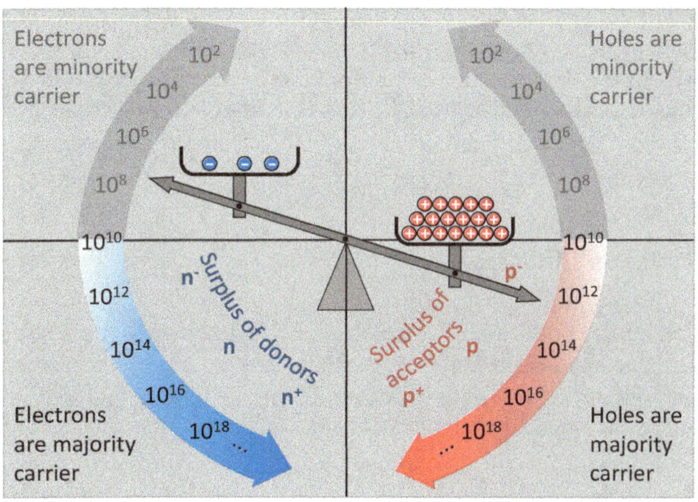

Fig. 2.29 Balance model for free electrons and free holes in a semiconductor. The scales indicate the density of these free carriers (per cm^3) which is determined by the surplus of the respective doping atoms (donors over acceptors, indicated by n$^-$...n$^+$, or acceptors over donors, indicated by p$^-$...p$^+$)

[11] Strictly speaking, the relationship only applies at thermodynamic equilibrium, i.e., when the generation and recombination of charge carriers are in equilibrium, which is what we assume here.

has become n-conductive and there is a conductive connection at the silicon surface between the n-conductive source and drain areas, called a *channel* (see Fig. 2.28). If a positive voltage V_{DS} is applied between drain (D) and source (S), current can now flow through this channel. The resistance of this channel can be regulated to a certain degree with the voltage V_{GB}.

The described effect is triggered by the electrical field of the capacitor formed by the gate and backgate. This is where the field-effect transistor gets its name. We are familiar with the formula

$$E = V/d \tag{2.2}$$

for an ideal parallel-plate capacitor, which states that, for a given voltage V, the field strength E is inversely proportional to the spacing d between the plates.

The MOS-FET is not an ideal parallel-plate capacitor, as the field spreads into the semiconductor. Nonetheless, the situation is near enough to the one we just described for us to make the following conclusion: the thinner the gate oxide, the lower the voltage needed at the gate for sufficient field strength for the field-effect.

There is another conclusion we make from Fig. 2.29. The NMOS backgate is p-doped, i.e., the scale in Fig. 2.29 tilts to the right for $V_{GB} = 0$ (i.e., without control). The resulting field-effect at $V_{GB} > 0$ metaphorically exerts a counter-clockwise torque on the scale. If this force is so big that the scale tilts to the left, the inversion is reached and the NMOS-FET conducts. It is clear from the figure without any further derivations that the inversion is more easily reached (i.e., that V_{th} is lower), with a smaller degree of tilt of the scale, which is defined by doping.

Summarizing we find that the field-effect rises—i.e., the threshold voltage V_{th} drops—as the thickness of the gate oxide decreases and as the backgate doping level sinks.

2.9.2 Process Options

In order to implement both NMOS- and PMOS-FETs, p-conductive and n-conductive bulk areas are needed. Different processes are available to achieve this, as shown in Fig. 2.30. A p-predoped substrate is depicted in the diagrams (colored in red). All statements apply analogously for n-doped substrates.

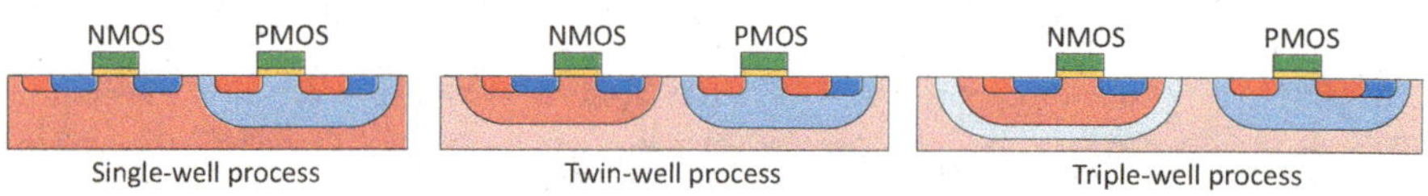

Fig. 2.30 Possible CMOS process options (all based on p-doped substrates)

Predoping the raw wafer as a common bulk for one transistor type means the bulk areas for the other type must be fabricated by selectively redoping the substrate. This is achieved by doping with diffusion (Sect. 2.6.2) or by implantation followed by diffusion (Sect. 2.6.3), whereby so-called *wells* are created. One variant of such a CMOS process is called a *single-well process* (Fig. 2.30, left).

The main issue with single-well processes is that the NMOS and PMOS transistors have different threshold voltage values (as long as no corrective measures have been taken). This becomes apparent if we recall the statements made in Sect. 2.6.4 (Subsect. "Vertical p–n Transitions Produced by Diffusion and Ion Implantation"), where we explained why redoped areas necessarily have a higher dopant concentration than before they were redoped. This means that the scale in Fig. 2.29 applied to an n-well created by redoping is tilted more (to the left), compared with the bulk formed by the p-predoped wafer (tilting to the right). Hence, the threshold voltage for the PMOS-FET is higher.

However, symmetrical threshold voltages should be used, if possible, for circuit design. This is generally corrected in single-well processes by performing a so-called "threshold-adjust" implantation across the whole wafer without a mask. In the example in Fig. 2.30, left, where the base is a p-doped wafer, just enough acceptors are implanted in a thin surface layer in this process that the PMOS- and NMOS-FET threshold voltages have the same value, i.e., the tilt angles of the "p–n scale" are the same in the end. The mobility of the holes in the PMOS-FET is reduced, however, by the increased number of dopant atoms.

Hence, it is advisable to begin with a weakly predoped raw wafer and to create the backgates for two transistor types as separate wells in the process (Fig. 2.30, center). The respective dopant concentrations can then be adjusted at will. These process options—called *twin-well processes*—are therefore most widely used despite the need for an extra mask.

Triple-well variants (Fig. 2.30, right) are available too. Here, one of the two backgate regions are produced by redoping twice, i.e., by a "well in a well". This creates p–n transitions to the raw-wafer substrate for the backgates of both transistor types. This has an additional degree of freedom, namely, that the backgates for the two transistor types can be set at any potential, by reverse biasing these p–n junctions. Applications requiring high voltages, such as automobile electronics, benefit from this option.

2.9.3 FEOL: Creating Devices

Recall that the front-end-of-line (FEOL) is the first portion of IC fabrication where the individual devices are patterned. We will describe a twin-well CMOS process below. We start on a p-conductive raw wafer, specifically one with a p^- substrate.[12]

We shall not depict the photolithographic steps in the diagrams for the sake of simplicity, rather we will show the status of each step after the photoresist has been developed. The mask openings can be gleaned from the photoresist structure. The five required masks are labeled with (1)–(5).

(1) "Nwell" Mask—Producing the n- and p-Wells

The substrate is first given a nitride layer, which is etched according to the Nwell mask (Fig. 2.31b). The remaining nitride acts as a mask for implanting the n-well regions with phosphor as donors (Fig. 2.31c). The thin pad oxide (needed for the adhesion of the nitride) now serves as a scattering oxide.[13] Following the removal of the photoresist, the implantation is driven into the depths of the substrate by diffusion. This causes an outdiffusion. A very thick thermal oxide is formed as well, which only grows outside the nitride layer, as the nitride masks the oxidation (Fig. 2.31d; cf. LOCOS, Sect. 2.5.4).

This oxide layer can be used as a mask for implanting the p-well regions when the nitride layer has been removed. The substrate is implanted with boron and does not require a separate mask, as it is self-regulating (Fig. 2.31e). This doping must be driven into the lower regions by diffusion, as well. Again, the pad oxide serves as a scattering oxide for the implantation.The entire wafer surface is now covered with n- and p-wells.

Before the next masking step, the wafer surface is planarized with CMP. The planarization is so deep that the entire oxide is removed (Fig. 2.31f).

(2) "STI" Mask—Producing the Field Oxide by Shallow Trench Isolation

As we know, active areas are laterally insulated with the field oxide. Very narrow oxide trenches can be created with the STI process (STI: shallow trench isolation) to produce a small field oxide at the n- and p-well interfaces. These narrow trenches ensure that the wafer surface is fully used.

Photoresist structured in a photolithographic process serves to mask the etching that produces the trenches (Fig. 2.32a, b). The trench depth is typically some hundred nm. The trenches are then filled with oxide insulating the n- and p-wells laterally with dielectric. The process is completed again by planarizing with CMP (Fig. 2.32c).

With some process options, a so-called *channel-stop* doping is implemented on the base of the trenches before they are filled with oxide. This suppresses the formation

[12] p-doped substrates are favored over n-doped substrates, as the die substrate is then at the lowest potential in the integrated circuit. If we define this potential as the reference potential (0 V), all our calculations will be based on positive voltages.

[13] A thin oxide, which deflects the ions and therefore prevents a parallel impact, is applied on top of the wafer surface.

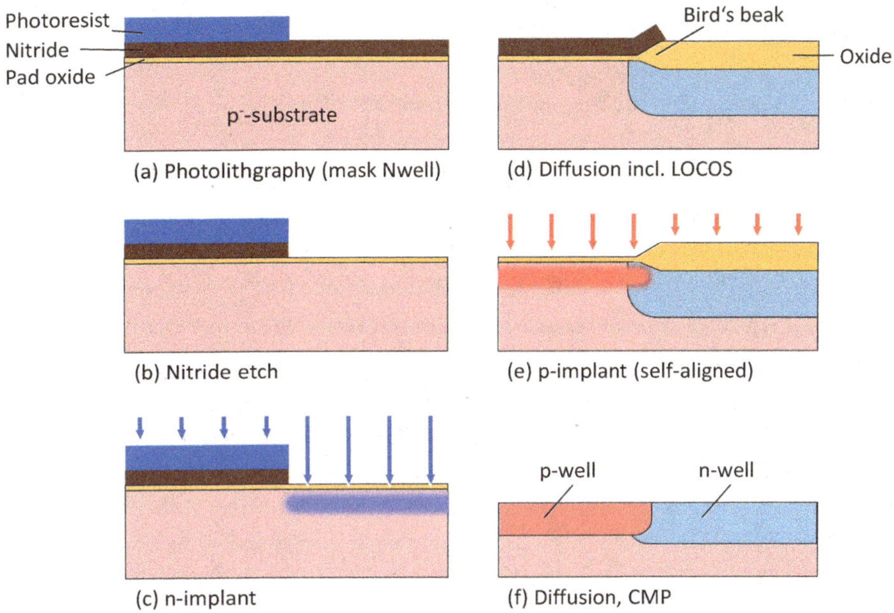

Fig. 2.31 Producing the n- and p-wells in a CMOS standard process

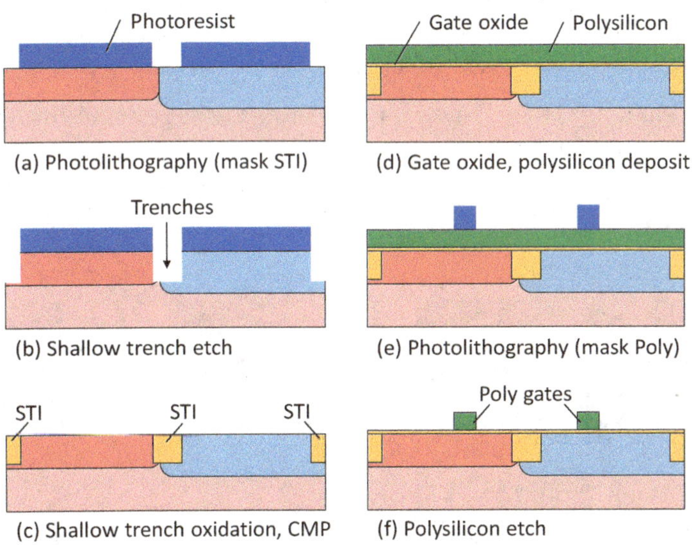

Fig. 2.32 Creating the shallow trench isolation (left) and the polysilicon structures (right)

of parasitic channels (Chap. 7, Sect. 7.2.1), but at the same time requires an additional masking step.

(3) "Poly" Mask—Producing the Polysilicon Structures

First, a very thin oxide layer is grown by dry oxidation on the cleaned silicon surface. This oxide layer will later be the gate oxide for the FETs (Fig. 2.32d). This process must be performed very carefully, as the purity and accuracy of the thickness of this layer are extremely important. Finally, the polysilicon is grown in heteroepitaxy (see Fig. 2.32d) and etched by a photolithographically structured mask (Fig. 2.32e, f).

(4) "NSD" Mask, (5) "PSD" Mask—Producing the Source and Drain Edges on the Channel Side

The left- and right-hand sides of Fig. 2.33 show two similar steps, where the source and drain regions of the two types of FET are lightly predoped. The gate oxide that is still in place is used as a scattering oxide for ion scattering. The implantation is not only masked by the resist, but also by the poly structure. The edges of the source and drain areas, which define the channel area, are set exactly by the edges of the polygate structure by leaving open the so-called *NSD* mask (NSD: NMOS, source, drain) over the NMOS gate and the so-called *PSD* mask (PSD: PMOS, source, drain) over the PMOS gate. As such, the edges arrange themselves (Fig. 2.33b, d).

The overlap and thus the unwanted parasitic capacitances C_{GS} and C_{GD} between gate and source and gate and drain are minimized by these "self-aligned" edges. The implantations also serve to dope the bulk contacts of the respective complementary FET type, (NSD for n-well, PSD for p-well).

(4) "NSD" Mask, (5) "PSD" Mask—Completing the Source and Drain Regions and the Backgate Connections

First, a thin oxide layer is produced across the entire surface by means of deposition. The process used here is the CVD (chemical vapor deposition) process, that ensures

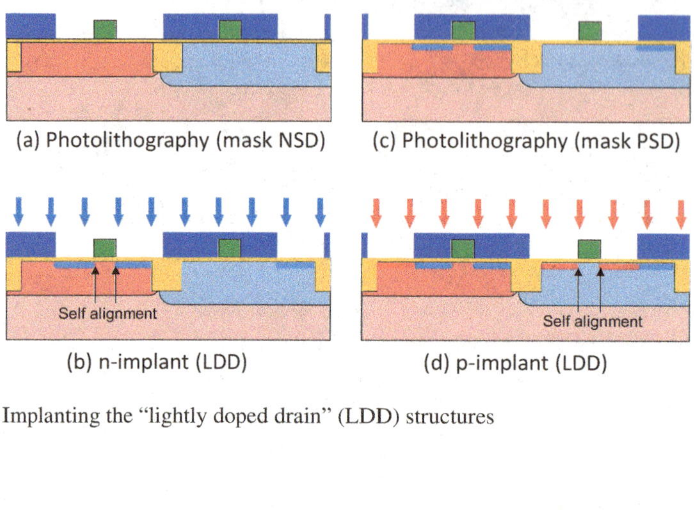

Fig. 2.33 Implanting the "lightly doped drain" (LDD) structures

that the oxide grows evenly on horizontal and vertical edges. The horizontal surfaces are then dry etched to fully remove the oxide from them. Almost all the oxide, however, remains on the vertical edges of the polystructures to form the so-called "spacers" (Fig. 2.34).

Following this preparation, the former process steps of masks (4) and (5) are repeated from the top using the same masks. The difference this time is that the substrate is heavily doped (i.e., n^+, p^+, Fig. 2.35). At the same time, the spacers ensure that the source and drain regions at the channel ends remain lightly doped. This configuration is called a *lightly doped drain* (*LDD*). It enhances the voltage capability of the transistor by increasing the breakdown voltage at the transition from the drain region to the channel. It is produced on both sides because each side can take on the role of a drain. We shall delve into the effect of this measure in Chap. 7 (Sect. 7.2.2).

Source and drain regions and the respective well connections are implemented in this way. The high degree of doping ensures a low-resistant contact to the metal that is

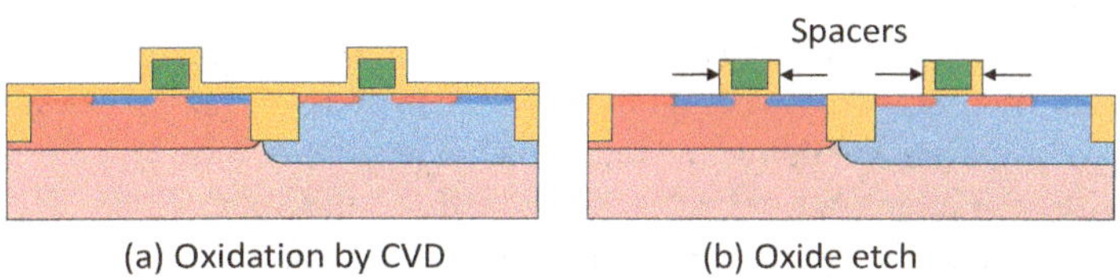

Fig. 2.34 Creating the spacers at the edges of the poly gates

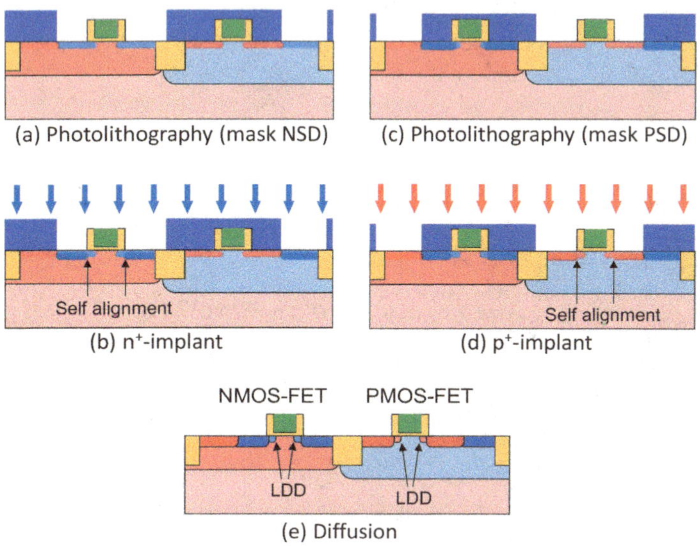

Fig. 2.35 High concentration doping of source and drain regions and well connections

later implemented. In the case of the source and drain, it also guarantees that enough majority charge carriers are available for the current flow through the transistors.

Finally, the surface is treated to prepare the contacts for the subsequent metallization. This concludes the front-end of line (FEOL), that is, all devices have been built.

Other features may also be patterned with the above process steps from which other (analog) devices, such as resistors, capacitors and diodes, can be formed [2]. These options will be covered in Chap. 6 (Sect. 6.3).

2.9.4 BEOL: Connecting Devices

Through contacts (i.e., contacts and vias) and interconnects (that connect the devices created in the FEOL process) are produced in the back-end-of-line (BEOL) steps by repeated application of the dual-Damascene process. As we have already discussed these process steps fully in Sect. 2.8.3, we will not go into them again here.

The structuring FEOL process steps in our exemplar CMOS process required five masks. For the subsequent BEOL metallization, two further masks per metal layer are required: a first mask to define the through contacts in an oxide layer and a second mask to define the interconnect layouts located above this oxide layer.

Chip fabrication is completed by the addition of a cap that especially protects all structures against penetrating moisture. This *passivation layer*—silicon nitride (Si_3N_4) is often used here—must be opened at the places through which the electrical chip connectors are routed. The so-called *bond pads* are located at these places. Small leads (aka *bond wires*) are attached to the pads here, or the pads are soldered directly onto a chip carrier, depending on the packaging used. These openings in the passivation layer are structured with an extra mask.

References

1. R.J. Baker, *CMOS: Circuit Design, Layout, and Simulation* (Wiley, 2010). ISBN 978-0-470-88132-3, 2010
2. A. Hastings, *The Art of Analog Layout*, 2nd edn. (Pearson, 2005). ISBN 978-0131464100
3. O. Kononchuk, B.-Y. Nguyen, *Silicon-On-Insulator (SOI) Technology: Manufacture and Applications* (Woodhead Publishing Series in Electronic and Optical Materials, Vol. 58) (Woodland Publishing, 2014). ISBN 978-0857095268
4. J. Lienig, M. Thiele, *Fundamentals of Electromigration-Aware Integrated Circuit Design* (Springer, 2018), ISBN 978-3-319-73557-3. https://doi.org/10.1007/978-3-319-73558-0
5. J.D. Plummer, M. Deal, P.D. Griffin, *Silicon VLSI Technology: Fundamentals, Practice, and Modeling* (Pearson, 2000). ISBN 978-0130850379
6. P. van Zant, *Microchip Fabrication: A Practical Guide to Semiconductor Processing* (McGraw-Hill Publ. Comp., 2004). ISBN 978-0071432412

Chapter 3
Bridges to Technology: Interfaces, Design Rules, and Libraries

Having presented fabrication technology for IC chips in Chap. 2, we now investigate in detail an important aspect of the physical design process: data interfaces. To be most effective, a layout designer should be aware of the links between physical design and the targeted technology—links that encompass layout and mask data, design rules, and libraries.

We introduce circuit, layout and mask data structures, that is, the main input and output data in the design steps, in this chapter. First, we explain the input to physical design—circuit data—while focusing on schematics and netlists (Sect. 3.1); we then discuss the output of the physical design step: layout data such as layers and polygons (Sect. 3.2). Mask data, which are the data required by the foundry and generated at the end of the design process, are described in Sect. 3.3. Here, we introduce "layout post processing", where amendments and additions to the chip layout data are performed in order to convert a physical layout into data for mask production.

Technology data, provided by the chip manufacturing foundry, are crucial for producing the physical design. An important portion of these data are technological constraints which are modeled in the geometrical design rules used in physical design. Essentially, geometrical design rules are constraints for physical design, whose compliance ensures the manufacturability of the layout results; for example, the minimum spacing between wires or components. Geometrical design rules are presented in detail in Sect. 3.4.

Technology data are organized in libraries. These libraries, which are extensively used in IC and printed circuit board (PCB) design, are covered in our final Sect. 3.5.

Figure 3.1 depicts the main design steps and how their interfaces bring together the different sections in this chapter.

J. Lienig and J. Scheible, *Fundamentals of Layout Design for Electronic Circuits*,
https://doi.org/10.1007/978-3-030-39284-0_3

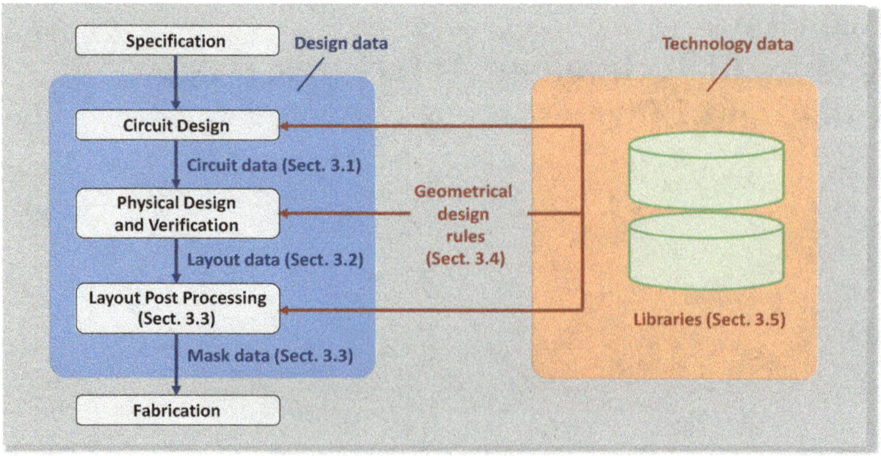

Fig. 3.1 Main design steps and their interfaces as covered in this chapter

3.1 Circuit Data: Schematics and Netlists

As depicted in Fig. 3.1, the circuit data are the input data for physical design. They are effectively a structural description of the circuit to be designed. First, we shall look at the general characteristics of a structural circuit description and then focus on the two typical representations used, the circuit schematic, also known as the circuit diagram, and the netlist.

3.1.1 Structural Description of a Circuit

The structural description of a circuit is a description of an electrical network. It contains information on the functional units in the circuit and their electrical connections.

Every functional unit has a certain number of connection points, often called *pins*. These pins are the functional units' interfaces, and they are connected electrically with pins on other functional units. These electrical connections are commonly known as *nets*, as they connect numerous (at least two) pins. A net is an ideal electrical connection, that is, a connection with zero impedance. Therefore, no potential differences can exist in a net. In other words, a net does not affect the electrical behavior of a circuit. This is why we talk about *potentials* and *nodes* in the context of nets. External connection points are also called *ports*.

The topology of an electrical network is shown in Fig. 3.2 (left). Here we can see, for example, that the "left-most" pin of functional unit C1 must be electrically connected to the left-most pin of functional unit C2, which is represented as net N1.

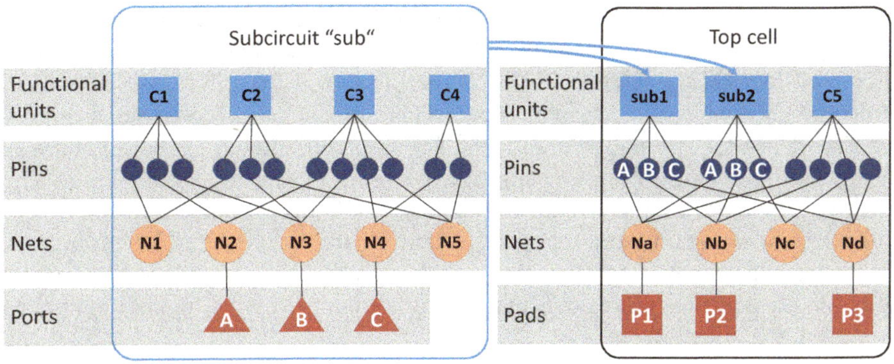

Fig. 3.2 Topology of a hierarchical electrical network

The functional units in a structural description are electrical devices from a fabrication process library. Parts of a circuit can be combined to form a functional unit in the structural description. In this case, we often call this type of functional unit a *function block* or a *subcircuit*. If this circuit is used as a function block in another structural description, its ports become the pins for this (higher level) function block. This is depicted in Fig. 3.2, where the (red-brown) ports A, B, and C on the left become the pins on sub1 and sub2 on the right.

If subcircuits (aka function blocks) are employed, the structural description becomes hierarchical. There can be (theoretically) any number of levels in the hierarchy. Subcircuits should be used when the same type of functionality occurs numerous times in a circuit; for example, a design that contains multiple adders, registers, or comparators. When subcircuits and hierarchical representations are utilized, the resulting structural description is easier to understand.

The top level in the hierarchy, often called the *top cell*, contains the complete system circuit; its ports (which connect to the next higher system level, e.g., the lead frame) are known as *pads* (shown in Fig. 3.2, right). As this is a term used in the layout implementation, we shall elaborate on it later (Sect. 3.4.4).

Analog IC Devices

Analog circuit devices in a structural description are the well-known passive and active basic electrical device types, such as resistors, capacitors, coils, diodes, transistors, and more complex active devices, like thyristors.

Digital IC Devices

These are the elemental and often small functional units used to design digital circuits, generally comprising logic gates and memory elements. These elements are small circuits made of transistors. They are designated as *devices* in the structural description of digital circuitry, as they will not be newly developed in the design flow, but are permanent, pre-designed units taken from a library (in the circuit design and in the design of the layout).

Same types of devices can often be implemented in different ways in a technology node, thus forming subtypes. For example, chips can contain both implanted resistors and polysilicon resistors. There can be as many as several dozen to over a hundred different subtypes of basic devices implemented in a state-of-the-art mixed-signal process, depending on the process technology options offered by the foundry. A range of different subtypes of devices are available as well in digital circuits. For every gate type, for example, there are variants with different numbers of inputs.

Every functional unit in a structural circuit description is called an *instance*. A device instance can thus be viewed as a copy of a device in the library. In reality, however, the device is often not actually copied from the library, rather a reference is made to the library element.

Each instance of a basic device is characterized by its electrical parameters: this is the capacitance in the case of a capacitor, for example, and the channel length and width for a field-effect transistor. Sometimes different driver strengths are available for logic gates; these different driver strengths correspond to different channel widths in the field-effect transistors making up the gates.

Summarizing the above statements, the complete structural description of a circuit contains the following information:

- A list of all devices and their properties. Each device instance should include:
  - The type (and subtype, if applicable) of device,
  - The instance designation (typically an instance name),
  - The dimensions for basic devices (electrical parameter settings).
- A list of all nets and their assignment to the device pins.

Printed Circuit Boards (PCBs)

The devices on a PCB are not manufactured with the board, but are externally sourced. The structural description of a PCB therefore contains a list of the devices to be mounted as physically separate (discrete) components. The electrical properties of these devices have no direct influence on the structural description of a PCB.

3.1.2 Idealizations in Circuit Descriptions

As mentioned above, nets are viewed as zero-impedance short-circuits, although they do not physically exist as such. Devices in a structural description are similarly idealized concepts, as well.

Analog Circuits

The basic devices in a structural description are so-called *lumped elements*. This implies that their electrical properties are not viewed as being physically distributed—they are, in other words, concentrated in one place. This idealized concept allows the conservation laws, namely, the *nodal rule* and the *mesh rule*, to be applied. These conservation laws are known as *Kirchhoff's laws*.

The nodal rule states that the sums of the currents flowing into and out of a node are equal. The mesh rule states that voltages occurring along a closed path in a network must cancel each other out. These two rules form the basis of network theory. Current flows and voltages can be calculated at every point in an electrical network using this theory; these calculations are performed in simulators by software tools, which we cover in more detail in Chap. 5.

We would like to point out that, although "lumped elements" is an idealized concept, we do not assume they are *ideal* elements. Parasitic properties, such as ohmic losses in a capacitor or in a coil, can be considered with lumped elements. This is done during simulation by using equivalent circuits for electrical devices, consisting of ideal elements and possibly controlled sources that model the parasitic properties and nonlinearities, respectively. The same applies to nets, where their parasitic properties are estimated or extracted from a layout (Sect. 5.4.6). We recommend [5] as a good treatise on the theory of circuit simulation.

Digital Circuits
Devices in digital circuit design are idealized to an even greater degree. In addition to the Boolean function, we are interested in the characteristics that impact timing in digital circuits. In the case of logic gates, this is the time delay between applying the input signals and the appearance of the output signal(s), called the *propagation delay*. Technology parameters, that are acquired by characterizing manufactured prototypes and assigned to the devices, are taken into consideration to determine these delays. Special attention should be given to signal delay times arising in the nets. These are considered in the design flow first by estimating them (circuit design), and finally by extracting them from the layout (physical verification). We recommend [6] as a useful resource on digital circuit design and its idealization.

3.1.3 Circuit Representation: Netlist and Schematic

The structural description of a circuit can be represented in text format as a *netlist* or in graphical format as a *circuit schematic*. A circuit schematic, also called a *circuit diagram*, is the pictorial representation of a circuit structure, in which the functional units are depicted as symbols and the nets as link lines. (We limit our discussion here to examples to be followed later with an introduction to schematics in Chap. 5, Sect. 5.2.)

The left-hand side of Fig. 3.3 contains the schematic of a simple digital circuit. Functional units are shown in green; their pins in red; and the nets (with the exception of the "Net1" net) in black. The functional units in this example are three logic gates. The three ports in the circuit are shown as red-brown colored triangles.

Instances in this example are designated according to a naming convention. Here, instance names are a combination of the designation for the device type with a number in square brackets appended to it. This defines the types of devices, as well. Ports can be given any names; they are called "A, B, C" in this example.

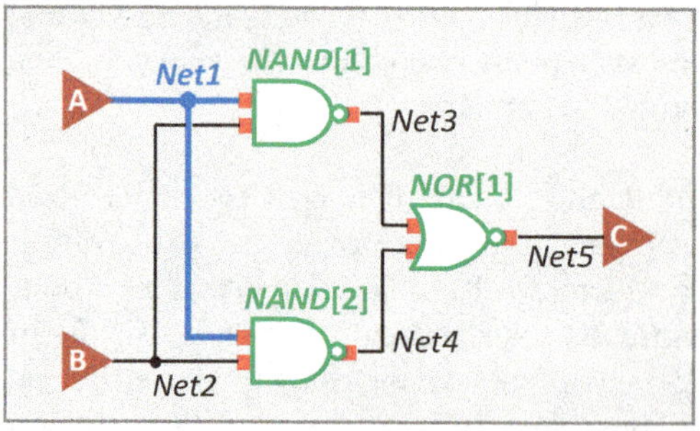

Pin-oriented netlist

(*A*: *Net1*)
(*B: Net2*)
(*C: Net5*)
(*NAND*[1]: *IN1 Net1*, *IN2 Net2*, *OUT Net3*)
(*NAND*[2]: *IN1 Net1*, *IN2 Net2*, *OUT Net4*)
(*NOR*[1]: *IN1 Net3*, *IN2 Net4*, *OUT Net5*)

Net-oriented netlist

(*Net1*: *A*, *NAND*[1].*IN1*, *NAND*[2].*IN1*)
(*Net2*: *B*, *NAND*[1].*IN2*, *NAND*[2].*IN2*)
(*Net3*: *NAND*[1].*OUT*, *NOR*[1].*IN1*)
(*Net4*: *NAND*[2].*OUT*, *NOR*[1].*IN2*)
(*Net5*: *NOR*[1].*OUT*, *C*)

Fig. 3.3 Circuit structural descriptions of a simple digital circuit; left as a circuit schematic, right as netlists. The latter can be differentiated into pin-oriented netlists, with each device having a list of associated nets, and net-oriented, where each net has a list of device pins assigned

Netlists are lists of functional units stored in a specific format in a file. There are two types of netlists, *pin-oriented* and *net-oriented*, depending on how they are sorted. Examples of both these netlists are shown on the right in Fig. 3.3. For syntax transparency, the "Net1" net is written in blue in the schematic and in the netlists. The syntax in these two simple examples is very similar to the EDIF standard (Electronic Design Interchange Format) [7].

The structural elements are sorted according to functional units in the pin-oriented netlists. Each functional unit is listed once and the nets connected to their pins are listed beside them (Fig. 3.3, top right). Nets appear several times in these lists.

In net-oriented netlists, the structural elements are sorted according to the nets. Each net is listed once and the functional elements connected by this net are listed with them (Fig. 3.3, bottom right). In this case, functional units appear a number of times in the list.

The assignment of the nets to the respective pins on the functional units must be unambiguous in each type of netlist. To assure correct assignment, the pins for every pre-defined functional unit must be designated in the technology data (i.e., in the library). Nets can thus be assigned to pins in the netlist by explicitly listing the pin names. This is the case for the netlists in Fig. 3.3. The pins on the logic gates are named "IN1, IN2, OUT" here.

After considering netlists, let us now discuss schematic representations. We motivate their use with examples from analog circuit design where they are heavily used. The quality of integrated analog circuits depends on several factors, an important one being the extent to which similar types of components have a symmetrical electrical response. These symmetry characteristics can be greatly impacted by how the devices are designed and laid out. Essentially, the devices need to be *matched* during the flow (which we discuss further in Chap. 6).

Matching issues and other requirements can be complex and challenging—so much so that in most cases this design task is still carried out manually by experienced layout engineers. Indeed, the layout designer must analyze the circuit in detail in order to select the right layout approach. If he/she had only a netlist to work with, it would be almost impossible for him/her to lay out the circuit to meet all these requirements. A pictorial image of a circuit helps engineers greatly when working with a circuit. This is in fact the main reason why circuit schematics are a must for the layout design of integrated analog circuitry.

Graphics need to be used sensibly, nonetheless. Standard design style rules should be followed by all designers to ensure circuit schematics are easy to read. Standard rules are in fact followed worldwide even though there are no hard and fast guidelines in place for their use. These de facto conventions mean that not only circuit design, but also developed circuits are more easily exchanged and their reuse promoted. (We elaborate on these rules in Sect. 5.2.2.)

We will explore several analog design concepts with an example circuit. A typical and widely used analog circuit is the "bandgap" circuit, which produces a temperature-compensated and supply-voltage independent reference voltage. The circuit schematic for our example is depicted in Fig. 3.4 (left). It contains basic devices and a so-called "Miller opamp" as a function block (subcircuit) which we depict in Fig. 3.5. The opamp is represented in the bandgap schematic as a function block by a (green) triangular schematic icon labeled "moa". The ports in this Miller opamp schematic (see Fig. 3.5, left) are the pins of the symbol in the bandgap schematic (see Fig. 3.4, left).

The basic device symbols are shown in green in our schematic examples. Assigned to each symbol are: an instance name (blue), device type designation (green), parameters with values for geometrical sizing (brown), and electrical parameters, if applicable. These details can be faded out in the editors to present a more uncluttered view. The net names are written in black beside the pins (red) in our schematic. Nets leading out through ports take their names from the ports (red-brown). Accordingly

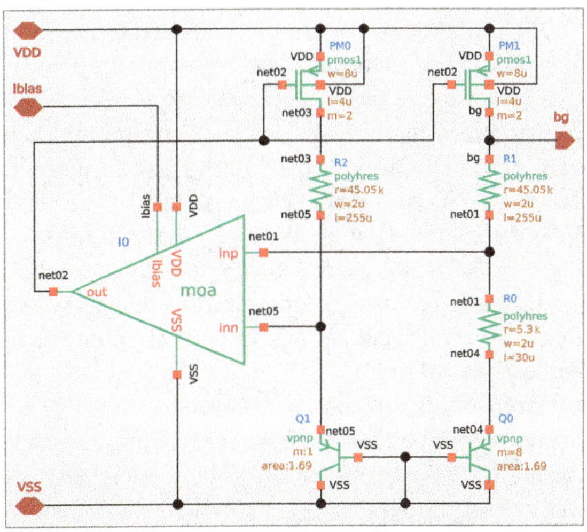


```
.SUBCKT bgap Ibias VDD VSS bg
QQ1 VSS VSS net05 vpnp M=1 EA=1.69
QQ0 VSS VSS net04 vpnp M=8 EA=1.69
XI0 Ibias VDD VSS net05 net01 net02 / moa
MPM1 bg net02 VDD VDD pmos W=8u L=4u M=2
MPM0 net03 net02 VDD VDD pmos W=8u L=4u M=2
RR2 net03 net05 45.05k polyhres W=2u L=255u
RR1 bg net01 45.05k polyhres W=2u L=255u
RR0 net01 net04 5.3k polyhres W=2u L=30u
.ENDS
```

Fig. 3.4 Schematic (left) and netlist (right) of a bandgap circuit

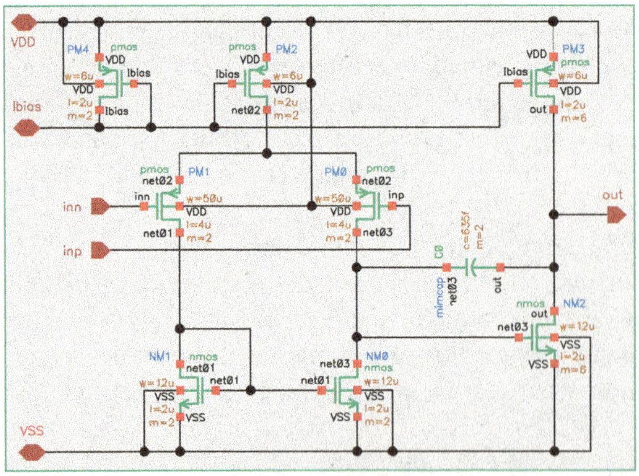

```
.SUBCKT moa Ibias VDD VSS inn inp out
MPM4 Ibias Ibias VDD VDD pmos W=6u L=2u M=2
MPM3 out Ibias VDD VDD pmos W=6u L=2u M=6
MPM0 net03 inp net02 VDD pmos W=50u L=4u M=2
MPM1 net01 inn net02 VDD pmos W=50u L=4u M=2
MPM2 net02 Ibias VDD VDD pmos W=6u L=2u M=2
MNM2 out net03 VSS VSS nmos W=12u L=2u M=6
MNM1 net01 net01 VSS VSS nmos W=12u L=2u M=2
MNM0 net03 net01 VSS VSS nmos W=12u L=2u M=2
CC0 net03 out 635f mimcap M=2
.ENDS
```

Fig. 3.5 Circuit schematic (left) and netlist (right) of a Miller opamp. Note that this Miller opamp is a function block, i.e., functional unit, in the bandgap schematic (see Fig. 3.4) and is represented there by a triangular schematic icon labeled "moa"

(and as mentioned earlier), the external ports in the circuit schematic for the Miller opamp (see Fig. 3.5) re-appear in the bandgap circuit schematic (see Fig. 3.4) as pins at the schematic symbol "moa".

Circuit schematics are drawn according to the style rules. Supply voltage and ground are shown as "VDD" and "VSS", respectively. The remaining ports are inputs and outputs. The transistors to be matched in the physical layout design are arranged horizontally beside each other in the schematics. This ensures that they can be easily located. Resistors need to be well matched as well. An experienced layout designer will immediately recognize these issues when examining the circuit's function by reading its schematic.

The netlists associated with the circuits in the right-hand side of Figs. 3.4 and 3.5 are given in the so-called *SPICE* (Simulation Program with Integrated Circuit Emphasis) netlist format [2]. This is a pin-oriented format. The corresponding information in both circuit schematics and netlists are in the same colors to ease clarity and comprehensibility. The first line contains the circuit name and the ports. Each of the following lines contains one functional unit with the following information: instance name (blue), connected nets (black), device type (green) and parameter settings (brown). The letters after the numbers designate the commonly used powers of 10 in physical units, e.g., "k", "u", "f" represent "kilo", "micro", "femto", respectively. The physical units themselves are associated with the device types, e.g., "ohm" in case of resistors, "farad" in case of capacitors. In case of geometrical sizing parameters, e.g., "*w*" or "*l*", the physical unit is "meter".

The pin names are not explicitly named in this format, i.e., the connections at the device pins are determined by the order in which the net names are listed.

The SPICE netlist format has an additional helpful feature. The instance names are preceded by an *identification letter* representing the type of device; we have highlighted these letters in violet in Figs. 3.4 and 3.5. The identification letters have the following meanings: M = MOSFET, C = capacitor, Q = bipolar junction transistor, R = resistor, X = subcircuit. For further information on the SPICE netlist format we recommend [2].

3.2 Layout Data: Layers and Polygons

As indicated in Fig. 3.1, the layout data are the result of physical design. These data are used not only to store a design result and to prepare this result for the fab; rather, a layout engineer works continuously with the layout data throughout the design process. Hence, we next look more closely at the structure of this data and the key graphics operations that the layout engineer may apply to such data.

3.2.1 Structure of Layout Data

We have already explored several of the most important aspects of the layout data in Chap. 1 (Sect. 1.3). We have seen that chip layout data are comprised only of graphical data, and that these graphics contain all the information necessary to produce masks. The same is true for PCB layout data, where the layout is described by polygon coordinates, accompanied by data containing diameters and positions for drilling the via holes and data for positioning the devices.

In general, a graphic can be represented as a raster or vector image. While raster graphics are bitmaps, i.e., a grid of individual pixels that collectively compose an image, vector images are mathematical calculations from one point to another that form lines and shapes.

Electronic layout data are saved only as vector images and processed as such. There are several reasons for this: (i) the data structure of vector graphics suits layout representations ("polygons") very well; (ii) they do not need much memory as compared to raster images; (iii) they can be processed more quickly and easily due to the information they contain; and (iv) they can be reshaped without loss of accuracy. The only disadvantage of the vector data structure is that it must be converted to raster data for presentation on computer screens. This is not a problem nowadays as state-of-the-art design environments are readily available with very efficient algorithms and high-performance hardware.

Layers

Graphics elements in layout data, for example the lateral structure of a doped region or an interconnect, are called *shapes*. Each shape is assigned to a unique *layer*. This layer association is an elemental attribute of every shape that enables it to be assigned to masks. It also forms the basis for graphical linking operations, as we shall see.

There is one important point we would like to make clear at this stage. When we talk about "layers" in the context of layout data, we are referring only to the above attribute in connection with the data structure. Such a layer often corresponds directly with a counterpart in the fabrication process. This counterpart could be a doped region in an "Nwell" layer, or "Metal1" for a metallic layer deposited on a wafer. This does not always have to be the case, however. Layout data also contain layers not directly associated with anything on a wafer, as we shall see (Sect. 3.3.4). The opposite can also be true: for example, the gate oxide layer on silicon is not modeled as a layer in the layout data.

We differentiate between the two types of "layers", as follows: we call layers used in the data structure as *drawn layers* and layers used in the fabrication process as *fabricated layers*. We shall only use this extended terminology where extra clarity is needed, i.e., in cases where the risk of misunderstanding is high if it is not used. In all other cases, and for the sake of simplicity, we will use the term "layers". Accordingly, all layers referred to in this chapter are "drawn layers" in the layout.

Shapes

The shapes in the layout data are always polygons. A polygon is a two-dimensional, continuous graphics element bounded by straight edges. This type of shape can be efficiently stored in the vector data structure as a list of successive corner coordinates. The resulting closed polyline determines the polygon; whether the polygon is to the left or right of the polyline must be defined, nevertheless. This varies from tool to tool. In some tools, the first coordinate in the list is appended at the end of the list; thus terminating the list.

Figure 3.6a shows an example of a general polygon with seven corners. Its coordinates are designated by C_i comprising two numerical values (x_i, y_i). By defining the smallest permissible grid (often called "manufacturing grid", "working grid", or simply "grid"), integer values, i.e., whole numbers, can be used for the (x_i, y_i) coordinates. Computer memory can thus be saved and the accuracy of the model is well-defined.

Donuts. Polygons with holes (also called *donuts*) can be modeled with this data structure. For efficiency in processing, this approach requires "dual" edge sequences in the data structure that "run" in two directions for this section. These superimposed edge pieces do not form a real polygon edge. An example is shown in Fig. 3.6b. The

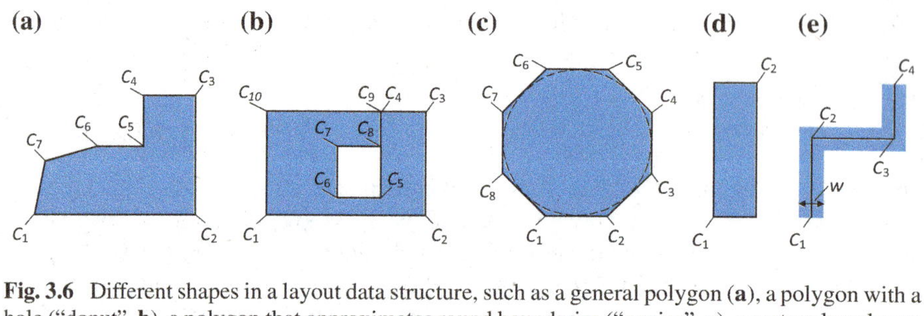

Fig. 3.6 Different shapes in a layout data structure, such as a general polygon (**a**), a polygon with a hole ("donut", **b**), a polygon that approximates round boundaries ("conics", **c**), a rectangle polygon (**d**) and a path polygon (**e**)

path (C_8–C_9) lies on the path (C_4–C_5) in this example. The coordinates C_4 and C_9 do not constitute real corners.

Conics. Graphics elements with round boundaries (some tool providers refer to these as *conics* in this context) cannot be precisely represented by vector images based on polygons. Round boundaries are always approximated by many straight-edge pieces; the level of approximation varies from tool to tool. For example, the number of edge pieces for a circle can be defined as a parameter. A circle approximated by a polygon with eight corners is shown in Fig. 3.6c.

Aside from general polygons, there are two other custom shapes: the *rectangle* and the so-called *path*. They are special cases of polygons, which can be more efficiently modeled in the data structure due to their special properties. Given that rectangles and paths are by far the most common graphics elements in a typical layout, this advantage is fully utilized in layout representations.

Rectangles. A rectangle is a polygon with four sides and four right angles. If the sides of the rectangle are parallel to the axes of the Cartesian coordinate system used for the design (which is almost always the case), a rectangle can be modeled with only the coordinates of two diagonally opposed corners (Fig. 3.6d). The data volume can thus be almost halved. This efficient data structure matches the way a rectangle is created in the graphics editor, that is, by the digitization of these two opposing-corner coordinates.

Paths. Paths are polylines to which a specific *width* w is assigned. These shapes are typically employed for the fabrication of interconnects to provide the electrical current with a continuous, constant interconnect cross-sectional area. The center line of the interconnect is digitized in the editor and the required path width is set as a parameter. The digitized coordinates, which in the example in Fig. 3.6e are the coordinates C_1 to C_4, are stored in the data structure along with the path width w. The data volume can thus be approximately halved compared to standard polygons. Paths stored in this way are also easier to modify.

In addition to the path shape in Fig. 3.6e, there are other non-standard shapes whose appearances can be manipulated to meet unorthodox technology constraints. A typical example is the expansion of the thickness with diagonal path segments (some tool providers call these paths segments "padded paths"). This thickness expansion enables the corners of the diagonal path produced by the polygon to lie on the grid. This is a means of preventing rounding errors from occurring when the mask features are produced. The beginning and end of paths with non-standard shapes can be automatically extended. Given that these non-standard path features are also tool-dependent, we shall not dwell further on them here; instead, we recommend the reader to refer to the relevant tool manual.

Edges. It is important to note that modern design tools manipulate shapes as well as parts of shapes. Take, for example, the individual edge segments (C_i, C_{i+1}): these are addressable data items for these tools. This means that individual edges and path segments can be selected in a layout editor, as well. Useful graphics operations can be performed with these options during layout processing, as we shall explain in Sect. 3.2.3.

Hierarchical Organization of Layout Data

As indicated earlier, layout data are organized in a hierarchy, and this hierarchical organization mirrors the corresponding structural description. Each function block in the structural description—and thus each schematic—is a self-contained subset of the complete layout, which is also called a *layout block*.

The hierarchical layout structure is illustrated as a tree in Fig. 3.7. A layout block (B) can contain components and other layout blocks. The components are described generally in the layout as *cells* (C). Basic components are also called *devices*, whose internal circuitry is typically designed in the *front-end-of-line* (FEOL) in the technology (FEOL is discussed in Chap. 1, Sect. 1.1.3, and demonstrated in Chap. 2, Sect. 2.9.3). The cell shapes (c) therefore are assigned to the layers in the FEOL in the case of devices.

Furthermore, a layout block contains the shapes that form the interconnect layouts produced in the *back-end-of-line* (BEOL) during fabrication (BEOL is introduced in Chap. 1, Sect. 1.1.3, and described in detail in Chap. 2, Sect. 2.8). These shapes are labeled "net shapes" (n) in Fig. 3.7. In contrast to cells and blocks, whose structures are to be found at lower levels in the tree, net shapes are always part of a block. They are said to be "flat" data within a block.

This tree structure of an entire layout exists in both the layout engineer's conception of the design as well as the organization of the design data in the design environment. Here, every layout block (and often every circuit schematic) is typically stored in a separate directory containing specific file formats. The exact data organization depends on the tool used.

While the shapes in the relevant layers are the main focus of interest when generating masks, the layout designer normally works with this layout tree structure,

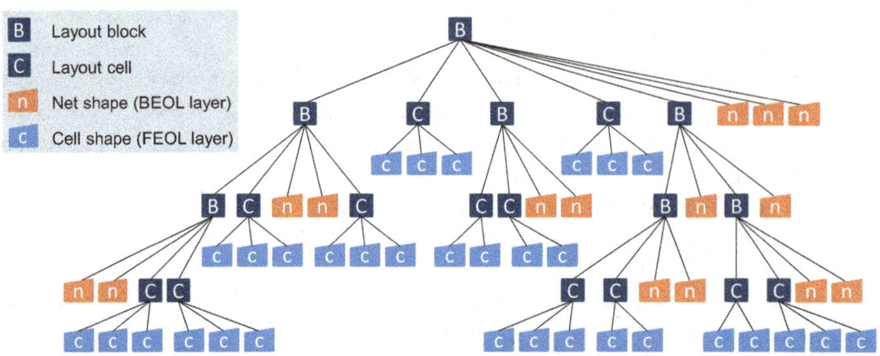

Fig. 3.7 Layout data structure derived from the hierarchical structural description. Layout blocks (B), which are a hierarchical subset of the complete layout, contain other blocks (B), cells (C) and net shapes (n)

and for good reasons. The physical design is considerably simplified by working at higher levels in the hierarchy, as it means the layout designer does not need to handle individual device shapes and parts of subblocks. What's more, thinking in functional units is supported by the hierarchy and the layout designer always has a clear insight into the circuit topology. This "global view" helps further the goal of an optimized final layout configuration.

Despite this facility, a layout designer must have a good grasp of the individual layers and the implications of combining them. He/she may also need to work on the polygonal level in certain cases (sometimes referred to as "polygon pushing"), or at least must take a closer look at it. We demonstrate this in the next section with some practice examples.

3.2.2 *How to Read a Layout View*

A small excerpt from a typical layout is shown in the top part of Fig. 3.8. We will next go through this example step by step, to learn how to "read" a layout.

The layout detail shown in the top of Fig. 3.8 is based on the CMOS standard process we discussed in Chap. 2 (Sect. 2.9). The bottom part of the diagram contains a sectional view of structures generated from this layout. (We use the same colors here as in Chap. 2, Sect. 2.9). If the layout view (Fig. 3.8, top) is a vertical view, the sectional view (Fig. 3.8, bottom) can be thought of as a "horizontal" view of the circuit, along the cutting line as identified in the layout (Fig. 3.8, top). Utilizing the

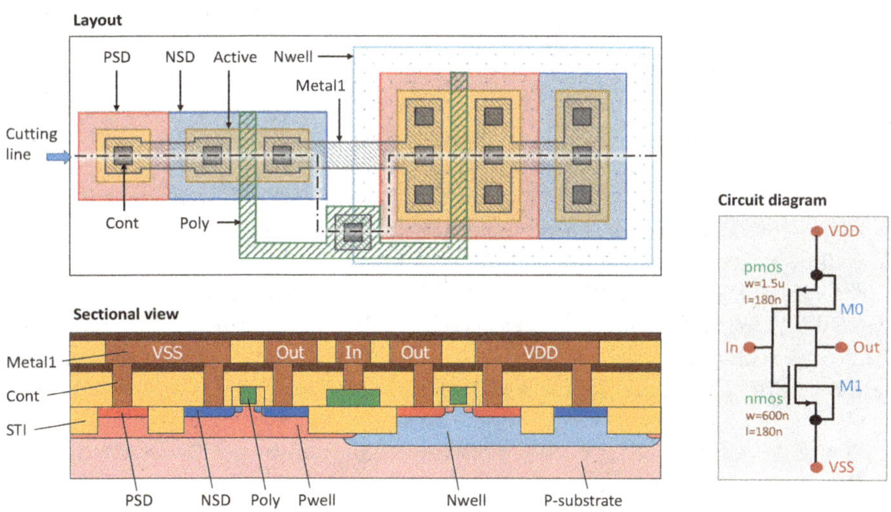

Fig. 3.8 The layout of a simple CMOS inverter (top), as shown in a typical layout editor, and the corresponding sectional view (bottom) and schematic (right)

circuit diagram (see Fig. 3.8, right), we see that the layout is a simple configuration comprising an NMOS transistor and a PMOS transistor. The sectional view shows that the connection points for these transistors on the silicon surface are in part interconnected and contacted with the first (bottom) metal layer (Metal1). This is, in effect, a "circuit" at this stage.

Using today's layout design tools, the engineer is only presented with a layout view (Fig. 3.8, top). As the tool does not generate or present the sectional view, he/she only sees the (two-dimensional) layout structure. Consequently, although the engineer has to "work" in two dimensions, it is useful—sometimes unavoidable—to "think" in three dimensions. "Reading" a layout means recognizing the devices and their electrical connections on the chip and imagining how they are shaped physically. Although this may initially seem a daunting task, there are techniques that can be learned that make it easier, as we discuss next.

As a first step, the devices need to be identified: this is done by examining the FEOL layers. We start by focusing on the drawn layer representing the "active" areas, which define sections of the chip surface without field oxide (shown ocher in Fig. 3.8, top). Normally, a drawn layer is assigned for these regions, but designations differ greatly from manufacturer to manufacturer. The drawn layer in question is called "Active" in our layout example in Fig. 3.8 (top). The mask "STI" (shallow trench isolation) is produced from this layer by negation, i.e., the shapes in "Active" define the regions that remain unaffected by STI.

In addition to these active surfaces, we are looking for polysilicon ("Poly", shaded green in Fig. 3.8, top) as both layers combined indicate a transistor. Specifically, wherever shapes from these two layers cross, there is a channel of a field-effect transistor (FET). It is often possible to identify most instances in a layout this way, given that FETs are by far the most common basic devices. The aforesaid applies to digital circuits and to most analog circuits.

Figure 3.8 shows that "Active" (ocher) and "Poly" (shaded green) cross in two places. We have therefore two FETs in our example.

Figure 3.9 depicts these two transistors separately (these transistors were introduced in Chap. 2, Sect. 2.9.3, see also Fig. 2.35e) by illustrating the layout and sectional views with labeled source (S), drain (D) and bulk (B) contacts. Bulk (aka *backgate*) contacts "belong" to the transistor layout as they define the potential of the well or the substrate, respectively.

The difference between NMOS-FETs and PMOS-FETs is that the bulk areas of PMOS-FETs are defined in processes with a p-substrate by a drawn "Nwell" layer (spotted pale blue, see Fig. 3.9, top right). The transistors outside Nwell areas are therefore NMOS-FETs. Bulks for NMOS-FETs are either the p-doped substrate of the wafer (for single-well processes) or the areas with a fabricated Pwell layer (twin-well processes). Either of these two scenarios could occur in our layout example, as the Pwell-doped areas could be derived from the drawn "Nwell" layer by negation (as we have seen in Chap. 2, Sect. 2.9.3) and would not then appear as a separate (drawn) layer in the layout.

Now that we recognize NMOS-FETs and PMOS-FETs, the dopant types (n or p) of the blue and red layers forming the source and drain areas should become clear

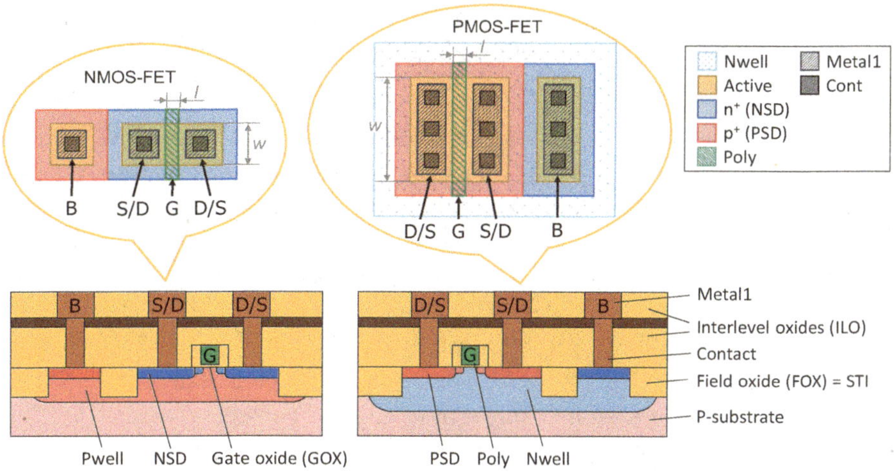

Fig. 3.9 The layout and the corresponding sectional views of the two transistors in Fig. 3.8 with marked contacts (D/S: drain/source, B: bulk) and gates (G). Transistors can be identified in any layout structure by focusing on crossings of the active areas (layer "Active", here depicted in ocher) and the polysilicon (layer "Poly", here shaded green)

as well. Additionally, n- and p-doping are reflected in the layer names: These n^+ and p^+ implanted layers are respectively labeled "NSD" and "PSD" in our example.

Finally, the BEOL layers, which connect the devices, are to be considered. Contacts and vias, for example, which are small, uniform squares in state-of-the-art processes, are generally easy to pick out. There are also metal features, which must always cover the contacts and vias and which form interconnects above the devices. In our layout example in Fig. 3.8, we have contact holes in the drawn "Cont" layer (dark gray) for contacting the source, drain and bulk regions. The layout of the interconnects are shown as the drawn layer "Metal1" (shaded bright gray). The gates, electrically connected by poly, have a common contact in metal, which is contacted to poly by the same "Cont" layer.

The two transistors are connected to form a logic inverter. The circuit schematic for the example is depicted on the right in Fig. 3.8.

3.2.3 *Graphics Operations*

A wide range of edit commands and graphics operators are available in modern layout editors. We shall only concern ourselves here with operators for manipulating and selecting shapes. Convenience commands for configuring a number of elements, such as the "Distribute", "Align", and "Compact" commands, are not dealt with here as they are well-known and intuitively understandable.

Interactive Shape Editing

Layout editors feature all of the standard graphics commands that we are familiar with in other drafting software. Shape commands that are generally available include "Add", "Delete", "Move", "Copy Paste", "Flip", and "Rotate". Layout editors also offer different user concepts specific to the layout process that make working with the tools easier. Data entry can be made with the mouse; numerical and text data can also be entered using the keyboard, and so on.

In addition to these standard functions, other commands are available for working with shapes when designing the layout:

- Stretching a shape by shifting a subset of its edges or corners ("Stretch"),
- Changing polygons by cutting out, truncating and attaching rectangles or more complex polygons (e.g., "Notch"),
- Merging overlapping shapes into one shape ("Merge"),
- Splitting polygons along (any) intersecting lines ("Split").

Logical Linking of Layers

Boolean operators from the field of mathematical algebra can also be applied to shapes in different layers. They are very important and powerful operators that "logically link" the "content" of these layers. While they are deployed sometimes in physical layout design, their main use is in the *design rule check* (*DRC*) to identify specific layout constellations for checking (Sect. 3.4 and Chap. 5, Sect. 5.4.5) and in the *layout post process* (Sect. 3.3) to produce mask data. We demonstrate the following, general standard logic operators in Fig. 3.10:

- OR: produces the geometrical union of two layers.
- AND: produces the geometrical intersection of two layers.
- XOR: produces the union minus the intersection of two layers.
- ANDNOT: generates the "geometrical difference" between two layers. Everything that is in the second layer is "punched" out of the content of the first layer.

The upper portion of the figure shows a simple sample layout. This layout consists of four rectangular shapes, two of which belong to a "red" layer and the other two to a "blue" layer. The results of the operations are written in a new layer "x", shown in gray at the bottom of Fig. 3.10.

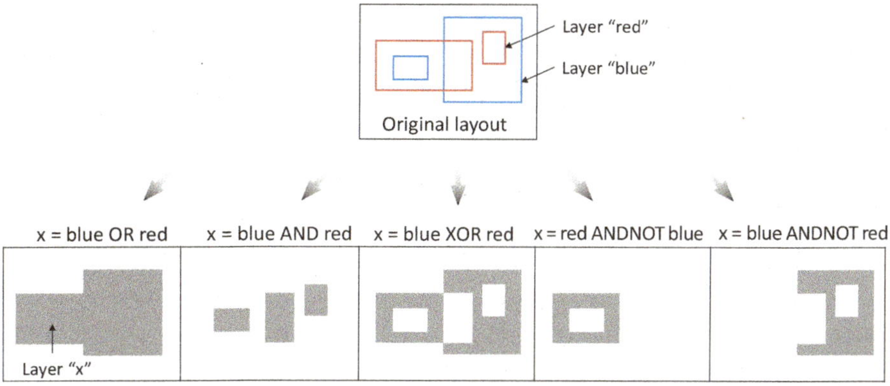

Fig. 3.10 Logical linking operations applied to four shapes on two layers (top) using the standard logic operators OR, AND, XOR, and ANDNOT (bottom, left to right)

Select Operations

Shapes that satisfy a specific criterion in a layer can be picked with select commands. In Fig. 3.11, we demonstrate some key selection criteria based on specific relationships between the shapes in the specified layers:

- INCLUDE: Selects shapes in a layer that overlap in any way with shapes in another layer.
- OUTSIDE: Selects shapes in one layer that do not overlap with shapes in another layer.
- INSIDE: Selects shapes in a layer that are fully covered by shapes in another layer.
- ENCLOSE: Selects shapes in a layer that fully cover shapes in another layer.
- CUT: Selects shapes in a layer that share a portion of their surface area (but not their entire surface area) with shapes in another layer.

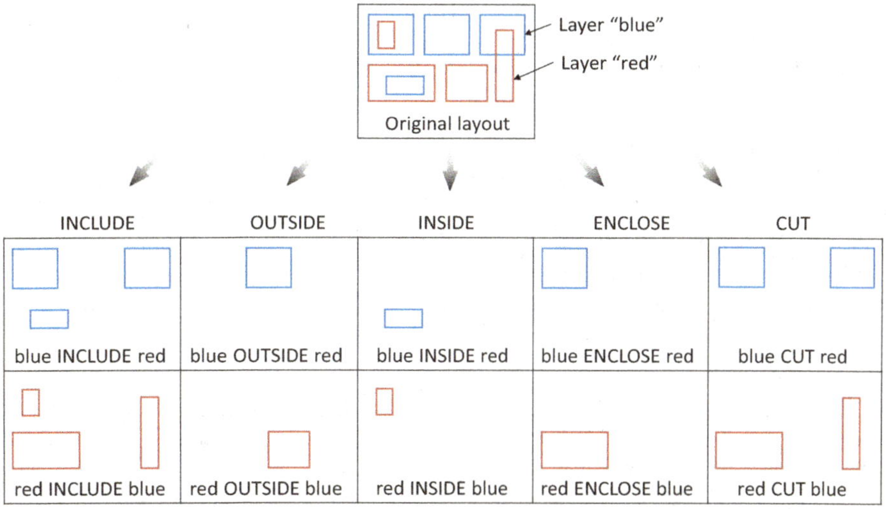

Fig. 3.11 Selection commands for filtering shapes from layers based on geometrical relationships

In contrast to the linking operations, no new geometries are created by the selection commands; instead, existing geometries that meet the criteria are selected (i.e., identified). The results can be saved in a new layer as required. The selection commands are of great interest for the DRC, as layer subsets of interest can be identified this way (Sect. 3.4.3, Example 1).

Sizing Operators

We introduced the *sizing* operator in Chap. 2 (Sect. 2.4.2) when we discussed preemptive edge shifts. (The structure's boundary lines are shifted outwards by a specific value or inwards in order to compensate for shrinking/enlarging effects that can occur in subsequent structuring process steps.) Sizing is also very useful in the DRC to check layouts for compliance with more complex design rules (Sect. 3.4.3, Example 2). Furthermore, sizing can be applied to "clean up" layouts, as we shall explain next.

A polygon is modified with the sizing operator by shifting all edges perpendicular to the edge alignment by a specific value. The polygon is enlarged if the edges are shifted by positive values; this operation is called *oversizing*. Whereas the polygon shrinks if the values are negative; this latter operation is called *undersizing*.

Sizing has several noteworthy properties that must be understood, as they can produce unfortunate and unexpected results if you are not aware of how they work. Having said that, you can also leverage these properties to produce specific effects that are helpful. We will take a closer look at these effects now.

Uneven growth. Oversizing by a value s causes the corners of a polygon to be shifted by a distance $v \cdot s$, where $v > 1$, i.e., the *corners* are always shifted from their original positions by *more* than the shift value s, e.g., $v = \sqrt{2}$ for right angles. For acute angles (angles $<90°$), the value v is greater than $\sqrt{2}$ and can theoretically be extremely large.[1] This effect is a generic defect in sizing, as "even" growth in all directions is the desired outcome in most cases (Fig. 3.12, left).

Ideally, we would like circular arcs at the corners (a circle defines a set of points with identical distances to the corner). But as we know, arcs cannot be modeled in the data structure. Oversizing can however be configured in some tools such that the corners can be "beveled" with additional edges to approximate a circular arc as per Fig. 3.6c. Two examples are shown in Fig. 3.12 (right).

Rounding error. Another difficulty with the sizing operator is that, in the case of inclined edges (see Fig. 3.12, bottom), the corners are not placed on the grid. This causes rounding errors because integer values are used for the coordinates. While these rounding errors can often be ignored, the angles w.r.t. the coordinate system may be altered (Fig. 3.13c), producing unwanted results in some scenarios. For example, design rules may be violated by this effect, which would not have occurred with mathematically correct sizing.

[1] Acute angles are often not allowed in layouts because the edges in question are treated as being "opposite" by DRC tools and are flagged as width rule violations. Even if this is not a problem for fabrication, these cases should be avoided to minimize the work involved in evaluating a DRC.

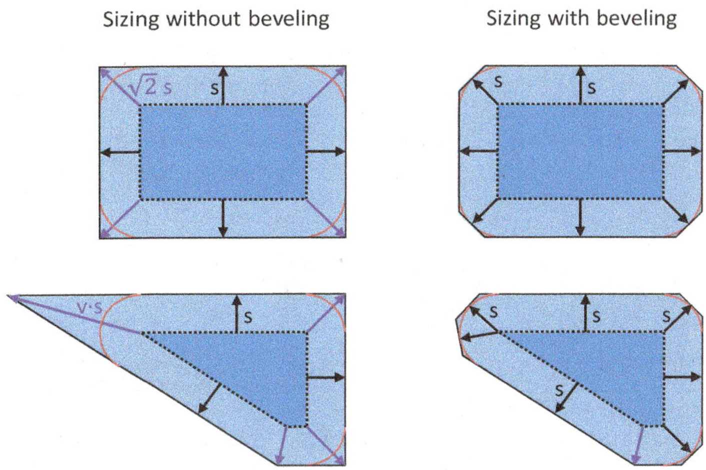

Fig. 3.12 Oversizing without (left) and with beveling corners (right)

Irreversibility. If two sizing operations are performed immediately one after another, with the same value but in opposite directions, the final result may not be the same as the original structure. There are a number of reasons for this, as illustrated in Fig. 3.13:

- Small polygons disappear during undersizing (also narrow ribs), Fig. 3.13a,
- Small holes in polygons disappear during oversizing, Fig. 3.13b,
- Rounding errors cause shape changes, Fig. 3.13c.

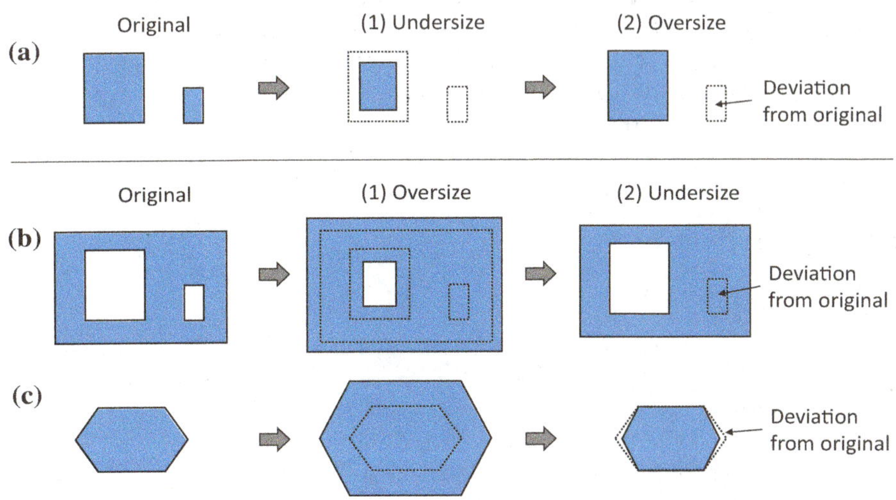

Fig. 3.13 The effects of two sizing operations with the same absolute value but different signs, i.e., sizing in opposite directions, that can produce polygons which deviate from the original shape

Cleaning up layouts. If sequences of sizing operations and logical links are used to create certain layout structures in the layout post process or in the DRC, these steps may produce unwanted shapes caused by rounding errors. These unwanted shapes are often very small. Hence, the depicted effects in Fig. 3.13a, b could be used to positive effect to eliminate such small artifacts.

3.3 Mask Data: Layout Post Processing

3.3.1 Overview

Following layout completion and final checking, there are still tasks that must be performed before the masks are produced. Some of the data need to be deleted; while other data must be altered; and new data has to be produced. These amendments and additions to the chip layout data are performed in *layout post processing*, which we divide into three stages:

(a) *Chip finishing* (Sect. 3.3.2),
(b) *Reticle layout* (Sect. 3.3.3), and
(c) *Layout-to-mask preparation* (Sect. 3.3.4).

These process stages and the data generated are summarized in Fig. 3.14. The numbering (1) to (8) indicates the order (steps) in which the contents are produced. The operations typically vary from company to company and can differ from process to process, too. The aforesaid applies to the choice of terminology, as well. Hence, our generalized description provides a path towards standardization in this regard.

In this treatise, we only describe typical operations with the key steps and data. In the real world, these processes will definitely be different in the certain cases. Compiling mask data from the layout data requires engineers with significant experience, and is carried out in all companies by specialists.

3.3.2 Chip Finishing

Product Structures

After the integrated circuit (i.e., the electrically active part of the chip) has been laid out, custom data that designates the product are introduced to the design (step 1 in Fig. 3.14). The content is normally integrated on the active chip surface; surface space must be available for this information. Chips generally contain an image of the company logo and a design designation, most often a chip name. Copyright information may also be included, to mark intellectual property. This information is typically integrated in the top metal layer so that it is easily visible. We have shown this data in brown in Fig. 3.15.

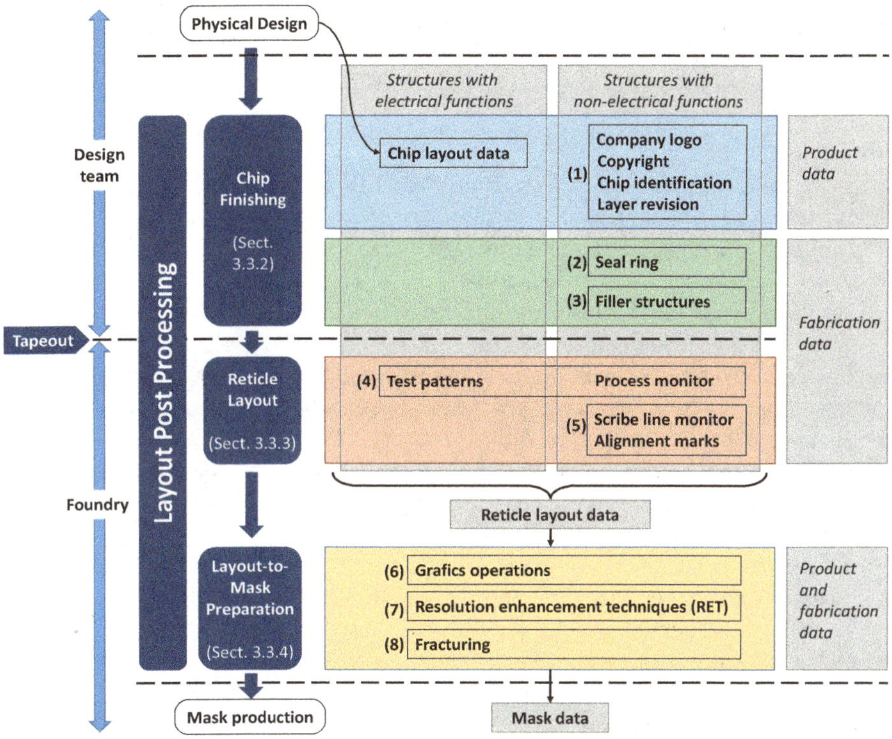

Fig. 3.14 Layout post processing for converting a chip layout into data for mask production, subdivided in three process phases (left) and eight process steps

Fig. 3.15 Custom designations (brown) for a design and the position of the seal ring (blue) on a chip

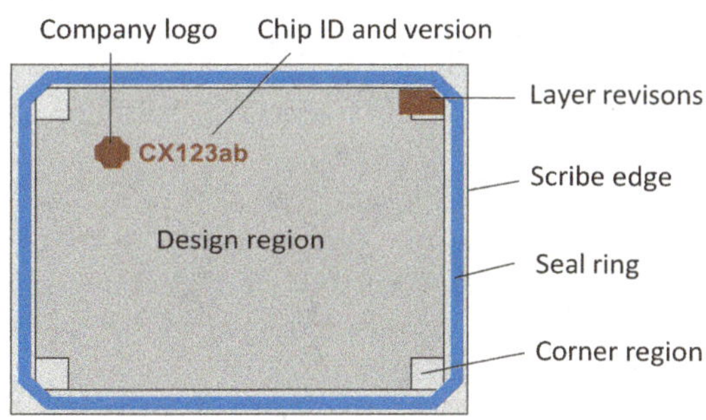

Structures with information on the design revision are generally implemented at the edge of the active surface area. They are either readable (e.g., numbers) or are encoded for a particular company. Structures are also entered at this stage that specify the revision level for every photomask. This is sensible as a new design revision does not typically mean all layers have been upgraded. Sometimes it is only necessary to change a few layers, such as layout changes in a subset of the routing-layer pairs (i.e., metal layer and corresponding via layer).

Fabrication Structures

Furthermore, additional structures are produced that are of significance for chip manufacturability. These measures are often labeled as *Design for Manufacturability* (*DfM*), which encompasses, among others, the creation of the *seal ring* and the *filler structures*—also called *dummy fill*.

The active part of the IC chip ("design region") is enclosed by the seal ring (step 2 in Fig. 3.14), shown schematically in Fig. 3.15. It ensures there is a clearance between active structures and the sawing trench and thus the later chip boundary. The chip boundary is very uneven as a result of a combination of sawing and breaking. It is therefore very susceptible to moisture penetration. The purpose of the seal ring is to protect the chip interior against moisture entering from the sides. It also protects the active structures against damage caused by sawing. This is why it is also called a *scribe seal*.

The internal design of a seal ring is very complex. It typically contains a stack of all processed metal layers. The layout structures in the layers are specified by the fab and, hence, are confidential.

Finally, filler structures are integrated in the entire chip area in a fully automated process (step 3 in Fig. 3.14) to improve the planarity achievable with the CMP process (chemical mechanical polishing). This concerns all layers that are planarized in CMP steps during fabrication. The material mix in these layers must be as homogeneously distributed as needed for the CMP process. If this is not the case, the material abrasion will be non-uniform, and may result in indentations and dents in the chip.

We explain these effects with an example of the Damascene process described in Chap. 2 (Sect. 2.8.3). In this example, excess copper is to be removed by the CMP process. The chemically active component in the slurry is selected so that the CMP removes copper only and not oxide, which we want to leave intact as far as possible. Some oxide is removed as well due to the mechanical CMP component. This phenomenon is called *erosion*. It is critical that the erosion depth be the same everywhere in order to maintain as planar a surface as possible.

The said effects are visualized in Fig. 3.16. Given that the erosion in regions with no copper is too weak (d), this effect is countered by introducing additional *dummy fill* copper structures. In other areas, where there is too much copper, too much pressure is applied to the oxide ribs, which are then excessively stripped (b). Consequently, care should be taken at the physical design stage that the copper density is not too high. In addition, slots must be added in the layout for very wide interconnects (a). This procedure is known as *slotting* and is equivalent to inserting oxide ribs. This measure impedes *dishing*—the removal of excessive copper (a).

The added "filler structures" serve no electrical function, and as such must be generated so that the electrically active structures nearby are not negatively affected. For example, the dummy fill structures inserted in the metal in our example should not cause a short-circuit anywhere. To ensure that no such defects have been introduced,

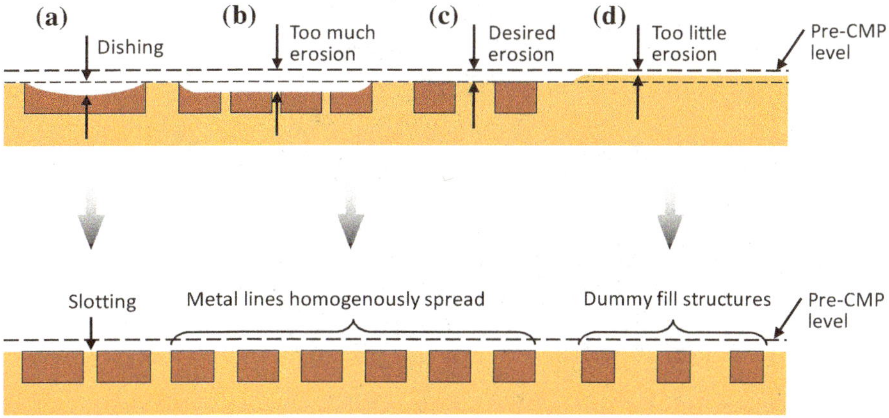

Fig. 3.16 Dishing and erosion caused by copper CMP (above) and remedial measures (bottom). Dummy fill structures are integrated during chip finishing

the complete design after chip finishing should be verified again before the layout is handed over to the fab. This verification encompasses a DRC and an LVS check (Chap. 5, Sect. 5.4).

It is important that data generated during chip finishing be stored separately from the layout data, i.e., data that describe the electrically relevant chip structures. We recommend storage in separate cells; typically, there are guidelines available on what type of data hierarchy should be used and how the cells should be named. This facilitates the use of automated processes in chip finishing. Furthermore, it simplifies redesigns, where filler structures must be removed prior to any layout modifications.

3.3.3 Reticle Layout

When a chip series is to be manufactured, the individual die is instantiated several times in the form of a matrix to produce what is termed the *reticle layout*. The reticle layout is comprised of vertical and horizontal scribe lines that separate the individual dies that have been placed in a matrix format, as shown in Fig. 3.17. The size of this matrix depends on the maximum exposable surface area based on the reticle size (cf. Fig. 2.3 in Chap. 2).

For a production run of only a few chip prototypes, the reticle surface can be used for different chip designs. The mask costs can thus be spread among the designs. A reticle layout with nine dies for a batch production run is pictured schematically in Fig. 3.17.

In order for the dies to be singulated after the wafer process, there must be sufficient spacing between neighboring dies. The dies are singulated with a diamond saw or with a laser beam. They cut notches between the dies that enable the individual dies

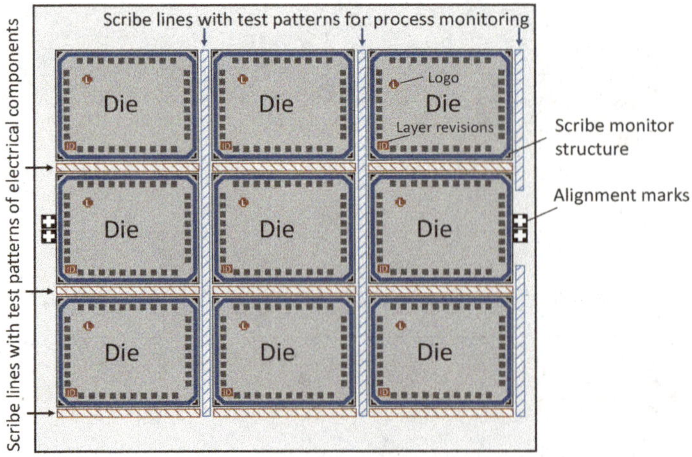

Fig. 3.17 Schematic of a reticle layout for batch production

to be broken apart. The spacing between dies ranges from 50 to 100 μm, depending on the saw or beam, and is called a "sawing trench", "saw street", or "scribe line".

Test patterns for use during wafer fabrication only are placed in the scribe line (step 4 in Fig. 3.14). There are two types of test patterns, the first of which consist of test patterns for monitoring the fabrication process (the blue hatched area in Fig. 3.17). These are features used to check individual process steps for compliance with manufacturing tolerances. This type of quality control could encompass optical checking of feature sizes or measuring electrical resistance values with needle tipped probes. If any faults are found, it may be possible to rectify them in some scenarios.

The second type of test patterns consist of complete electrical basic devices or cells. They can only be analyzed when the FEOL and the first metal layer have been completed (brown hatched area in Fig. 3.17). For positioning, the test patterns are often sorted according to their types. One type is assigned to the vertical and the other type to the horizontal scribe lines. Tests using these patterns will be performed using the entire wafer, before it has been cut into individual dies.

Sometimes all of the desired test patterns cannot be entered in the scribe lines in the case of very large dies because of the lack of scribe line space. There are two ways of overcoming this issue. (i) A sawing trench may be widened to create space for extra test patterns; unfortunately, this uses up wafer surface. (ii) Alternatively, some test patterns may be purposely omitted to save wafer surface, i.e., to increase the chip yield per wafer.

After scribe lines and test patterns have been defined, alignment marks are also placed in the scribe lines on opposite sides of the reticle (step 5 in Fig. 3.14). These alignment marks are required to align mask structures and wafer structures during exposure. We have discussed this operation in detail in Chap. 2 (Sect. 2.3.4).

Finally, markers for the diamond saw or the laser beam are introduced. They ensure the blade/beam always cuts exactly in the sawing trench.

3.3.4 Layout-to-Mask Preparation

Once the reticle layout has been produced (Sect. 3.3.3), it must be converted to mask data, which will be used by the fabrication systems to generate the masks. The mask data are generated in a fully automated layout post process comprised of three stages (Fig. 3.14, steps 6–8), summarized below and described in detail in the paragraphs that follow.

(a) First, the layout data are amended with *graphics operations* (Fig. 3.14, step 6).
(b) The graphical data are then subjected to measures, such as *resolution enhancement techniques* (*RET*), which enhance the optical resolution (Fig. 3.14, step 7).
(c) Finally, a *fracturing* step is performed that adapts the data for the mask production devices (Fig. 3.14, step 8).

Graphics Operations

To better understand the graphics operations, we first compare the input data (reticle layout) with the output data for the layout post-processing step:

- The input data are first converted to a *flat* data structure.
- The output data will include extra layers not contained in the input data. We call these additional layers *derived layers*.
- The input data also contains layers that will not be present in the output data, as they will be deleted. We call these (input layout) layers that will be deleted *logical layers*. The remaining layers used for generating masks are called *physical layers*.
- The shapes in the output data will be altered in some physical layers.

Some individual content in the reticle layout may be ready for the second stage (b) (Fig. 3.14, step 7) and even for the third stage (c) (Fig. 3.14, step 8) in mask preparation, depending on the fabrication process. This content can be excluded from the modifications of the graphics operations. The layout data from the chip layout is always subjected to graphics operations.

When the layout data have been converted to mask data, functional layout data structures, such as blocks and cells, are no longer used. Instead, only the layer association of the geometrical structures is relevant for fabricating the photomasks. The data hierarchy is removed for this reason and all shapes are sorted according to their layers. Now that the hierarchy levels have been removed, the resulting data are called “flat” data.

Derived layers. For some layers, mask geometries can be derived from the data in other layers, as we describe in the example below. In such cases the layout designer can ignore this type of layer in the physical design, which reduces the workload. Derived layers are automatically generated at the graphics operations step during the layout-to-mask preparation stage. We use an example next to illustrate derived layers.

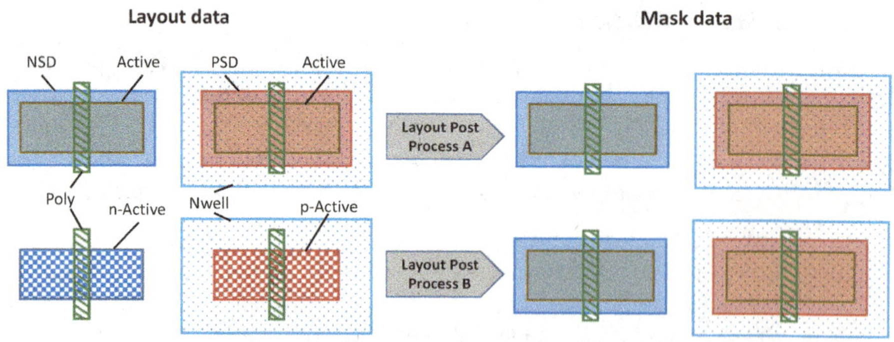

Fig. 3.18 Producing mask data of NMOS- and PMOS-FETs. Layout post process B (bottom) illustrates a layout simplification by using graphics operations where the three mask layers "Active", "NSD" and "PSD" are derived with graphics operations from the two layout layers "n-Active" and "p-Active". The shown layers are not processed in layout post process A (top)

Figure 3.18 shows a typical example of how the active areas in an NMOS- and a PMOS-FET are produced. In process A (top), all layers are used in the layout as they should appear later on the mask. (This process is similar to the one shown earlier in Fig. 3.8.) In Fig. 3.18, process B (bottom), the three mask layers "Active" (opening the field oxide), "NSD" (n-implantation) and "PSD" (p-implantation) are derived with graphics operations from the two layout layers "n-Active" and "p-Active". Here, we are using the logical linking operations (illustrated previously in Fig. 3.10) and sizing operations (illustrated in Fig. 3.12):

- "Active" is derived from the logical operation "Active = n-Active OR p-Active".
- "NSD" and "PSD" are derived from "n-Active" and "p-Active", respectively, by oversizing by a value k with the commands: "NSD = SIZE (n-Active, k)" and "PSD = SIZE (p-Active, k)".

The same result is achieved by process B with one layer less than process A and with fewer shapes in the physical layout design. The value k is based on the maximum overlay error for the process (i.e., the tolerance of photomask alignment, Chap. 2, Sect. 2.4.1).

Logical layers. Additional layers in the layout, which are not needed for generating the masks, are required for the automatic geometrical design rule checks (we will deal with these next in Sect. 3.4). Modern fabrication processes are very complex and thus have complex design rules. Rules can be based on the electrical function of the given structure, and therefore may apply to individual layers. Hence, this function must be made identifiable for the DRC; identification is provided by means of logical layers.

For example, a structure in the "Poly" layer can serve as a gate, a resistor, or a capacitor electrode. Upon careful consideration given the relatively high layer resistance, it may also be used as an interconnect. Let us assume the structure is a

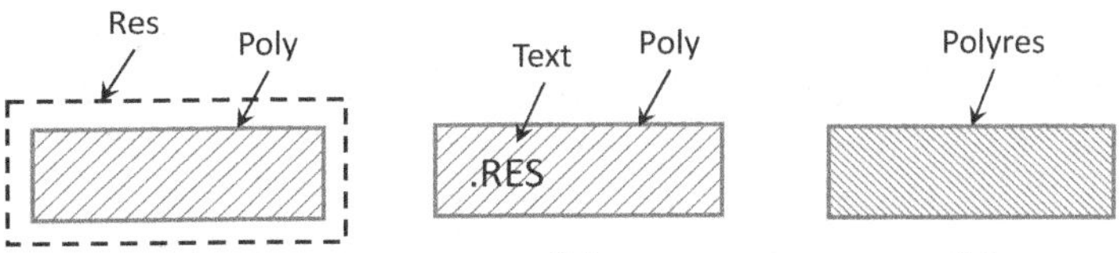

Fig. 3.19 Options for designating the electrical function of a layout structure with logical layers "Res" (**a**), "Text" (**b**), or with a designated physical layer "Polyres" (**c**)

resistor. Figure 3.19 depicts three typical ways this electrical function of the Poly shape can be made automatically identifiable:

(a) We define a layer "Res" and draw another shape that encloses the Poly structure.
(b) We use a layer "Text" to introduce written instructions in the layout. We can place an explanatory text string "RES" over the Poly structure. (Texts by definition are not shapes and cannot appear in the mask data. In effect, they take up no space.)
(c) We set up a separate drawn layer (in this case, "Polyres") for every electrical function that can use a structure in a fabricated layer, and draw the structure in this layer.

The type of designation selected depends on what is supported by the deployed tools or rule files.

The "Res" and "Text" layers in examples (a) and (b) are called *logical* layers. As mentioned previously, these layers are not used for generating the masks. Mask-generating layers are called *physical layers*. "Polyres" in our example (c) is a physical layer. The fabricated layer "Poly" in our example can be created in the post process by the union of all drawn physical layers that contain Poly structures, representing resistors, gates, capacitors and Poly interconnects, with the logic operation:

"Poly = (Polyres OR Polygate OR Polycap OR Polyline)".

In addition to the electrical function, design rules often depend on the electrical potential applied to a structure (this topic will be discussed in more detail in Chap. 6, Sect. 6.2). This is another example of how additional layers are used in the layout to inform the DRC of the electrical potential applied to a structure. Typically, voltage classes that are marked with appropriate logical layers are used in these cases (similar to the example above).

Altered layers (pre-sizing). In Chap. 2, we saw how shapes on a mask were enlarged or reduced in size and mapped to structures on a wafer during fabrication. This resizing can be described as an outward edge shift (oversize) or inward edge shift (undersize), based on the polygonal shape. This edge shift by a value k during fabrication can be compensated by the process graphics operators by shifting the shape edges in the layout data by an equivalent negative value $-k$. We called this a "pre-sizing" operation in Chap. 2 (Sect. 2.4.2) and provided an example of how it works. Pre-sizing aims to present the structures in the layout as they appear later on the processed wafer.

Resolution Enhancement Techniques

Nowadays, the exposure wavelength in cutting-edge processes is greater than the smallest structures being exposed. To overcome this reversal (and "physical contradiction"), sophisticated *resolution enhancement techniques* (*RET*, also referred to as *resolution enhancement technologies*) are required which ensure that an optical resolution is obtained nevertheless (step 7 in Fig. 3.14). One of these techniques is *optical proximity correction* (*OPC*), whereby the layout structures are altered so that the intensity distribution of the light striking the photoresist approximates as closely as possible the structures drawn in the layout. Two OPC measures have been demonstrated in detail in Chap. 2 (Sect. 2.4.3).

Numerous other RET are available for increasing the optical resolution. Options that are frequently deployed industrially include the use of phase-shift masks and multiple exposure techniques (Chap. 5, Sect. 5.5). Another development is the usage of extreme short-wave UV radiation (EUV) which requires reflecting masks and mirrors instead of lenses. The necessary specific manipulations of graphics data are to be performed in the layout-to-mask preparation step, as well.

Detailed descriptions of RET are outside the focus of this book; a good reference for such techniques is [8].

Fracturing

In fracturing (step 8 in Fig. 3.14), graphics data are converted from the polygonal shape used in the physical layout design (as explained in Sect. 3.2.1, see Fig. 3.6) to a shape required by the hardware for mask fabrication. Masks are produced using photolithography, in a manner similar to how wafers are structured, where the radiopaque chrome layer is structured by means of coating, exposing, developing and etching. Two primary techniques are available in the process for the masked exposure: (i) deploying adjustable apertures, or (ii) having the system "write" directly with a focused electron beam.

Two adjustable apertures aligned orthogonally to one another form the basis of the aperture system (i). Polygons must be converted to simple rectangles to serve as input in this procedure. The apertures are adjusted according to the lengths of the sides of the rectangles.

In the direct writing technique (ii), the electron beam sweeps the mask to be structured one line at a time; it is switched on and off in the process. The polygons in the graphics data must be converted to scan lines to guide this technique.

3.4 Geometrical Design Rules

3.4.1 Technological Constraints and Geometrical Design Rules

We refer to the limitations and capabilities of a fabrication approach as the *technological constraints*. The purpose of *geometrical design rules* is to *model* these technological constraints for use in the design and verification of the layout. (The aforesaid also applies to PCB design.) Hence, the geometrical design rules are constraints for the physical layout design, whose compliance assures the manufacturability of the layout results (see Fig. 3.1).

Compliance with the geometrical design rules (we shall refer to them as *design rules* or *rules* throughout the rest of this section) is mandatory in physical design. While optimization goals (e.g., aiming for the smallest possible chip surface area) have no fixed values and are thus seen as "soft" criteria, the design rules have defined boundary values, and are thus "hard" criteria.

A design rule check (DRC) verifies a layout for compliance with geometrical design rules. If no defects are found, the layout is marked as *error-free*. This verification step is crucial in electronics manufacturing. Applied to IC design, a positive result means the layout data can proceed to the tape-out phase (cf. Fig. 3.14), meaning it is ready to be manufactured w.r.t. technological constraints.

There are several issues with this definition of "error-free". If, for example, a specific configuration of layout structures cannot be fabricated as intended, despite having completed a positive DRC, this is by definition not a layout error. This means that the layout is not the cause of the problem, rather than the (not met) technological constraint which has not been correctly modeled in the corresponding design rule. This design rule is called *non-robust*. Hence, a design rule that models a technological constraint correctly is *robust*. Furthermore, we call a layout *robust* when it can be safely produced in spite of manufacturing tolerances. From this, we can see that for a layout to achieve *robustness*, it must be error-free *and* the design rules must be robust. We will demonstrate this finding with some examples in the following Sect. 3.4.2.

On the other hand, an infringement of a design rule remains a layout error even if fabrication can proceed despite this rule violation. There are several scenarios where this type of event cannot be avoided (Chap. 5, Sect. 5.4.1).

3.4.2 Basic Geometrical Design Rules

A permissible value range is defined in a simple design rule for the dimension (often referred to as the *structure size*) of a specific layout feature. The following range options are available:

(a) $structure\_size \geq min\_value$,
(b) $structure\_size \leq max\_value$,
(c) $structure\_size = exact\_value$.

The constants *min_value*, *max_value* and *exact_value* are technology-specific. Design rules typically define minimum values (case a), such as a minimum wire width. In rare cases, a maximum value may be prescribed (case b) or even an exact value (case c). A value range can be specified as well by design rules by combining case (a) and case (b).

Generally speaking, *structure_size* represents the spacing between opposing polygon edges. Consequently, this parameter can be combined in multiple ways: the edges can belong to the same polygon or to different polygons. In the latter case, the polygons can belong to the same layer or to different layers. Finally, the dimension specified can refer to the inside or the outside of the polygon in question. Basic design-rule groups are derived from these combinations, which are given in Table 3.1. Their names describe the type of layout arrangement addressed by the rule.

Several typical examples of the basic design rules noted in Table 3.1 are illustrated in Fig. 3.20. The polygon edges referred to by the rules are colored red. The case when two edges are "opposite" needs to be further defined for these rules to be applied meaningfully. Edges are said to be opposite if the angle between them is smaller than 90° for the area being measured (shown by arrows in Fig. 3.20). In other words, they are considered opposite if they are parallel to one another or the angle between them is acute. If this is the case, the relevant rule is applied and checked (Fig. 3.20, top); otherwise not (Fig. 3.20, bottom).

If opposite edges are in contact (which can only happen with acute angles), this typically signals a fault with width and spacing rules, as both rules require a minimum distance of the edges in consideration. These cases are illustrated by dark blue shapes in Fig. 3.20.

Cases where opposite edges at acute angles to each other do not touch, also require attention. Width and spacing measurements for these layout features depend on whether a rule requires a minimum or maximum value. The latter is indicated by dash-line arrows in Fig. 3.20.

It is important to note that geometrical design rules that are related to only *one layer*, i.e., width and spacing rules, are derived from the resolution capability of

Table 3.1 Basic design-rule groups (left) cover various edge, shape, and layer combinations

Design Rule	Relationship between edges	Number of shapes	Number of layers
Width	Inside/inside	1	1
Spacing	Outside/outside	(a) 1 or 2 (b) 2	(a) 1 (b) 2
Extension	Inside/outside	2	2
Intrusion	Inside/inside	2	2
Enclosure	Outside/inside	2	2

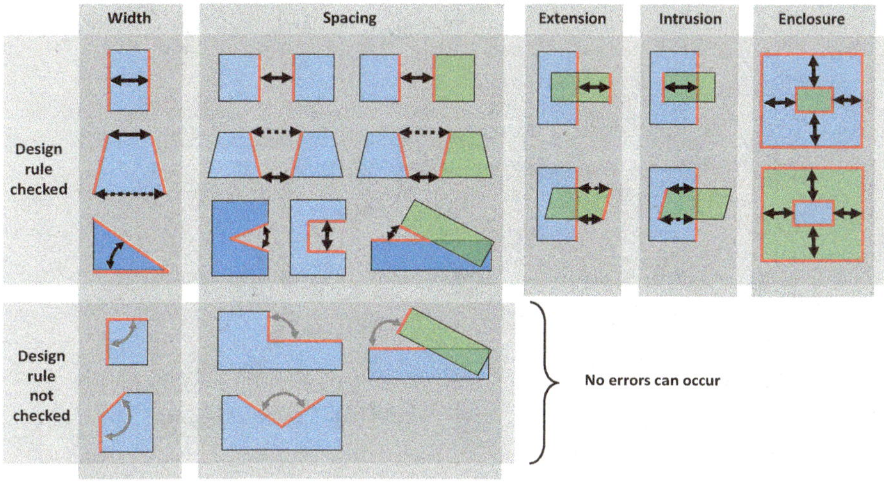

Fig. 3.20 Illustrations of basic geometrical design rules (width, spacing, extension, intrusion, and enclosure). While width rules relate to one layer, spacing rules can consider one or two layers. Extension, intrusion and enclosure rules (the first two sometimes loosely defined as "overlap rules") relate two different layers

the semiconductor process. Specifically, these rules are determined by the smallest manufacturable feature sizes.

Example 1

Let us now visualize these one-layer rules with an example. A simple layout instance of a metal layer is shown in Fig. 3.21. The layout comprises two parallel interconnects. Two different layouts are drawn on the left; both are error-free. In both cases, the interconnects are generated with minimum width and minimum spacing according to the respective design rules. In the upper case, the design rules are "non-robust", i.e., the technological constraints have not been correctly modeled in the corresponding design rules. In the lower variant, the design rules are "robust", i.e., they model the technological constraints such that the layout can be safely produced in spite of manufacturing tolerances.

The two variants clearly demonstrate the consequences of unavoidable manufacturing tolerances, which can lead to deviations in "two directions". These consequences are illustrated (i) with interconnects in the center column of Fig. 3.21 that are smaller than the nominal size and (ii) interconnects that are wider than the nominal size in the right column. Applying non-robust design rules risks a break in the first case, and can cause short-circuits in the second case. These critical faults are avoided by using robust design rules as shown in the lower variant.

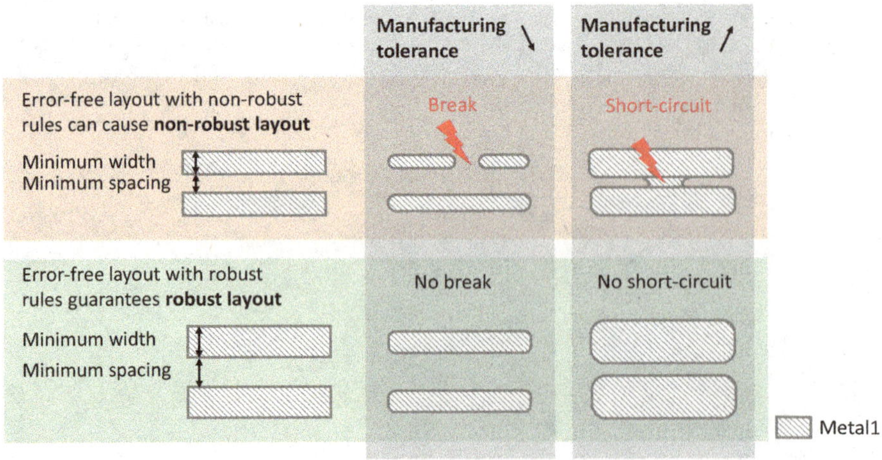

Fig. 3.21 Layout example for design rules that apply to one layer. The consequences of tolerances, which can lead to smaller (middle) or wider wire widths (right), depend on the robustness of the imposed design rules (left). A robust layout can be safely produced in spite of manufacturing tolerances

Example 2

The design rules that involve *two layers* relate to the maximum overlay error that can occur (Chap. 2, Sect. 2.4.1). As such, these rules are based on the tolerances for mask alignment and exposure.

An example of an NMOS-FET layout is shown in Fig. 3.22. Two different layouts are pictured on the left. Again, we assume both layouts to be error-free (i.e., they comply with the respective rules). The top layout is non-robust due to non-robust rules. In the bottom case, the layout is robust because the "Enclosure" and "Extension" overlap rules, which are critical for overlay errors, are formulated in a robust manner. The results for three different overlay errors, i.e., variously misaligned masks, are depicted in the three columns. Different faults associated with the top variant are shown in the figure. Again, these errors can be avoided by adhering to the robust rules.

The examples in Figs. 3.21 and 3.22 all relate to design rules that require a minimum value. However, single layer design rules in particular can also have maximum values. These include constraints relating to the CMP process, as previously discussed (and illustrated in Fig. 3.16). Restricting the width of metal structures to avoid the dishing effect is a prime example. Upper limits are often applied to the spacing in the "Active" layer, given the embedded STI structure produces mechanical stress in the silicon crystal. This in turn impacts charge carrier mobility and, hence, the conductivity of the doped regions.

In Fig. 3.13, we demonstrated a procedure for eliminating "small" shapes by undersizing and oversizing by the same constant in series. This is an elegant solution for checking for maximum widths. The shapes of interest can thus be very easily localized (and then corrected) by setting the sizing value to *max_value*/2 and by performing an "undersize" followed by an "oversize" in the respective layer.

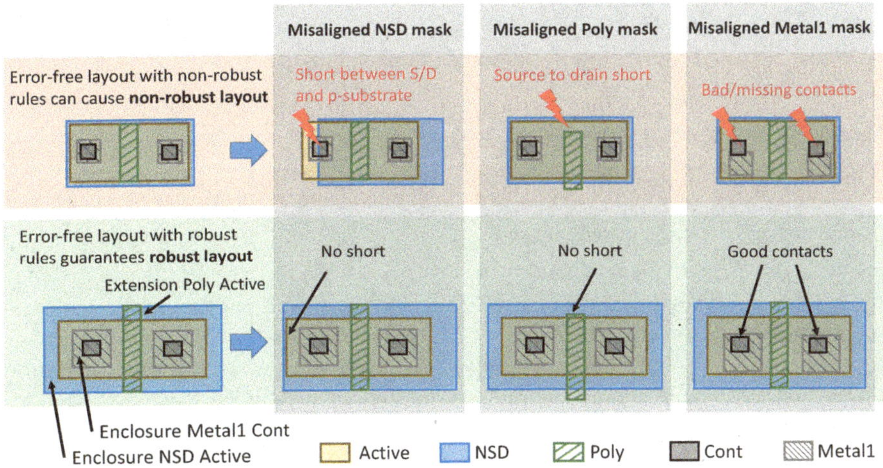

Fig. 3.22 Layout example for rules that apply to two layers. The consequences of misaligned masks are effectively mitigated by using robust "Enclosure" and "Extension" overlap rules, i.e., these rules model the respective technological constraints correctly (bottom)

Contacts and vias in state-of-the-art processes are typically defined by design rules that are not based on an inequality (i.e., parameter ranges), but that define an *exact* width. The reason for this is that the structuring process for the intermediate oxides are very precisely optimized for this size. If a larger cross-sectional area is needed for the current flow between two metal layers, this can only be achieved by using more vias, also called "multiple vias" or "via arrays" [4]. The same applies for contacts.

3.4.3 Programmed Geometrical Design Rules

In addition to the layout arrangements that have been effectively determined by the "basic" design rules we have discussed up to now, there are often many layout constellations (i.e., specific combinations of shapes in multiple layers) whose technological constraints require a more complex description. In these cases, the layout constellations of interest are first extracted by the graphics operations described in Sect. 3.2.3, such as "select", "sizing", and "logical linking operations". Any number of intermediate results may need to be produced until the geometrical scenario to be verified is present in a so-called "calculated layer". One of the basic geometrical design rules (Sect. 3.4.2) is then "run" on this result, to identify any design rule violations.

Design rules are said to be "programmed" in such cases. Modern tools have many functionalities for programming design rules. We illustrate two simple examples next.

Example 1
Let us examine the inverter layout in Fig. 3.8. For the PMOS-FET (on the right in the image), the n-well must enclose the shapes in Active. This is the only way to ensure that the active region is fully embedded in the n-well (the backgate Nwell) while considering the process overlay. On the other hand, there must be a safe distance between the active region of the NMOS-FET and the n-well for this transistor to be fully embedded in the right backgate (Pwell). To meet both constraints by means of design rules, the two Active regions must be differentiated w.r.t. their association with the two different transistors. This can be achieved by defining the Active regions with the following "select" commands:

(a) PMOS-Active = Active INSIDE Nwell
(b) NMOS-Active = Active OUTSIDE Nwell.

The same result can be achieved in this example when we substitute in these commands the "select" operators "INSIDE" and "OUTSIDE" by the Boolean operators "AND" and "ANDNOT", respectively. Having generated the layers "PMOS-Active" and "NMOS-Active", the two cases of interest can now be separately addressed in the following simple design rules:

(a) ENCLOSURE (PMOS-Active, Nwell, min_value_2)
(b) SPACING (NMOS-Active, Nwell, min_value_1).

Example 2
Source and drain areas should always be connected electrically with as many contacts as possible, as otherwise the current flow from source to drain would be inhomogeneously distributed, resulting in a degraded FET response in the "ON" state. Figure 3.23 shows an NMOS-FET where this constraint is not met.

This constraint can be checked with the instruction sequence illustrated in the example in Fig. 3.23. In the first step, the source and drain regions are extracted and written in layer "X1". In the second step, the contacts are punched out and the result is written in layer "X2". An undersize and oversize sequence is then executed with a *clear* value in the third and fourth steps. By carefully setting the *clear* value, the calculated layer "Result" indicates sufficient space for placing (additional) contacts in the source/drain portions.

3.4.4 *Rules for Die Assembly*

For an IC to be used in an electronic system, the chip's die must be (i) securely mounted in the intended system's device and (ii) electrically connected (through the chip's external connectors) to the system.

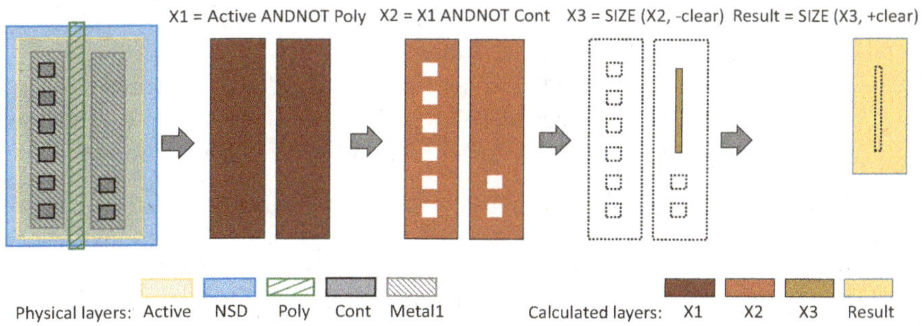

Fig. 3.23 Example of design rule programming where, with the help of calculated layers (X1, X2, X3 and Result) and an instruction sequence (e.g., undersize and oversize), the source/drain regions with missing contacts (yellow, right) are localized. Calculated layers such as the "Result" layer can serve as an error layer in a DRC, visualizing design rule violations

These external connectors on a layout area are often called *pads*. In chips, however, pads are more than just simple electrical contact points. Typically, they are included in so-called *pad cells* that contain complete function blocks, which not only contain a metallic surface area for external contacts, but also have circuitry to protect the chip core against damage from surge voltages, i.e., electrostatic discharge (ESD, Chap. 7, Sect. 7.4.1). Pad cells often also contain driver circuitry.

Many options are available today for implementing the technological interface between the chip's die and the (next higher level) electronic system [3]. To ensure these options can be implemented, further criteria and constraints must be taken into account during the IC physical design. They are known as *rules for die assembly*.

A common method of assembly is to place the chip's die in an enclosure, called a *chip package*. A sectional view of a die in a chip package is depicted in Fig. 3.24. The die is mounted on a metallic substrate, called the *lead frame*, and the die pads are then connected to the contact points on the lead frame. These contact points later become the external connections on the packaged chip (also called the *leads* of a package) that can be soldered to a PCB, for example. *Bonding wires* are often used to make these connections, in which case the pads are also called *bond pads*.[2]

The entire configuration is then encapsulated in plastic, the *mold compound*. During this procedure, the package leads on the lead frames are punched out so that

Fig. 3.24 Chip packaging in sectional view

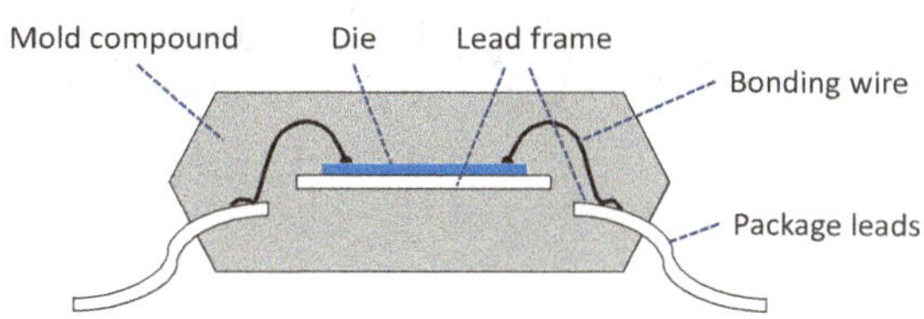

[2]Besides making contact within the enclosure via bonding wires, other contacting methods are also available. We will not dwell on these other options at this point in order to keep our treatise simple.

they are electrically isolated from one another. Using this enclosure method, the chip is mechanically stabilized by the mold compound.

In order to make the chip suitable for this type of assembly operation in batch production, rules for die assembly must be followed during layout design. As illustrated by the example in Fig. 3.25, some typical die assembly rules are as follows:

- The die must fit on the lead frame. An enclosure rule is generally defined for this purpose. This rule defines by how much the lead-frame rectangle must be larger than the die.
- Width and spacing rules apply to the bond pads on the chip. The values to be observed depend on the thickness of the bond wire and the bonding facility.
- There must be a minimum spacing between bonding wires everywhere in the plan view.
- Bonding wire angles must not be arbitrarily acute, i.e., there is a minimum size for the angles of the bonding wires (blue angle in Fig. 3.25).
- There should not be any bond pads in the die corners (colored red in Fig. 3.25). This is to avoid the risk of the die breaking during bonding.
- The die must not be much smaller than the lead frame because the lengths of the bond wires must not exceed a maximum value. This is to avoid the risk of neighboring bond wires making contact with one another (and thus forming a short circuit) during the molding process.

The *bond diagram* depicted in Fig. 3.25 shows how the bond pads are electrically assigned to the contact points on the lead frame—so-called *pinning*. Other details are indicated to illustrate the above-mentioned rules for die assembly.

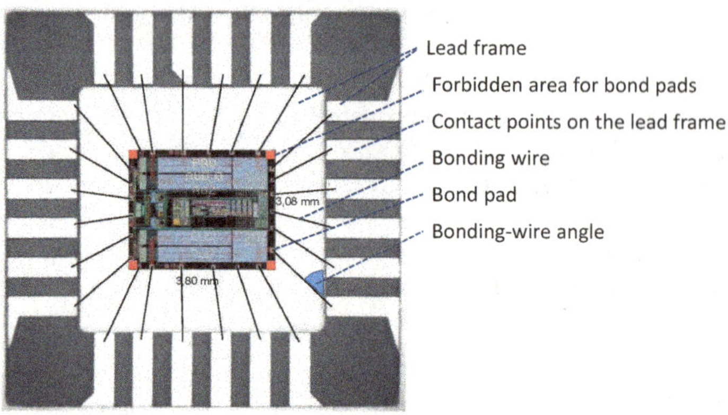

Fig. 3.25 Bond diagram for a chip for a 24-pin package with bonding wires between die pads and contact points on the lead frame. This top view complements the sectional view in Fig. 3.24

3.5 Libraries

When we create a schematic, we use device symbols from a library and connect them according to the design requirements. Simulating the design requires a library of device models in addition to the netlist and the input stimuli. A netlist, a technology file and device layouts (the latter, again, provided by a library in case of digital devices) are needed to generate the layout.

Every design flow, therefore, is inherently coupled to *libraries*. Libraries contain relevant design information, such as design rules, and pre-designed layouts for components such as macros and standard cells.

For each component, the library must provide three aspects: (i) a symbol that represents its type and interfaces, (ii) a model to describe its behavior, and (iii) a layout to characterize its geometry (Fig. 3.26). This not only applies to IC cells, but also to discrete devices, such as transistors, resistors and capacitors, arranged on a PCB, which have a symbol view, behavioral description, and housing geometry attached as well [1].

We next investigate these important constituents of all design flows by introducing libraries used in IC design (Sects. 3.5.1 and 3.5.2) and PCB-design libraries (Sect. 3.5.3).

3.5.1 Process Design Kits and Primitive Device Libraries

The technology that will be used for fabrication must be known when the chip is being designed. The available primitive devices, their electrical properties and the design rules depend on this technology. Hence, every foundry provides a *process design kit (PDK)* for each of their technologies. The properties of the fabrication process that are of interest during the design are represented by a PDK, which essentially forms

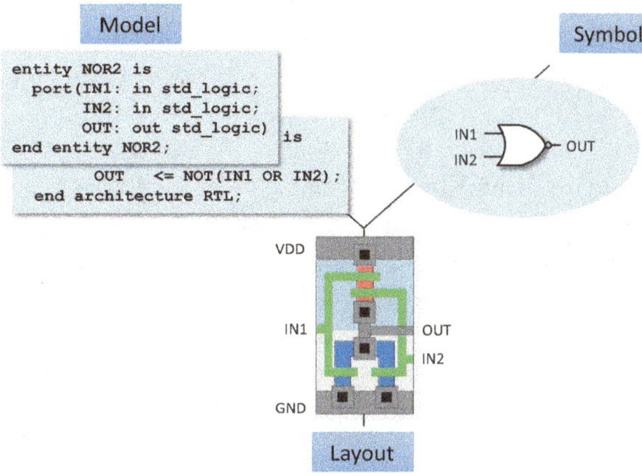

Fig. 3.26 The three aspects of a library component, such as a NOR gate, can be represented by the views in the Gajski-Kuhn Y-chart (Chap. 4, Sect. 4.2.2). These aspects or views come into play at various design stages, such as circuit design (symbol), simulation (model) and physical design (layout)

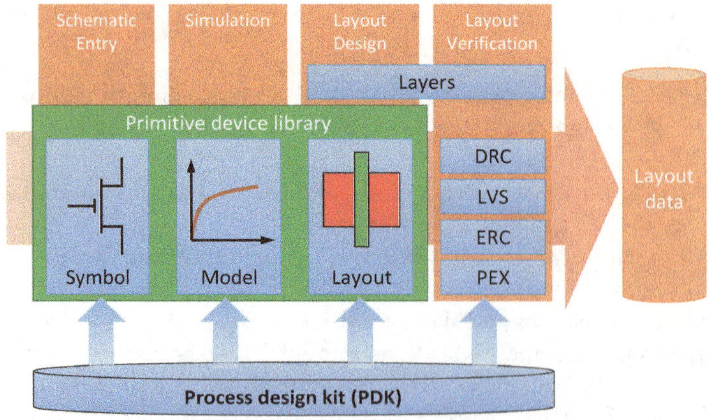

Fig. 3.27 Elements of a process design kit (PDK, shown in blue) and their relation to the various design steps. The PDK contains, among others, the primitive device library with basic devices such as transistors, resistors, and capacitors

the basis for circuit design, simulation, physical layout design and verification in this technology (Fig. 3.27). Every PDK contains, among others, the following:

- A primitive device library containing device symbols and models, and parameterized cells (PCells),
- Verification decks for DRC, LVS and other design verification steps, and
- Technology data, such as layers, sheet resistance, and routing rules.

In short, the PDK is used to design, simulate, draw and verify the design before handing it over to the foundry for chip production.

Foundries offer proprietary PDKs that support different (commercial) design, simulation and verification tools.

Devices supported by a technology are stored in the *primitive device library* of the PDK. This library contains different transistors (MOSFETs, bipolar junction transistors), resistors, capacitors, diodes and I/O cells that can be constructed in the technology. For schematic design, each of these devices contains a *symbol* and a list of device parameters, such as a transistor's width and length, for example.

The circuit function can be verified with simulation based on simulation *models* that are also contained in the PDK. In addition to DC operating point analysis, AC analysis and transient analysis, SPICE device models also typically support best-/worst-case corner simulations and Monte-Carlo simulations. With the latter, process parameter variations and the statistical mismatches between identical device parameters can be simulated, as well. Estimates of fabrication yields can be derived from the results of Monte-Carlo simulations.

For physical design, the PDK contains a layout generator, known as a *parameterized cell* (abbreviated as *PCell*) for every device. These are programs that can automatically create a layout based on parameters passed to them. For example, a transistor PCell produces a transistor layout using the width, length and other parameters defined during the schematic and layout design of this transistor (Chap. 6, Sect. 6.4).

Essentially, devices can be used in circuit schematics, simulations and when creating the layout by means of the above-mentioned symbols, models and PCells, which also represent our three views (see Fig. 3.26). It is only later during device generation, placement, and routing that technology layers must be taken into consideration. For this, a range of attributes are specified by the PDK for each layer: the layer name and its graphical representation in the layout editor, technological properties, such as sheet resistances (Chap. 6, Sect. 6.1) and sheet/line capacitances, as well as routing rules, such as preferred direction and pitch. Layouts for different via patterns (e.g., stacks, arrays) to connect neighboring metal layers are often supplied as well.

As mentioned before, the manufacturability of an integrated (sub)circuit is assured and design flaws are detected by performing layout verification during and after physical design. Therefore, each PDK contains rule decks for different verification steps and tools, such as the design rule check (DRC).

The PDK contains rules for the layout versus schematic (LVS) check, as well. An LVS extracts a netlist from the layout and compares it with the netlist of the corresponding schematic.

The extracted netlist can also be used together with the layout for a parasitic extraction (PEX). The PEX yields a result similar to that of the LVS extracted netlist, but it also contains parasitic resistors, capacitances and inductances (i.e., parasitic devices) occurring in the interconnect layers and between interconnect layers and the silicon substrate. This "parasitics-extended" netlist is then simulated to verify that the circuit functions properly despite the presence of these parasitic devices. In order to achieve this, the PEX refers to a detailed description in the PDK, consisting of the vertical layer structure and the configuration of the supported verification tools.

A PDK can contain rule checks for other verification steps as well, such as the *antenna rule check* and the *electrical rule check* (*ERC*). These and the aforementioned verification steps will be discussed in detail in Chap. 5 (Sect. 5.4).

3.5.2 Cell Libraries

Standard Cell Libraries

The design of digital (sub)circuits requires a *standard cell library*. A standard cell performs a low-level logic function; it is constructed from multiple transistors that are connected by interconnect structures. All standard cells have a variable width, but have either the same or one of a small number of fixed heights (Chap. 4, Sect. 4.3.1); the cell heights are thus restricted to one or a small number of "standard" sizes, hence the name. Some relevant standard cell examples are combinatorial elements,

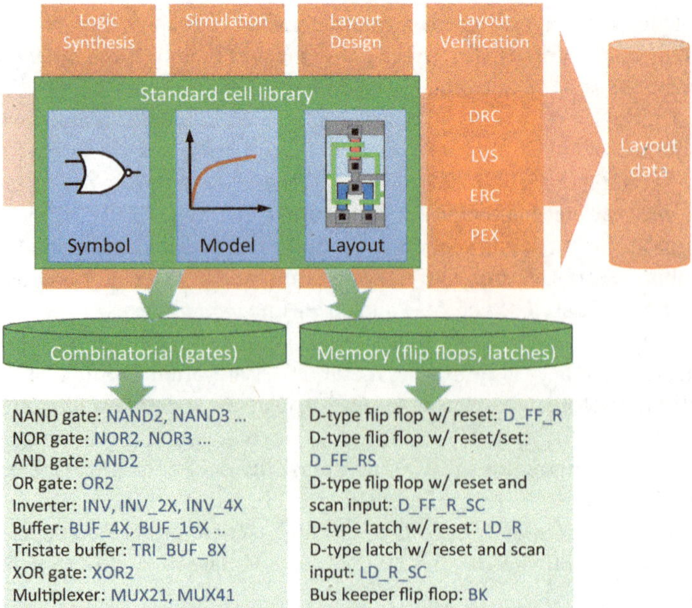

Fig. 3.28 Elements in a standard cell library, which consists of basic combinatorial gates and memory elements in the form of flip flops and latches

such as NAND or NOR gates, and memory elements such as flip-flops and latches[3] (Fig. 3.28).

Each standard cell in the library is assigned a symbol, model and layout similar to primitive devices (Sect. 3.5.1). Models, however, are typically available as VHDL or Verilog code and are assigned additional timing information for digital simulation. Layout generators are not used, given that each standard cell function already has an optimized and thus fixed layout.

Standard cell libraries are directly associated with a specific IC fabrication process, and as such are often available from the manufacturer, i.e., the fab. As these fabs closely guard their intellectual property (IP), detailed design information on the internal layout structure of the standard cells is sometimes kept secret. Thus, only the external interface and shape information is available in the library for use by the contractor. The hidden internal structures are only inserted during mask generation in the fab [1].

[3]Flip flops and latches are circuit elements with a memory/storage function. This memory function is needed in most digital systems because the output often not only depends on the current inputs (as in combinatorial networks), but also on the previous input(s) and output(s) which determine the *state* of a system.

Pad Cell Libraries

Pad cells provide the means of connecting an IC core to the surrounding world, i.e., to the next higher level in a hierarchy. Their tasks are: (i) buffering input signals, (ii) driving of external loads and adaption to external logic levels, (iii) providing supply voltage, and (iv) shielding the core against electrostatic discharge (ESD), wrong polarity, and transients.

Pad cells are typically provided with protective guard structures. These guard structures consist of integrated diodes connected to the pad supply rings, which protect the internal core against destructive ESD events (Chap. 7, Sect. 7.4.1).

Macro Libraries

Macro libraries contain macro cells, often referred to simply as *macros*, which are pre-designed combinations of basic cells that perform extended functions. They can be differentiated into (i) *hard macros* that have a fixed shape and, hence, are already fully placed and routed, and (ii) *soft macros* that have flexible shapes, as their internal placement and routing has not yet been defined.

3.5.3 Libraries for Printed Circuit Board Design

In PCB design, a library is also a file or a database for organizing elements that can be re-used many times during physical design. PCB libraries include *symbol libraries* that hold schematic design symbols; *footprint libraries* that contain device footprints (land patterns), and *model libraries* with models for the simulation programs. While these three aspects of a library element are combined in a single technology library in IC chip design, they are typically stored in separate (technology) libraries in PCB design.

The libraries can typically be searched with a library management system. When the PCB design is being drawn, copies of the required elements are placed in the circuit schematic or layout. Here too, this procedure is referred to as *instantiation*, as with chip design.

Within the PCB library, different library elements can reference one another. PCB libraries are often tied to specific tools and tool providers; as a result, symbol, footprint and model libraries for PCB design are often not interchangeable between different design tools. These libraries are stored in either ASCII or binary formats, depending on the tool provider.

Symbol Libraries

Symbol libraries contain the symbols needed to create the circuit diagram (Fig. 3.29). They can be searched, viewed, selected and instantiated in the library management system. The symbol in a symbol library typically references to one or more suitable footprints for the associated physical device. Other information, such as the manufacturer, the vendor or the link to electrical models can generally be stored as well.

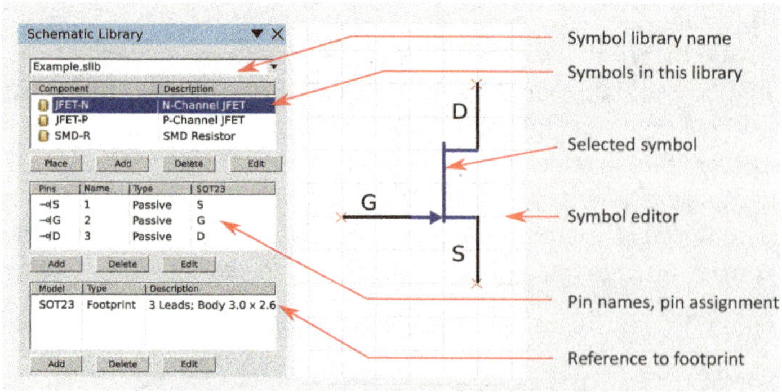

Fig. 3.29 Example of a schematic library for an n-channel junction FET

If the libraries being used do not contain a required symbol, design tools generally allow the designer to create a new symbol with a symbol editor and to store it in the symbol library. The new symbol should have all the necessary information for the design process, such as the link to the footprint, the pin assignment and, if necessary, the link to the electrical model.

Footprint Libraries

While the term "footprint" (also known as "land pattern") is often used to describe the arrangement of pads or through-holes used to physically attach and electrically connect a component to a PCB, it is also applied in the context of libraries. Here, the term is used more broadly and encompasses all physical information for a device required for PCB design. Aside from the geometrical description of the individual layers (routing layers, solder-resist layer, solder-paste layer, silk screen layer, mechanical layer, drill holes, cut-outs, etc.), the footprint may contain three-dimensional (3D) physical device models, as well.

Footprints are managed in footprint libraries, an example of which is depicted in Fig. 3.30.

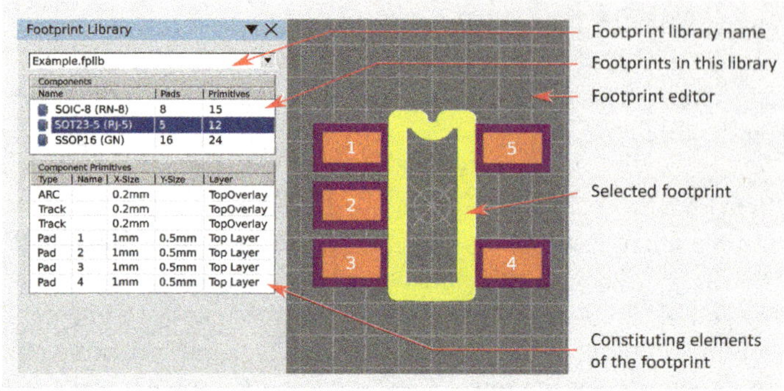

Fig. 3.30 Example of a footprint library for a 5-pin SMD package

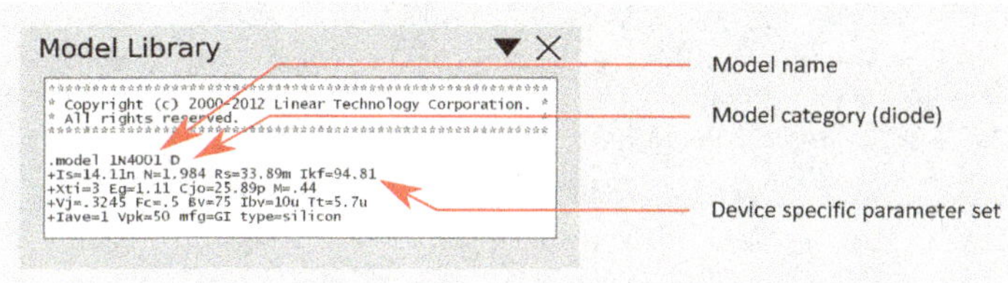

Fig. 3.31 Example of a model library for a diode with parameters

Every device symbol on the schematic refers to one (or more) associated footprint(s) in the footprint library. This relationship is used during the transition from the circuit diagram to the PCB layout design to derive the associated footprints from the footprint library, which are then available for placement on the PCB.

New footprints can be designed in the footprint editor and stored in a library; however, care should be taken that the footprint is suitable for the prospective fabrication technology. If the PCB is needed as a 3D model in other fabrication steps, the matching 3D physical model must be included for the device.

Model Libraries

Electrical models for simulation programs, such as *PSpice*™, *LTSpice*™ or *SIMetrix*™, are arranged and stored in model libraries. They contain the written description of the device's electrical response (Fig. 3.31).

The model describes the electrical behavior of basic elements, such as resistors, capacitors, and inductances. Models of semiconductor diodes and transistors are adapted to the device in question by means of a number of characteristic parameters. More complex devices are modeled by connecting a number of basic elements.

In order to protect their intellectual property, some vendor models are only available as encrypted entities, which do not allow any insight into the inner workings of the simulation models.

References

1. D. Jansen et al., *The Electronic Design Automation Handbook* (Springer, 2003). ISBN 978-1-402-07502-5. https://doi.org/10.1007/978-0-387-73543-6
2. K. Kundert, *The Designer's Guide to Spice and Spectre* (Springer, 1995). ISBN 978-0-792-39571-3. https://doi.org/10.1007/b101824
3. J. Lienig, H. Bruemmer, *Fundamentals of Electronic Systems Design* (Springer, 2017). ISBN 978-3-319-55839-4. https://doi.org/10.1007/978-3-319-55840-0
4. J. Lienig, M. Thiele, *Fundamentals of Electromigration-Aware Integrated Circuit Design* (Springer, 2018). ISBN 978-3-319-73557-3. https://doi.org/10.1007/978-3-319-73558-0
5. F.N. Najm, *Circuit Simulation* (Wiley, 2010). ISBN 978-0-470-53871-5
6. J.M. Rabaey, A. Chandrakasan, B. Nicolic, *Digital Integrated Circuits—A Design Perspective* (Pearson Education India, 2017). ISBN 978-933-257392-5

7. S.M. Rubin, *Computer Aids for VLSI Design—Appendix D: Electronic Design Interchange Format* (Addison-Wesley Publishing Company, 1987). ISBN 0-201-05824-3. https://www.rulabinsky.com/cavd/index.html
8. A.K-K. Wong, *Resolution Enhancement Techniques in Optical Lithography* (SPIE Digital Library, 2001). PDF ISBN 978-081947881-8, Print ISBN 978-081943995-6. https://doi.org/10.1117/3.401208

Chapter 4
Methodologies for Physical Design: Models, Styles, Tasks, and Flows

In Chap. 2 we covered technologies and in Chap. 3 we saw how these technologies interface with physical design. Here in this chapter we now provide an end-to-end overview of the physical design process, namely how to physically construct the layout of an electronic circuit. In this chapter we present the fundamental knowledge an engineer must possess to carry out this task. In Chap. 5 we then discuss each of the specific physical design steps in further detail.

We begin the chapter by introducing the design flow (Sect. 4.1), design models (Sect. 4.2) and design styles (Sect. 4.3). Next, we investigate various design tasks and related tools (Sect. 4.4), before discussing optimization goals and design constraints (Sect. 4.5). Up to this point our treatise has focused mainly on the digital design flow. In Sect. 4.6 we introduce the characteristics of, and differences between, analog, digital, and mixed-signal design flows. Looking toward the future, we conclude the chapter by presenting two different yet complementary visions for analog-design automation to overcome the analog-digital design gap (Sect. 4.7).

4.1 Design Flow

We can divide the design flow for an electronic circuit—or any industrial product for that matter—into different stages (Fig. 4.1). Each stage in this sequential process focuses on a specific aspect of the circuit, and the output of one stage generally becomes the input of the following stage.

A set of documents, typically electronic files and artifacts, is produced one after another during the design stage of this process. These documents are based on an input specification of the circuit being designed; they must define it completely and correctly. The circuit must be described in such a manner that it can be manufactured (during the fabrication stage) in a way that meets all requirements in the specification.

Millions or even billions of items must be effectively managed when designing electronic circuits. A task of such complexity requires a completely systematic approach. Although a fully automated design flow that is guaranteed to be flawless

J. Lienig and J. Scheible, *Fundamentals of Layout Design for Electronic Circuits*,
https://doi.org/10.1007/978-3-030-39284-0_4

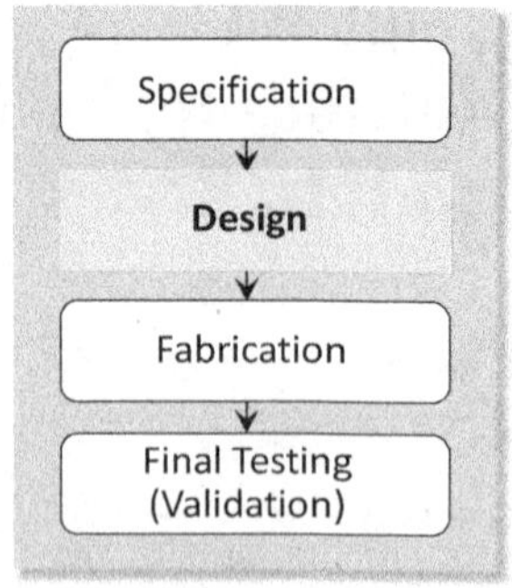

Fig. 4.1 The primary stages in a product development flow

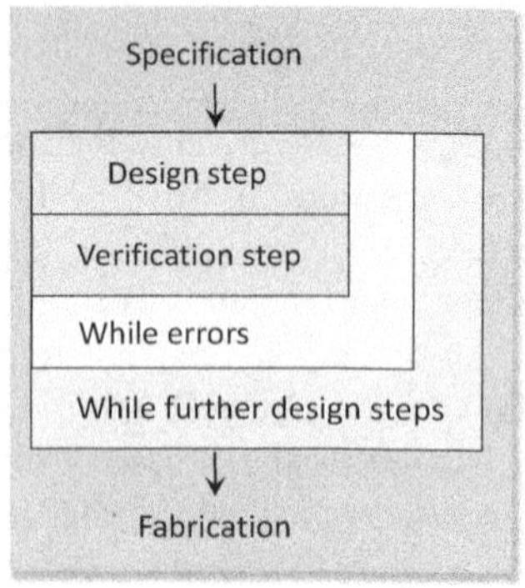

Fig. 4.2 The general design flow illustrated as a (looped) sequence of design and verification steps

is still out of reach today, certain design strategies have proven themselves useful in practice for implementing the design stage in Fig. 4.1:

- Splitting the overall design process into individual design steps, and
- Appending a verification step to each design step.

As we will see throughout this chapter, a useful way of viewing the design process is as a continuous sequence of design and verification steps, as outlined in Fig. 4.2.

The need to split the design process into multiple design steps first arose with the advent of highly complex integrated circuits, which coincided with the evolution of digital systems according to Moore's law (Chap. 1, Fig. 1.11). Figure 4.3 depicts this step-wise design approach, illustrating a design flow for a modern very large-scale integrated (VLSI) circuit.[1] As will be shown below, steps performed at the beginning of this flow are more abstract than steps towards the end. At the end of the process, and prior to fabrication, detailed information on each circuit element's geometric shape and electrical properties are known. The expansion in the right-hand side of Fig. 4.3 illustrates the processing for the physical design step.

The various stages of the design flow in Fig. 4.3 are discussed in detail next. While our subsequent description focuses primarily on the (digital) VLSI design flow, these

[1]We use the term "VLSI circuit" to denote pure *digital* ICs, which represent the leading edge in circuit complexity.

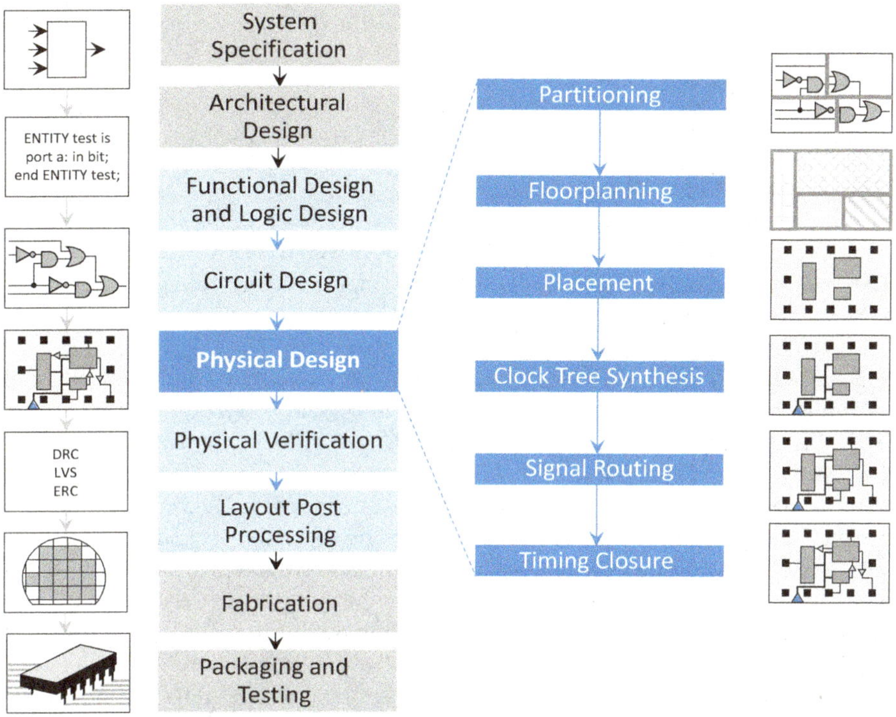

Fig. 4.3 The major steps in the VLSI circuit design flow, with a focus on physical design [2]

steps can be similarly applied to any electronic circuit, such as an analog circuit or a printed circuit board.

System Specification

The overarching goals and high-level requirements of the system are collectively defined by chip architects, circuit designers, product marketers, and operations managers. In the case of an application-specific integrated circuit (ASIC), which is typically built to order, the customer is additionally involved in this initial design phase. These goals and requirements cover functionality, performance, physical dimensions, and fabrication technology [2].

Architectural Design

A basic architecture must be designed to meet system specifications, as captured in the preceding step. Some of the decisions involved are [2]:

- Partitioning in hardware and software,
- Integration of analog and mixed-signal blocks,
- Memory management (serial or parallel) and the addressing scheme,
- Number and types of computational cores, such as processors and digital signal processing (DSP) units—and specific DSP algorithms,
- Internal and external communication, support for standard protocols, etc.,

- Usage of hard and soft intellectual-property (IP) blocks,
- Pinout, packaging, and the die-package interface,
- Power requirements,
- Choice of process technology and layer stacks.

Functional and Logic Design

Once the architecture of the entire system has been agreed upon, the functionality and connectivity for each module (such as a processor core) must be defined. During functional design, only the high-level response (input, output, timing behavior) for each module is determined.

Logic design is performed at the *register-transfer level* (*RTL*) using a *hardware description language* (*HDL*) by means of software programs that define the functional and timing behavior of a chip, or of modules within a chip. Two common HDLs are *VHDL* (Very High-Speed Integrated Circuit HDL, established by U.S. Department of Defense, approx. 1983) and *Verilog* (portmanteau of the words "verification" and "logic", established by Gateway Design Automation, approx. 1985). HDL modules are essentially software programs that are written by the designer; and which are run during functional and logic design as part of a simulation of the chip's behavior. They implement the functionality, input, output, and timing behavior of the "hardware" module they represent. HDL modules must be thoroughly simulated and verified for correctness before being used for logic synthesis, as described next.

Logic synthesis tools automate the process of converting HDL into low-level circuit elements. The HDL synthesis tool essentially translates (i.e., converts) the programming language-like HDL statements into a corresponding representation of logic gates and low-level circuit elements. In this way, a logic synthesis tool takes as input a Verilog or VHDL description and a technology library, and maps the described functionality to a list of signal nets, or *netlist*, and specific circuit elements, such as standard cells and transistors. In this case, the next step "circuit design" can be omitted as these netlists are used in physical design.

Circuit Design

In addition to analog units, several critical, low-level digital elements must be designed at the transistor level; this process is referred to as *circuit design* [2]. Among the elements designed at the circuit level are analog circuits, static RAM blocks, I/O, high-speed functions (multipliers), and electrostatic discharge (ESD) protection circuits. The correctness of circuit-level design is predominantly verified by circuit simulation tools, such as SPICE.

Physical Design

During physical design, the input netlist (consisting of logic- and circuit-level design components such as macros, cells, gates, transistors, etc.) is instantiated with geometric representations. In other words, all macros, cells, gates, transistors, etc. are defined using shapes and sizes per fabrication layer, are assigned spatial locations on the chip (placement), and have appropriate routing connections (routing) completed in metal layers. The result of physical design is a set of manufacturing specifications that must subsequently be verified.

Physical design is rules-based: these rules reflect the physical limitations of the fabrication medium. For instance, the minimum distances between wires and their respective minimum widths must be predefined, in accordance with the targeted fabrication medium. In addition, the design layout must be recreated in (*migrated* to) each new manufacturing technology, so that it obeys the design rules of the new technology.

In this book, we emphasize that physical design is a vital step in chip design as it directly impacts circuit performance, area, reliability, power, and manufacturing yield.

Due to its highly complex nature, physical design of digital ICs is itself split into the following individual key steps (see Fig. 4.3) [2]. (These steps are described in detail in Chap. 5.)

- *Partitioning* divides a circuit into smaller subcircuits or modules that can each be designed or analyzed individually.
- *Floorplanning* (aka *chip planning*) determines the shapes and layout of subcircuits or modules, as well as the locations of external ports and IP or macro blocks. It includes *power and ground routing* which distributes power (VDD, PWR) and ground (VSS, GND) nets throughout the chip.
- *Placement* determines the spatial locations and orientations of all cells within each block.
- *Clock tree synthesis* determines the buffering, gating (e.g., for power management) and routing of the clock signal to meet prescribed skew and delay requirements.
- *Signal routing*, consisting of
  - *Global routing* allocates routing resources that are used for connections; example resources include routing tracks in global routing cells (gcells).
  - *Detailed routing* assigns routes to specific metal layers and routing tracks within the global routing resources.
- *Timing closure* optimizes circuit performance by specialized placement and routing techniques.

After detailed routing, the layout is accurately optimized electrically at a small scale [2]. Parasitic resistances (R), capacitances (C) and inductances (L) are extracted from the completed layout and passed to timing analysis tools to check the chip's functional behavior. If the analyses reveal erroneous behavior or an insufficient design margin (*guardband*) against possible manufacturing and environmental variations, then the design is updated incrementally to alleviate such shortcomings and to optimize the design.

As described later in detail (Sects. 4.6, 4.7 and Chap. 6), physical design for analog circuits uses a very different approach. Here, geometric representations of analog circuit elements are created in real time using *layout generators* or by manually drawing them rather than selecting predesigned cells from a library, as is the case in digital layout design. These generators use circuit elements with specified electrical parameters, such as the resistance of a resistor, and accordingly generate the appropriate geometric representation, e.g., a resistor layout with automatically calculated length and width to produce the required resistance. Additional requirements, such as *matching* the devices for symmetrical behavior, must be considered in the placement step, and interconnect wire widths need to be adapted in the routing step to provide sufficient cross-sectional area for currents.

Physical Verification

After the physical design is complete, the layout must be fully verified to ensure correct electrical and logical functionality. Some problems flagged during physical verification can be tolerated if their impact on chip yield is negligible. In other cases, the layout must be changed, but these changes must be minimal and should not introduce new problems. Therefore, a layout is typically modified manually by experienced design engineers at this stage. The main physical verification methods are presented in detail in Chap. 5 (Sect. 5.4) and are summarized here as follows:

- *Design rule checking* (*DRC*) verifies that the layout meets all technology-imposed constraints. These design rules include the following categories.
  - Typical geometrical design rules that require minimum or maximum values for widths, spacings, extensions, intrusions, and enclosures for layout polygons. These rules make sure that the layout structures can be correctly generated on the silicon due to process accuracy.
  - Layer density rules to ensure that chemical-mechanical polishing (CMP) produces a desired planarity.
  - *Antenna rule checking* that seeks to prevent the *antenna effect*, which may damage the thin transistor gate oxide during plasma etching through accumulation of excess charge on metal and polysilicon wires.
- *Layout versus schematic* (*LVS*) checking verifies the electrical compliance of the layout with the input netlist. A netlist is derived from the layout and compared with the original netlist produced from logic synthesis or circuit design.
- *Parasitic extraction* (*PEX*) derives electrical parameters for the layout elements from their geometrical representations; these are used in conjunction with the netlist to verify the circuit's electrical characteristics.
- *Electrical rule checking* (*ERC*) covers additional requirements to ensure design functionality. ERC verifies, for example, the correctness of power and ground connections, and that signal transition times (*slew*), capacitive loads and fanouts are appropriately bounded.

Layout Post Processing

The final DRC-/LVS-/PEX-/ERC-clean layout data then undergoes *layout post-processing* where amendments and additions to the chip layout data are performed in order to convert the physical layout into data for mask production (mask data). This post processing can be divided into three steps: (1) *chip finishing*, (2) producing a *reticle layout*, and (3) *layout-to-mask preparation* (which we all cover in detail in Chap. 3, Sect. 3.3.).

Fabrication

The mask data is then used for manufacturing at a dedicated silicon foundry (*fab*, short for *fabrication*). The design handoff to the manufacturing process is called *tapeout,* a historical term that refers to the large file that is produced and which was, in previous generations, delivered using a magnetic-tape medium. Generating the data for manufacturing is also sometimes referred to as *streaming out* [2].

At the fab, the design is patterned onto different layers using photolithographic processes. (We discuss the fabrication technologies for IC chips in Chap. 2.) Photomasks are used such that only certain silicon patterns, specified by the layout, are exposed to a laser light source. Many masks are used, one after another, to create the microscopic structures on the silicon. If the design is changed, some or all the masks must be changed as well.

Chips are fabricated on round silicon wafers. The chips are then tested and labeled as either *functional* or *defective.* They are sometimes sorted into *bins*, depending on the functional or parametric (speed, power, etc.) tests. The wafer is cut into smaller pieces, called dies, at the end of the fabrication process.

Packaging and Testing

After dicing, functional chips are typically packaged. Packaging is configured early in the design process and reflects the anticipated environment and usage for the application, along with cost and form-factor requirements. Package types include *dual in-line packages* (*DIPs*), *pin grid arrays* (*PGAs*), and *ball grid arrays* (*BGAs*). After a die is positioned in the package cavity, its pins are connected to the package's pins, e.g., by *wire bonding* or with solder bumps (*flip-chip*). The package is then sealed.

With integration based on multi-chip modules (MCMs), chips are usually not packaged individually; rather, they are integrated as bare dies into the MCM, which is packaged separately at a later point.

The finished product is often tested after packaging to ensure the fully assembled unit meets design requirements, such as functionality (input/output relations), timing, and power dissipation.

4.2 Design Models

Design models are a somewhat abstract concept that are used to systematically link the available design styles together. They provide an overview of the various degrees of abstraction of a design and combine it with key parameters to be considered at this level. For example, the designer can select one of the abstraction levels and then switch from one view to another.

We will now introduce two design models. The first one is a three-dimensional (3D) design space that links the dimensions "hierarchy", "version" and "view". We will then present the Gajski-Kuhn Y-chart. Here, the aforementioned "view" dimension is split into three different domains (behavioral, structural, physical) arranged in three axes, and the "hierarchy" dimension is mapped to concentric abstraction levels.

4.2.1 Three-Dimensional Design Space

The different models describing an integrated circuit design, or its constituent parts, can be systematically split into numerous dimensions. This section describes the *3D model* where the design space is represented in three dimensions, as shown in Fig. 4.4.

The *hierarchy* dimension is instrumental in mastering the structural complexity of an electronic circuit. In it is specified the process starting with the general circuit representation, through more particular intermediate steps, to the completely detailed circuit description. We assign these intermediary steps to different hierarchal levels. By moving from one hierarchical level to the next lower level, problems are split into subproblems, which now can be solved by multiple experts in parallel. This enables better and faster solutions for the problem on the higher level and is a classic application of "divide and conquer".

Design clarity is also a benefit of this approach. For example, prior to considering how to realize a (high-level) logic gate using various (low-level) transistor options, the entire circuit's task is first subdivided into smaller, more manageable subtasks. Only after these subtasks are small enough to comprise a few gates, does the designer consider the options for building the gate's transistors in a custom fabrication process.

This approach also allows the reuse of many of the design decisions at the upper levels in the hierarchy when transitioning to another implementation or fabrication technology.

Obviously, the design task is simplified by this hierarchal organization by splitting a complex task recursively into several smaller tasks. We subdivide a microprocessor for example into logically independent subtasks, such as control unit, RAM, ROM, ALU. The ALU is further divided into a multiplier, shifter, logical operations, etc. These are interconnected via bus systems that control the data and instruction transfers between the components. The individual subtasks are themselves further subdivided. The multiplier is organized such that its adders can be accessed separately.

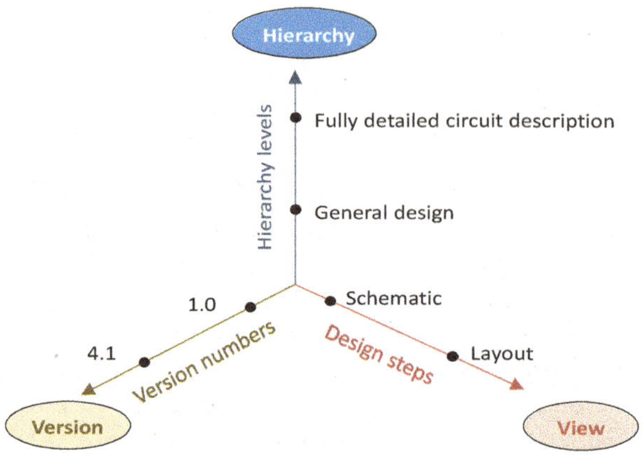

Fig. 4.4 Three-dimensional (3D) design space with its main components

Again, the data and instruction streams must enable multiplication and summation. The adder itself is then constructed with simple logic gates.

Figure 4.5 shows how an analog circuit can be broken down into a tree of different hierarchical levels. As we shall see later, the levels of such a breakdown can only be defined for digital circuits in a way that it is universally applicable. There is still no standard classification of hierarchy levels available for analog circuits like the one shown in the figure. That is, a different analog circuit would likely require a different set of hierarchical levels than those illustrated in Fig. 4.5.

There are benefits to using the hierarchical structure, and drawbacks as well. For example, a small part of the circuit can be reused many times and must only be designed once, which can save on design time; however, the design quality may unfortunately suffer because options for (layout) optimization might remain unused.

The *version* dimension of the 3D-model describes the design history, that is, as the design develops over time. Figure 4.6 shows an example of the version history of a circuit as a tree. Managing and monitoring design versions (i.e., consistency) has long been neglected in EDA systems. They are a major source of errors in the design process and their proper usage is an absolute must in modern design environments.

Figure 4.7 visualizes the third design-space dimension, the so-called *view* axis. Here, a circuit is represented in different domains (aka views), such as the schematic (circuit diagram) and the layout. Both representations describe the same design, but in completely different ways. The number of different design views can be high, and depends at least indirectly on the attributes of the underlying design system.

Many attempts have been made to systematically combine the "hierarchy" and "view" dimensions in a single design model. One of the most widely used models in this regard is the Gajski and Kuhn Y-chart for digital circuits introduced in 1983. We will describe this model in the next section. It is a simple and elegant chart that helps to visualize the design styles concisely.

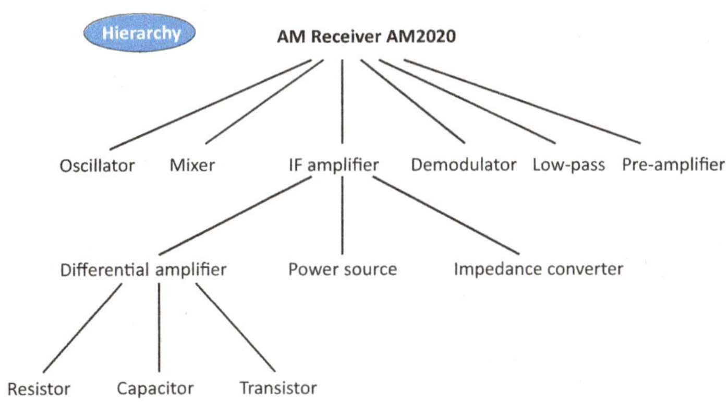

Fig. 4.5 Example of an analog circuit broken down into different hierarchical levels

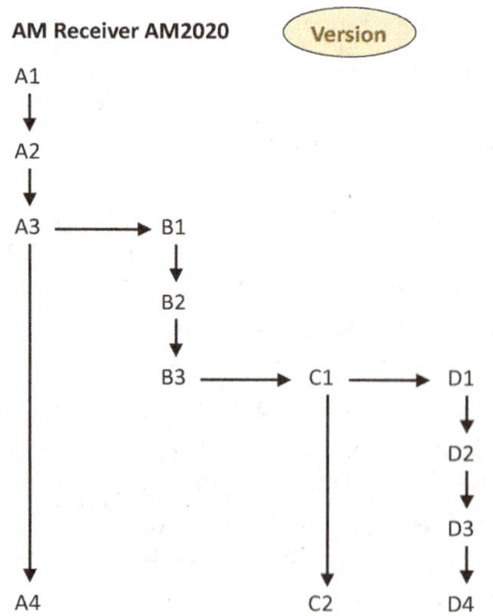

Fig. 4.6 Versions describing the trajectory of design variants over time

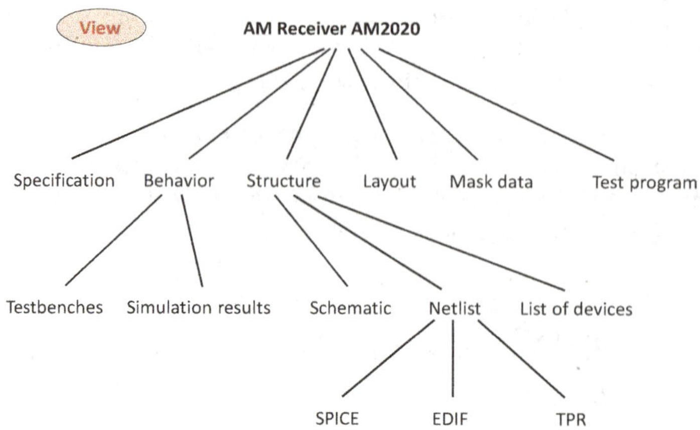

Fig. 4.7 A design's views are different (design-stage dependent) representations of the same circuit

4.2.2 The Gajski-Kuhn Y-Chart

A circuit's three domains, i.e., its behavioral, structural and geometrical/physical views, were first arranged by Gajski and Kuhn in the so-called *Gajski-Kuhn Y-Chart* [1]. This chart is also known as the *Y-chart* because of its three arms (Fig. 4.8). In this model, the "hierarchy" dimension of the 3D model (Sect. 4.2.1) is mapped to concentric abstraction levels. The "view" dimension is split into three different domains (behavioral, structural, physical) arranged in three axes.

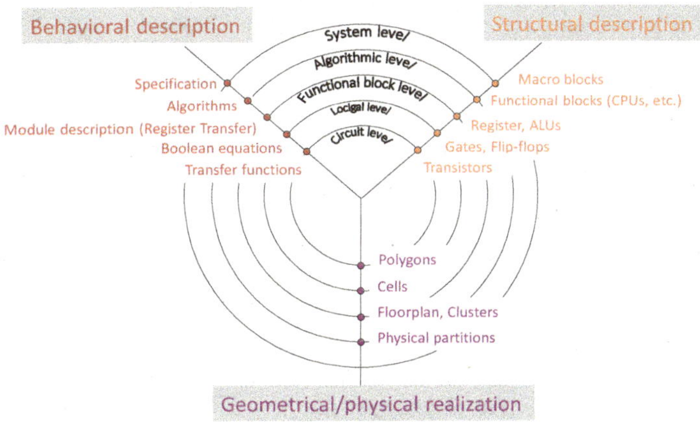

Fig. 4.8 The Gajski-Kuhn Y-chart for digital circuits is composed of three circuit domains "behavioral description" (functional view), "structural description" (structural view), and "geometrical realization" (physical view), with different abstraction levels, which become more specific towards the center. Traversing along the branches, we refine or abstract designs, while a move to another branch along one of the concentric circles changes the representation

Any circuit development is considered from the perspective of three domains ("views") depicted as three axes. The *abstraction levels* describing the degree of abstraction (also considered as *hierarchy* or *design levels*) are aligned along these axes. The outer shells are generalizations, the inner cells incremental refinements of the same aspect.

Each of the three axes represents a different design perspective, in other words a different view or domain. Each view describes certain attributes that characterize the element being designed. All relevant aspects of the subject of the design, such as a NAND gate, are completely described by the collective views (Fig. 4.9).

The *behavioral view* (aka *behavioral description*) describes all aspects of the design subject's behavior. It contains operations performed by the element being designed and its dynamic response. It comprises equations, functions and algorithms.

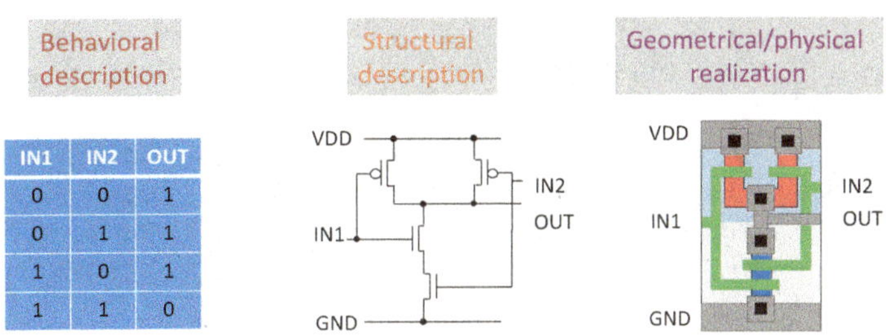

Fig. 4.9 A NAND gate depicted in its behavioral, structural, and physical domain

The *structural view* (aka *structural description*) designates the logical structure—in other words, the abstract implementation of the subject of the design—as the topographical configuration of the components and their connections. The *physical view* (aka *geometrical/physical realization*) describes how the design object is implemented, that is, how the (structural) components are built with real physical objects. This latter view contains the complete and precise geometrical layout and configuration of all components and connectivity architectures.

In each of the three views, the design is defined by several documents that describe the unit being designed in different degrees of abstraction. The degrees of abstraction of the descriptions are characterized by the respective abstraction levels, sometimes also referred to as design levels. In other words, the distance to the center of the Y is a measure of the degree of abstraction at each level.

In the Y-chart, there are five different abstraction levels starting at the outermost circle and ending at the innermost circle, as follows:

- The *system (architectural) level* is where the global properties of an electronic system are specified. Block diagrams containing abstractions of signals and their transient responses are part of the behavioral description. Block symbols for CPUs, memory chips, etc. are used in the structural view (structural description) at this level.
- The *algorithmic level* is where concurrent algorithms, such as signals, loops, variables, and assignments, are defined. Blocks like ALUs are part of the structural view.
- The *functional block level* (*register transfer*) is a more detailed abstraction level where the interactions between communicating registers and logic units are described. Here, data structures and data flows are defined. The floorplan design step is part of the geometrical view (physical implementation) at this level.
- The *logical level* is described in the behavioral perspective by Boolean equations. This level in the structural view comprises gates and flip-flops. In the geometrical view, the logical level is described by standard cells.
- The inner-most *circuit level* is modeled mathematically using differential equations. This is the actual hardware level; it consists of transistors and capacitors all the way "down" to crystal lattices.

As depicted in Fig. 4.10, the Y-chart can be used to illustrate the physical design terms relating to synthesis and generators (left) as well as individual design steps (right).

The "hierarchy" design dimension from the 3D design space (Sect. 4.2.1) complements the abstraction-level concept deployed here (Y-chart, Sect. 4.2.2) by allowing us to define a separate hierarchy for each of the three domains (views). In the Y-chart, an abstraction level is derived from a given modeling concept for the domain level in question, while a hierarchy in the 3D model embodies the composition/decomposition concept ("dividing in smaller pieces" while staying on a domain branch of the Y-chart).

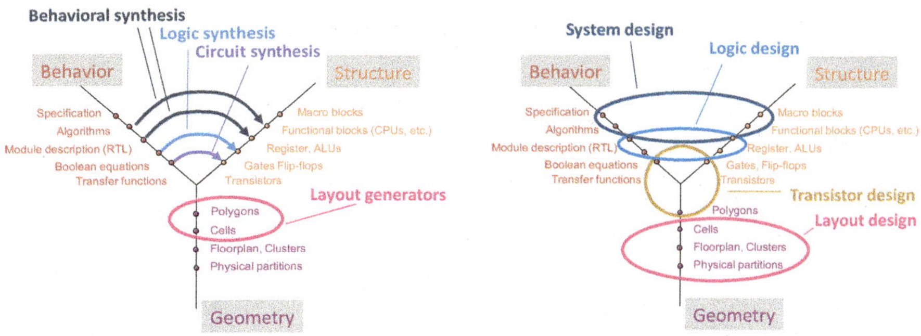

Fig. 4.10 Using the Y-chart to visualize the terms used in synthesizing EDA tools (left) and the individual design steps (right)

4.3 Design Styles

A design step is a mapping between two design states. A *design style* is characterized by a sequence of design states, terminating in a geometrical representation of the design (i.e., a layout).

Any design style can be represented by the Y-model covered in the previous section. Each design state can be characterized by the point of intersection between one of the axes and one of the concentric circles in the Y-model. The starting point in a design flow, the idea of an electronic system, is normally represented in this model by the design state "Behavioral view at the system level".

We shall now investigate in more detail examples of design styles and types of integrated circuits associated with them.

4.3.1 Full-Custom and Semi-Custom Design

Selecting an appropriate circuit-design style is very important as it has a bearing on time-to-market and design cost. There are two types of (digital) VLSI design styles: *full-custom* and *semi-custom*. Full-custom design is primarily used for extremely high-volume circuits, where the high design costs are amortized over large production volumes. Semi-custom design is the more standard approach as it simplifies the design process, and hence reduces time-to-market as well as overall costs.[2] Semi-custom design can be divided into two methodologies:

- *Cell-based*: typically using standard cells and macro cells, this design has many *pre-designed* elements, such as logic gates that are copied from libraries.

[2]Analog circuits (which we will consider in Sects. 4.6 and 4.7) are always full-custom designed. Analog design requires many more degrees of freedom in order to deal with its vast diversity of constraints.

- *Array-based*: either gate arrays or FPGAs, the design has some *pre-fabricated* elements, such as transistors.

We next discuss full-custom design briefly, followed by more in-depth discussion of several semi-custom design styles.

Full-Custom Design

Among available design styles, a full-custom design style is subject to the fewest design-methodology constraints during layout generation. For example, blocks can be placed anywhere on the chip without restriction. The result is a very compact chip with highly optimized electrical properties. This design is however laborious, time-consuming, and can be prone to error due to the low level of automation.

Full-custom design is primarily useful for microprocessors, where the high cost of design effort is amortized over large production volumes. It is also well suited for analog circuits for technical reasons (and not for economic ones) as extreme care must be taken to achieve matched layout and adherence to a multitude of stringent electrical performance specifications. These types of circuits can only be designed by experienced layout designers—and, hence, the design style is mainly manual.

Standard-Cell Design

A digital standard cell is a predefined element that has a fixed size in layout and that realizes a standard Boolean function. Figure 4.11 shows three examples with their schematic symbols, truth tables and layout representations (corresponding to the structural, behavioral, and physical views of the Y-chart, respectively). Standard cells are distributed in cell *libraries*, which are often provided at no cost by foundries and are pre-qualified for manufacturing.

Standard cells are designed in multiples of a fixed cell height, with fixed locations for *power* (*VDD*) and *ground* (*GND*) ports. Cell widths vary depending on the

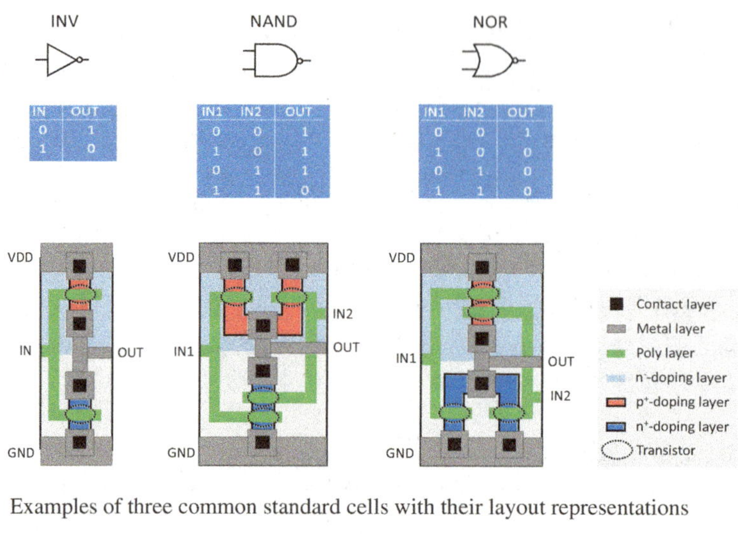

Fig. 4.11 Examples of three common standard cells with their layout representations

transistor network implemented. Because of this restricted layout style, all cells are placeable in rows such that power and ground supply nets are distributed by (horizontal) abutment (Figs. 4.12 and 4.13). The cells' signal ports are located throughout the cell area.

Since standard-cell placement is more restrictive, the complexity of this design methodology is greatly reduced. This enables a fully automated layout design flow

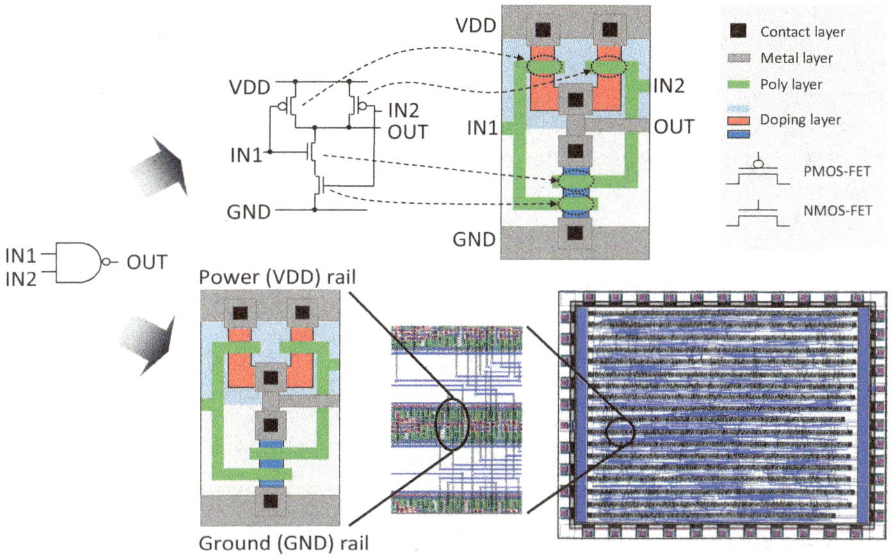

Fig. 4.12 Schematic and layout of a NAND gate using CMOS technology (top) and its implementation in a standard-cell design (bottom) [2]

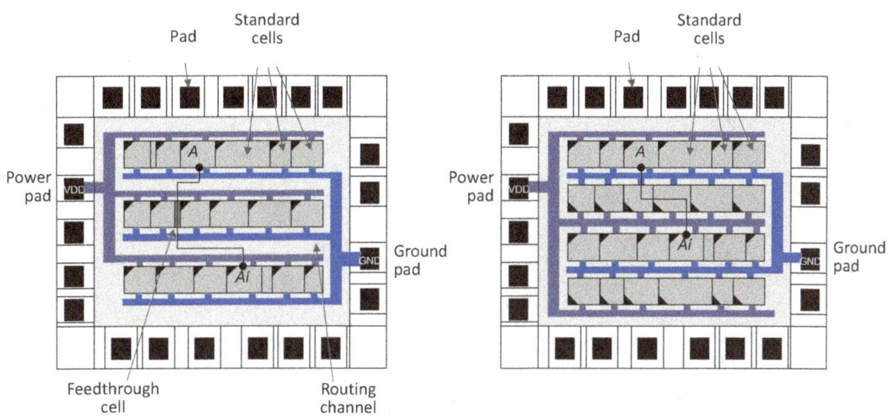

Fig. 4.13 Typical standard-cell layouts with each row having its own power and ground rails (left) and shared power and ground rails (right) which requires cell orientations to alternate [2]

including placement and routing. Time-to-market can thus be decreased compared to full-custom designs at the cost of factors such as power efficiency, layout density, or operating frequency. Hence, different market segments are served by standard cell-based designs, e.g., ASICs, than full-custom designs, e.g., microprocessors. Developing the cell library to qualify it for manufacturing and design requires considerable initial effort. This "up front" expenditure of effort (to build the cell library) is done once; the use of these library cells across many subsequent designs results in significant cost savings and efficiencies in these designs.

Macro Cells

Macro cells are usually large pieces of logic with standard functionality that can be reused. They can be simple cells that contain a couple of standard cells, or they can be highly complex cells containing entire subcircuits such as an embedded processor or memory block. They have a wide range of different shapes and sizes, as well. Macro cells can usually be placed anywhere in the layout area to optimize routing distance or the electrical properties of the design.

Due to the increasing popularity of reusing optimized modules, macro cells, such as adders and multipliers, have become popular. In some cases, almost the entire functionality of a design can be assembled from pre-existing macros; this calls for *top-level assembly*, through which various subcircuits, e.g., analog blocks, standard-cell blocks, and "glue" logic, are combined with individual cells, e.g., buffers, to form the highest hierarchical level of a complex circuit (Fig. 4.14).

Gate Arrays

Gate arrays are silicon chips containing elements with basic standard logic functionality, e.g., NAND and NOR, but with no connections (Fig. 4.15). The *interconnect* (routing) layers are added later after the chip-specific requirements are known. Thus, gate arrays are not initially customized and can be mass produced. The time-to-market for designs based on gate arrays is mostly constrained by the fabrication of

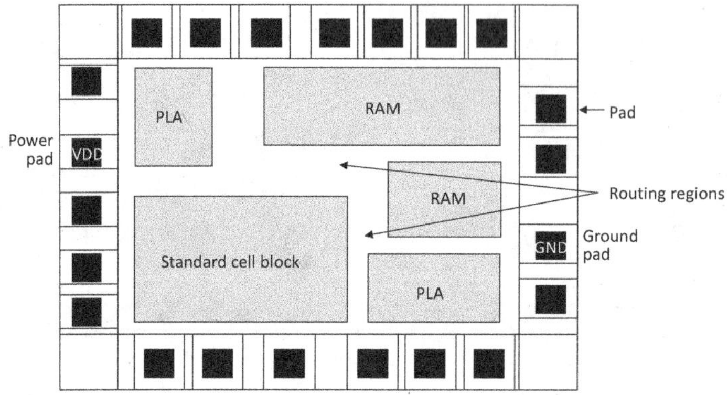

Fig. 4.14 Typical macro-cell layout that might include standard-cell blocks and other pre-designed modules [2]

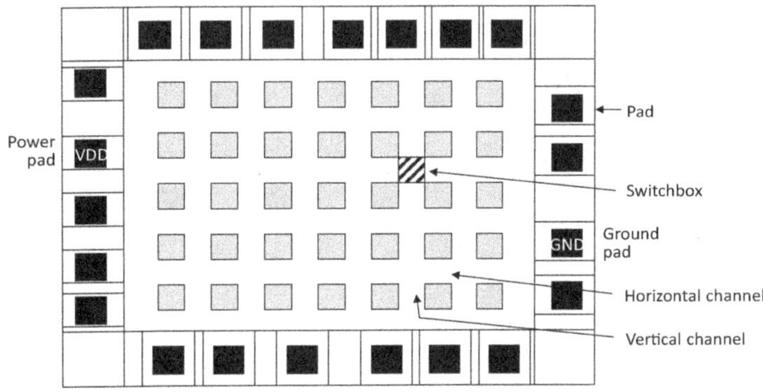

Fig. 4.15 A gate array layout

interconnects. Hence, gate array-based designs can be produced cheaper and faster than standard cell-based or macro cell-based designs, especially for low-volume production.

The layout of gate arrays is greatly restricted to simplify modeling and design. This limitation means that wire-routing algorithms can be very straightforward. Only the following two tasks are needed [2]:

- *Intracell* routing: creating a cell (logic block) by connecting certain transistors to implement a NAND gate, for example. Common patterns of gate connections are typically located in cell libraries.
- *Intercell* routing: connecting the logic blocks to form nets according to the netlist.

When physically designing gate arrays, cells are selected from what is available on the chip. A bad placement (which is strictly speaking an assignment) however may cause failures at the routing stage, since the demand for routing resources depends on the placement configuration.

Field-Programmable Gate Arrays (FPGAs)

Logic elements *and* interconnects in an FPGA are prefabricated and can be configured by users with switches realized by transistors (Fig. 4.16). *Lookup tables (LUTs)* are used to implement logic elements. Each logic element can represent any k-input Boolean function, e.g., $k = 4$ or $k = 5$ [5]. *Switchboxes* that join wires in adjacent routing channels are used to configure interconnects. LUT and switchbox configurations are read from external storage and stored locally in memory cells.

The main advantage of FPGAs is their customization without the involvement of a fabrication facility. This dramatically reduces up-front investment and time-to-market. Design costs are also lower because the design flow effectively ends at "netlist level". However, FPGAs typically run much slower and dissipate more power than ASICs, and thus may not be suitable for applications requiring the highest performance or the lowest power consumption. Above certain production volumes, e.g., millions of chips, FPGAs become more expensive than ASICs since the non-recurring design and manufacturing costs of ASICs are amortized.

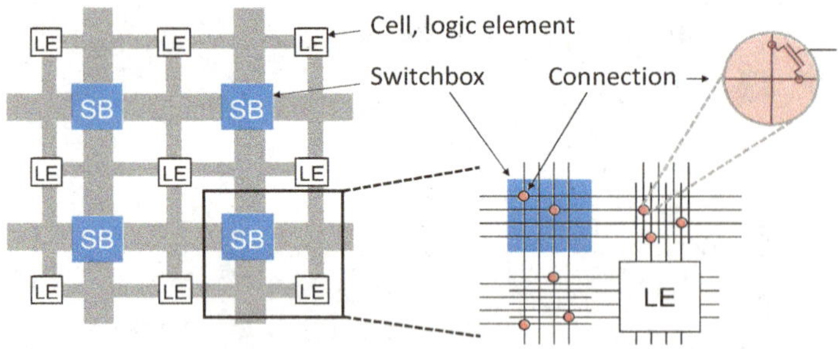

Fig. 4.16 A field-programmable gate array (FPGA) consists of logic elements (LE) and switchboxes (SB) that form a programmable routing network

Table 4.1 Comparison of design styles. *Design costs* are determined by the design effort to be spent, *mask costs* by the number of masks needed. *Fab costs* relate to the circuit's area consumption imposed by the design style, *performance* expresses the electrical performance of the circuit and *economic volume* indicates the production volume needed to reach cost amortization

Design style	Design cost	Mask cost	Fab cost	Performance	Economic volume
Full custom	High	High	Low	High	High
Standard cell	Low	High	Medium	Medium	Wide range
Macro cell	High (low if reuse)	High	Low	High	Wide range
Gate array	Low	Medium	High	Low	Low
FPGA	Very low	None	High	Low	Low

The major characteristics of the above-mentioned design styles are summarized in Table 4.1. These options represent a broad range in terms of cost, performance, and schedule (e.g., time to market) requirements that the design team must consider when evaluating and selecting a design style.

4.3.2 Top-Down, Bottom-Up and Meet-in-the-Middle Design

In the prior discussions about the different design styles, we assumed that the design process starts at high design levels and gradually progresses towards lower design levels. This type of design is generally known as a *top-down design flow*. While this approach is very effective in dealing with complexity, available degrees of freedom cannot be fully leveraged with it, as they cannot be used at the higher levels. These issues have heretofore tended to favor the use of a *bottom-up style*. In this style, elements designed at low levels based on geometrical design rules and electrical transistor properties are used to design elements at the next higher level.

The following styles, based on the direction of the design process, are available today:

Top-down. In the top-down design style, the design is progressively refined from the specification to the finished layout.

Bottom-up. With a bottom-updesign style, a very small circuit (such as a simple current mirror containing two active devices) is designed starting at the electrical level in the Y-chart. A library of different types of circuits is thus compiled. Macro blocks of these circuits are then created at the next higher level. These macro blocks can then be stored for re-use. This process is repeated until a circuit is produced as per the specification.

Meet-in-the-middle. This design style is a combination of the above two styles. The bottom-up design style is deployed at the low levels, while the top-down style is used at the higher levels.

Modern digital circuits are designed using standard elements (e.g., standard cells). These standard elements are created with a bottom-up strategy and subsequently stored in a library. This happens prior to and independent of the design of a circuit. When designing a digital circuit, these standard elements are selected from the library as fixed cells, i.e., they are instantiated in the design where they remain unchanged. Thus, the actual design of the circuit is a fully top-down design flow.

The meet-in-the-middle design style is the commonly used design method for analog circuits. Here, all basic elements including primitive devices are created during the design phase, i.e., in real time, in a bottom-up manner by hand or using layout generators which have to be guided by the layout designer through parameters. The entire (analog or mixed-signal) system is designed top-down considering the analog functional entities as building blocks. This also is illustrated in the Y-chart in Fig. 4.17 with the special case of a top-down structural decomposition combined with a bottom-up layout generation.

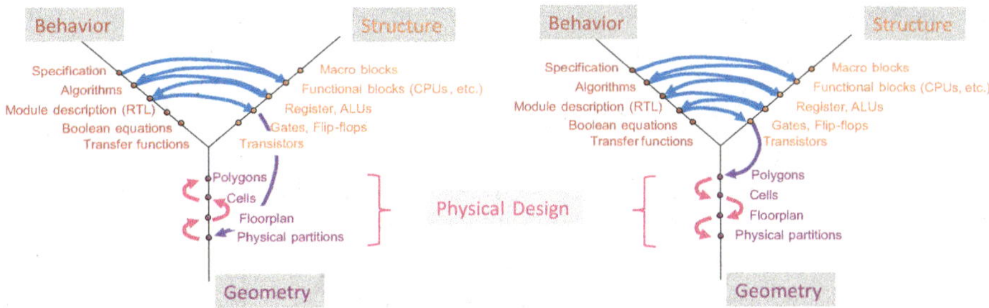

Fig. 4.17 The classical (digital) top-down design flow (left) and a top-down structural decomposition combined with a bottom-up layout generation as applied in analog design (right)

4.4 Design Tasks and Tools

Having introduced design models and design styles, we now wish to examine possible design tasks and related tools. This involves investigating in more detail the tasks performed in a single design step.

As mentioned before, a design step is a mapping between two design states. If we consider a design step towards the design goal (a geometric circuit representation, i.e., a layout), the associated mapping is called a *synthesis step*. The opposite direction is then labeled an *analysis step*.

4.4.1 Creating: Synthesis

A synthesis step brings the design subject to a state that is nearer to the implementation goal, i.e., the final design layout. Such a step is typically characterized by a decline in the degree of abstraction and a greater level of detail in the description. New information is thus introduced into the design description that was not explicitly there previously. The synthesis step is therefore creative.

The next design state in each case must be produced from the data and documents previously generated. Design states may be produced manually, automatically or semi-automatically. Designs can be produced automatically, for example, with place and route software, which generate cell layouts in the geometry domain from gate circuits in the structure domain, or with *silicon compilers*, which seamlessly compile a behavioral description from any abstraction level directly into a layout

Some placement and routing tools enable semi-automatic design. With these software packages, the designer can intervene interactively with the program or make pre-placements. Other packages analyze an algorithmic description and make prioritized proposals for architectures, which the designer can then select.

In general, we will call EDA tools generative ("creating"), if the view or both the view and the design level change. We differentiate between *synthesis tools* and *generators* in this regard. Synthesis tools perform a design step following general rules that are independent of the specific problem, whereas generators produce their result following a problem-specific course of action without the ability to optimize. A tool that turns a behavioral description (such as Boolean equations) into a gate circuit is thus a logic synthesis tool because it can perform this design step for all behavioral descriptions regardless of which specific function they realize. In contrast, a layout generator uses specified (electrical) parameters and generates an appropriate geometric representation of a cell or a circuit. An example is a *cell generator* that generates the layout of a resistor based on the parameter value defining the desired resistance. As the generator is limited to a specific problem, it must be re-written when the design problem changes, such as would occur for changes in the constraints or the related course of action, or if additional input parameters are introduced.

4.4.2 Checking: Analysis

An analysis step comprises an abstraction/extraction and operates in the opposite direction to the design direction (and thus to the synthesis operation). Details from a given detailed design description are summarized and generalized to generate information. (Note the difference to synthesis, where new information is introduced into the design description.)

These checking steps are commonly used to verify a synthesis step. An analysis is required if a synthesis step is performed manually or by a process that has not been formally verified itself, such that the "correctness by construction" cannot be guaranteed.

The only case where the result must not be checked is when the design step is performed automatically, and the correctness of the result is guaranteed by the tool used. As the process is extremely error-prone in all other scenarios, design methodologies must include "verifiers". There are two different types available: (i) tools that verify the correctness of the result and (ii) tools that verify the correctness of the design step.

Examples of the first category of tools are: design-rule checkers (DRC) that verify the geometrical rules for mask layout; syntax checkers for descriptions of functions in high-level description languages; and simulators especially for the different abstraction levels. The second set of tools comprises netlist extractors combined with netlist comparators that check the step from the netlist to the mask layout (layout verification). Programs that compare the results from simulations at different abstraction levels also belong to the second set of tools, as well as formal verification, insofar as it can be applied.

We also need to explain the difference between *validation* and *verification* (Fig. 4.18). Validation is the process of acquiring enough evidence that the design goals have been reached. Validation answers the question: "Has the correct circuit been designed?" The answer assigns a "value" to the object under consideration.

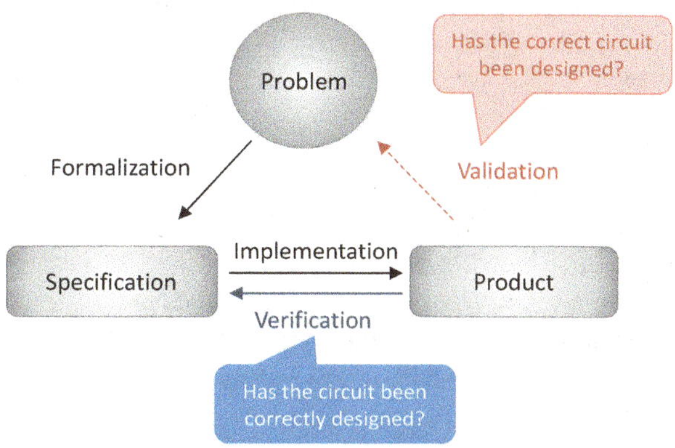

Fig. 4.18 Visualizing the difference between the terms "validation" and "verification". While validation checks the overall design goals from a customer prospective, verification confirms compliance with the formalized specification by design rule checking, for example

Verification, on the other hand, is the comparison with a standard known to be error-free to confirm compliance with explicitly given requirements. These can originate from the project (notably the specification) or from the technology (i.e., all types of design rules). Verification answers the question: "Has the circuit been correctly designed?" The answer "yes" or "no" is obtained by a formal examination based on mathematical models and operations.

4.4.3 Eliminating Deficiencies: Optimization

If the point reached in the Y-chart does not meet all requirements, an attempt is made to eliminate any deficiencies by means of (local) optimization. This is not a step towards or away from the design goal (w.r.t. our initial definition of a design step), as in the case of a synthesis or analysis step, rather the current design state is optimized here.

Examples are compacters that reduce the required surface area of the mask layout, logic optimizers that reduce the number of required gates, and architectural modifications for performance improvement.

4.5 Physical Design Optimization and Constraints

4.5.1 Optimization Goals

Physical design is a complex *optimization problem* with several different objectives, such as minimum chip area, minimum wirelength, and a minimum number of vias. Improving circuit performance, reliability, etc. are some common optimization goals. How well the optimization goals are met determines the quality of the layout.

Different optimization goals may be difficult to incorporate in algorithms and may conflict with each other. However, tradeoffs among multiple goals can often be expressed concisely by an *objective function*. For instance, we can optimize wire routing with the formula:

$$F = w_1 \cdot A + w_2 \cdot L, \tag{4.1}$$

where A is the chip area, L is the total wirelength, and w_1 and w_2 are *weights* that represent the relative importance of A and L.

Hence, the weights determine the impact of each objective goal on the overall cost function. In practice, $0 \leq w_1 \leq 1$, $0 \leq w_2 \leq 1$, and $w_1 + w_2 = 1$.

4.5.2 *Constraint Categories*

When optimizing the layout, *constraints* need to be considered. Compliance with these constraints is mandatory for the proper functioning and implementation of the layout (methodological design constraints, which we will introduce later, are an exception here).

While missed optimization goals limit the quality of the circuit (and not its overall functionality), constraints are critical "boundary values" that, if not complied with, the circuit layout is rendered useless. Hence, constraints are either met or violated; the degree to which they are met or not, as the case may be, is immaterial (again, note the difference to optimization goals).

The constraints (aka boundary conditions) for a layout design can be divided into three categories, as follows (Fig. 4.19).

Technological Constraints

Compliance with the technological constraints assures the manufacturability of the IC. The technological constraints are derived from the fabrication technology and their boundary values and are converted into geometrical design rules (Chap. 3, Sect. 3.4.2). These rules can be divided into the five basic groups—width, spacing, extension, intrusion, and enclosure rules (cf. Fig. 3.20 in Chap. 3).

Geometrical design rules are stored in a technology file that is part of the design suite for a given technology (process design kit, PDK, Chap. 3, Sect. 3.5.1). They are defined by the fab, hence, they are given to the circuit and layout designer prior to the actual design.

Functional Constraints aka Electrical Constraints

Compliance with these constraints assures the targeted electrical response (and thus the functionality) of the circuit. They also consider the required circuit reliability. These rules arise from the project or from the technology.

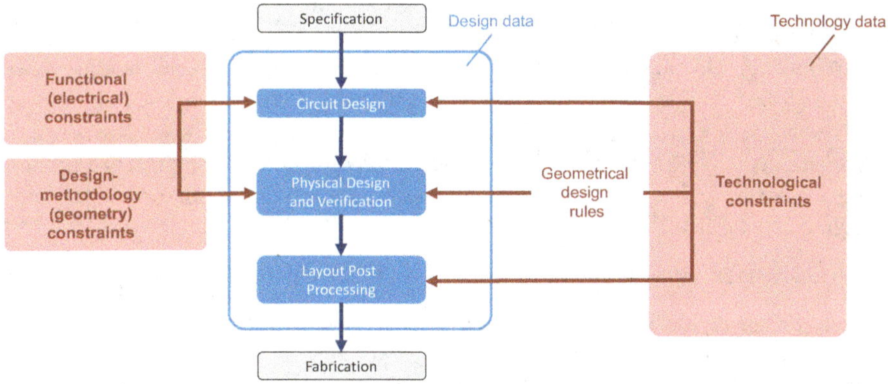

Fig. 4.19 The three constraint categories and their relations to the main design steps

Examples of project-specific functional constraints are upper bounds for mutual couplings to prevent disturbances to signals in various circuit lines. Examples of technology-specific functional constraints are upper limits for the current density in metal lines to guard the circuit against degradation by the electromigration effect.

Functional constraints are defined prior to layout design: they are the result of simulation and are forwarded to the layout designer, e.g., in a Standard Delay Format (SDF) file. They are verified during or after layout design, e.g., by calculating signal delays based on actual interconnect geometries.

Design-Methodology Constraints aka Geometry Constraints
These constraints are applied to reduce the complexity and degree of difficulty associated with a design. Their purpose is to enable the layout task to be performed with the aid of EDA tools.

Degrees of freedom that are theoretically available are artificially restricted by these constraints. They have made algorithmic solutions possible especially in digital design. Examples of these constraints are layer-dependent preferred routing directions or row assignments for standard cells with pre-defined power and ground connections.

Design-methodology constraints are either defined prior to physical design when choosing a design style such as standard cells, for instance, or during physical design when assigning certain optimization requirements, such as layer-specific preferred routing directions.

4.5.3 Physical Design Optimization

When considering optimizing goals and constraints in physical design, several challenges arise:

- Optimization goals may conflict with each other. For example, reducing wirelength too much can cause congestion and increase the number of vias.
- Constraints often cause discontinuous, qualitative effects even when objective functions remain continuous. The floorplan design, for example, might allow only some of the bits of a 64-bit bus to be routed with short wires, while the remaining bits must be detoured.
- Constraints, arising from scaling and tough interconnect requirements, are becoming more restrictive, and new types of constraints are being added in each new technology node.

These challenges motivate the following physical design characteristics [2]:

- Each design style (Sect. 4.3) requires its own custom flow. That is, there is no universal design tool that supports all design styles.
- Imposing design-methodology constraints (geometry constraints) during chip design can potentially ease the problem at the expense of layout optimization.

For instance, a row-based standard-cell design is much easier to implement than a full-custom layout, but the latter could produce significantly better electrical characteristics.

- In order to handle the extreme complexity of design tasks, the design process is divided into sequential steps. For example, placement and routing are performed separately, each with specific optimization goals and constraints that are evaluated independently. However, there is a caveat here: this is not a global optimization as each step is optimized on its own.

4.6 Analog and Digital Design Flows

4.6.1 *The Different Worlds of Analog and Digital Design*

Comparing analog and digital circuits, we see that they make use of the same physical principles, the same materials, and the same basic elements. In addition, in both cases these elements—mostly transistors—are electrically connected to form larger circuits. Obviously, analog and digital circuits are grounded on the same construction principle.

However, there are many differences between analog and digital: the design flows and tools are different in both cases. Furthermore, extremely different automation levels are encountered in both flows. In fact, analog and digital look like two completely different worlds.

Investigating this further, we find that there is a fundamental difference between the two technologies, which is the root cause for the genesis of these two different worlds. If we take the circuit in Fig. 4.20 as an example, we see that the output is a continuous *curve* (Fig. 4.20, right). In digital design, however, we are only interested in the initial and final voltages; this is a binary function that is either on ("1") when the input signal is off ("0") or off ("0") when the input signal is on ("1") (Fig. 4.20, left). This is why this circuit is called an "inverter" by digital designers. Here, we only need to define thresholds on the voltage axis and rise and fall times on the time axis acting as "forbidden" ranges (yellow shaded regions) in order to control correct voltage and timing. The real transient response, which may vary, is irrelevant as long as signals are outside the forbidden voltage range—which can be achieved by reading them outside the forbidden time range.

Obviously, an *abstraction* is all that is needed in digital designs. In analog design, on the other hand, there is no abstraction: all signal characteristics must be considered.

Hence, digital is the world of discrete signals, analog that of continuous signals. Consequently, the *smallest design entity* is different in both circuits:

- The smallest entities to be considered in digital designs are logic gates. These gates are predesigned subcircuits, which are not changed during the design process.

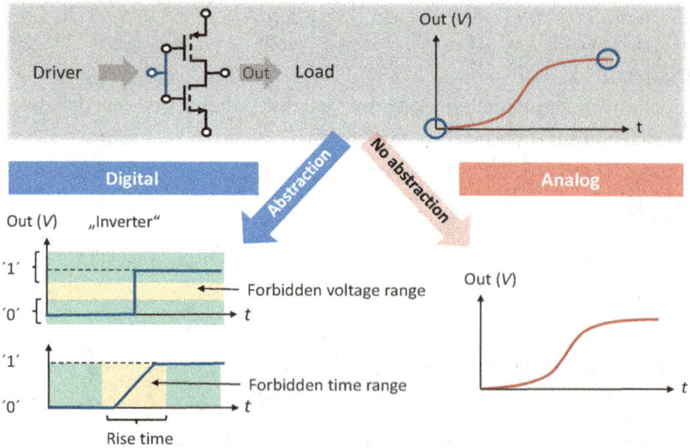

Fig. 4.20 An illustration of the differences between digital (left) and analog (right) circuits. While the former focuses on discrete values at discrete points in time, analog circuits, and thus analog designs, require the exact continuous signal response to be taken into account. Thus, digital designs deal with logic states, while analog designs require the actual transistor parameters to be considered

- In analog flows, however, the basic design entity must be the transistor as its properties and response must be specified in detail in order to build the analog circuit. The transistor "width" and "length" parameters must be carefully selected during the design, for example.

Figure 4.21 illustrates these characteristics as they are also reflected in the Y charts for both digital and analog circuit designs. Obviously, different design methodologies are required in physical design to deal with the differences between analog and digital design discussed above.

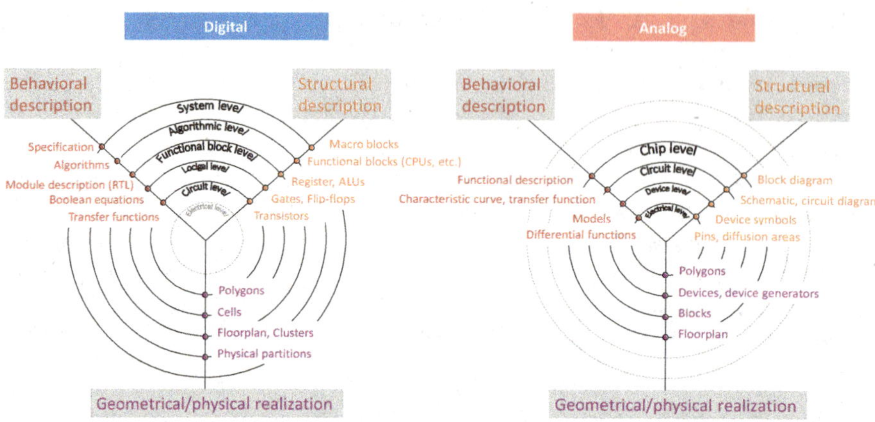

Fig. 4.21 Comparing the Gajski-Kuhn Y-charts for digital (left) and analog circuits (right)

In general, there are three different types of integrated circuit: (i) digital, (ii) pure analog, and (iii) mixed-signal; mixed-signal comprises both digital and analog subcircuits. Let us now examine the different design approaches for analog and digital (sub-)circuits in detail.

Analog circuits are generally less complex in terms of transistor count and are designed in a manual fashion. As many functional constraints must be considered here, the distinct lack of design automation and analog synthesis means that manual design is still widespread today.

Digital circuits, by contrast, are notable for their large number of nets and cells and their fully automated design flows. While the numerical complexity of analog circuits generally does not exceed a few thousand nets, digital circuits typically comprise millions of nets.

The design steps for digital circuits are mostly discrete and are performed sequentially. Analog design steps, on the other hand, typically overlap and several steps are performed quasi simultaneously. In analog, device generation, module placement and routing are usually executed together (Fig. 4.22).

You might recall that, in general, we move inwards in the Y-chart when designing a circuit. In other words, the closer we are to the center, the more specific the model and our design becomes. This also means that any given design problem is solved by sequentially removing its degrees of *design freedom*.

Functional models with many degrees of design freedom are successively transformed into equivalent ones with fewer degrees of freedom. Take for example a functional specification: it is first transformed into a netlist, then into a floorplan, a placement order, a wired layout and finally into a physical mask layout with no further degrees of design freedom. Hence, the design freedom in a design flow is restricted

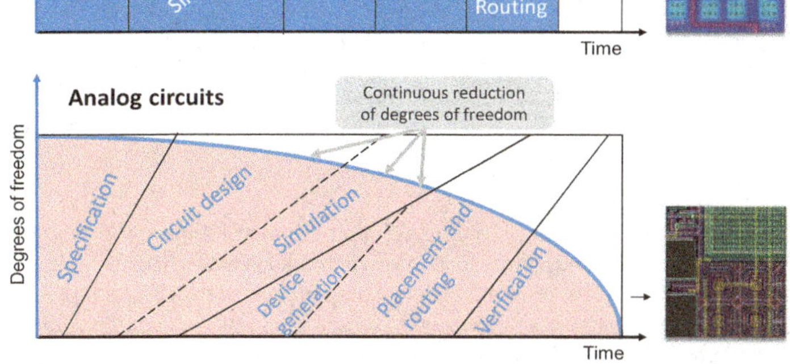

Fig. 4.22 Simplified design flows for digital (above) and analog circuits (below). In the analog flow, design steps typically overlap and are tightly linked [8]. Both flows are also characterized by a reduction in the "design freedom" throughout the process

over time, however, it is curtailed step by step in digital designs and gradually in analog circuit designs (see Fig. 4.22).

While there have been significant improvements recently in physical design automation for analog IC design, it has not advanced at anything close to the rate of its digital counterpart. (We discuss the related gap between analog and digital design productivity in Chap. 1, Sect. 1.2.3, cf. Fig. 1.11). For many years, efforts for solving the analog layout problem have focused on techniques like those successfully applied in the digital space. These techniques are mostly based on optimization algorithms that require a mathematical modelling of the design problem, which is always in conjunction with a certain degree of abstraction or, in other words, with a reduction of the degrees of freedom. On the one hand, the model may be constructed to be at a low level of abstraction due to the attempt to closely mirror physical reality. In this case the solution space becomes very complex and optimization algorithms are generally too weak to find sufficiently "optimal" solutions in reasonable time. On the other hand, there may be an emphasis on the use of efficient algorithms to the detriment of model quality. In this case, optimizers can easily find a formally "optimal" solution, but its quality suffers from the fact that the model is too far away from physical reality.

Figure 4.23 illustrates this dilemma. The efficiency of optimization algorithms, often measured in speed and the ability to find a global optimum, is generally inversely proportional to the accuracy of the underlying mathematical model, which is derived

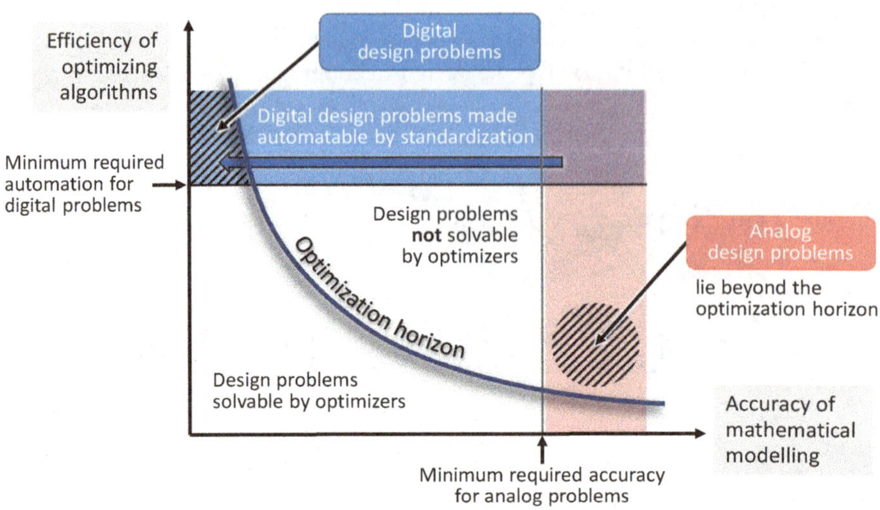

Fig. 4.23 Illustration of the efficiency of optimization algorithms, which is generally inversely proportional to the accuracy of the underlying mathematical model visualized as an "optimization horizon" (blue curve). Analog designs of high qualitative complexity demand high modeling accuracy. Digital designs that are usually of high quantitative complexity require high algorithmic efficiency. Only design problems located below the curve can be satisfactorily solved by optimizers—which thus excludes most analog design problems

from the physical world. This interrelation is visualized by the *optimization horizon* in Fig. 4.23 (round curve). Only design problems located below the optimization horizon can be satisfactory solved by optimizers.

Analog design problems are highly complex *qualitatively* and therefore need a high degree of modeling accuracy. As such, they lie in the red shaded region of Fig. 4.23. The situation is different for digital design problems, which are highly complex *quantitatively*, and therefore require a high degree of algorithmic efficiency. Consequently, they lie in the blue shaded region of Fig. 4.23.

In contrast to digital design problems, where a high level of abstraction (i.e., a standardization) had paved the way for design automation based on optimizers (see Fig. 4.23, top left), in analog design an abstraction from physical reality is unacceptable. In other words, analog designers cannot abstain from using the degrees of freedom. Quite the contrary, they extensively exploit the degrees of freedom because they are key to fulfilling all the above described requirements in order to achieve a desired circuit quality. Thus, unfortunately, analog design problems lie beyond the optimization horizon (i.e., above the curve).

The impossibility of abstraction is a severe obstacle for the application of optimization-based automation methods to analog design problems. In our opinion, this is the main reason for the observation that, despite occasional successes, optimization-based approaches have not yet been able to establish themselves in the analog domain.

4.6.2 Analog Design Flow

Analog circuits are designed in single modules, with a verified layout produced for each module. As indicated above, the design process itself is only minimally automated; starting with primitive devices, which can be created with the aid of layout generators, an engineer typically drafts the layout for an analog circuit (and analog parts in mixed-signal designs) almost completely manually using a graphic editor.

State-of-the-art design flows for analog circuits are characterized by two different design styles—top-down and bottom-up [8]. While so-called *optimizers* perform layout generation top-down, the procedural approaches (*procedures*) generate the final layout with the bottom-up style.

As illustrated in Fig. 4.24 (left), the top-down approach makes use of optimization-based tools similar to conventional digital flows. Their overall structure is given by an exploration engine generating solution candidates by exploring a defined solution space and an evaluation engine selecting the "best" candidates based on design objectives in a loop-wise manner [7]. An optimizer can produce new (genuine) design *solutions*.

In contrast, bottom-up procedures re-use "expert knowledge" with the *result* of solutions previously conceived and captured in a procedural description by a human expert, thus imitating the expert's decisions in a straight-forward manner (Fig. 4.24,

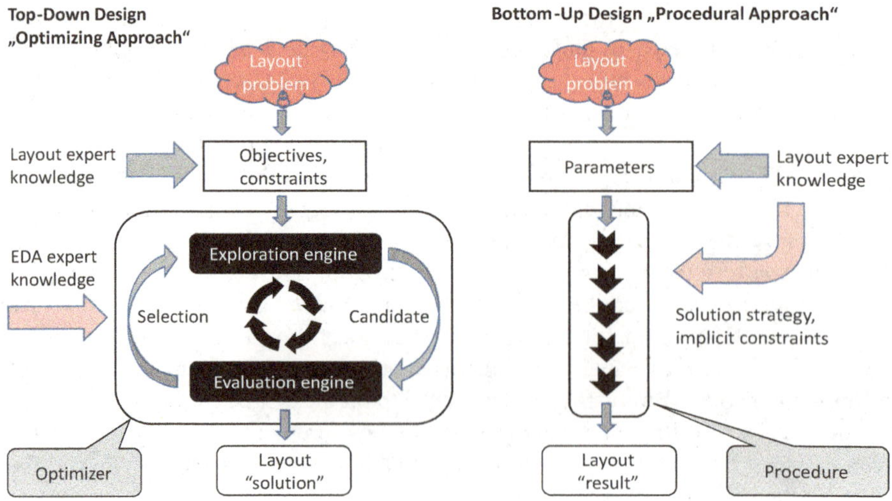

Fig. 4.24 Top-down optimization versus bottom-up procedures in analog design [8]. The top-down approach (left) makes use of optimization-based tools similar to conventional digital flows. The bottom-up automation method (right) reproduces a design solution previously conceived and captured in a procedural description by a human expert. The gray arrows indicate the data flow of the layout design process. Whereas optimizers are built by EDA tool experts, procedures are built by layout experts (pink arrows)

right). Cell generators, such as *PCells* (which stands for *parameterized cells*), are an example. They are widely used procedures that process the required electrical or geometrical device properties as input parameters and automatically create a correct layout cell for a specific technology.

As already mentioned, analog design is characterized by a large number of degrees of freedom, influencing factors, and functional constraints. Circuit topography, component parameters, arrangement of matching components, and application of special routing patterns are some of the degrees of freedom in analog design. Influencing factors that negatively impact the performance and robustness of an analog circuit are, among others, nonlinearities, parasitic components, electromagnetic coupling, temperature, and electromigration. Constraints, as mentioned before, describe boundaries to a circuit's performance and robustness, and they absolutely must be fulfilled for a successful design. Unfortunately, a formal description of the influencing factors and constraints is often missing in analog design. This lack of a complete problem definition is the primary reason why the analog design process has not been automated yet [3, 6, 8]. (We discussed this and further reasons in Sect. 4.6.1.)

Analog "expert knowledge" is an important "ingredient" in the design of analog circuits which, as a human resource, cannot be properly translated into formal expressions of high-level, abstract design requirements (constraints). Since procedures contain the "expert knowledge" implicitly and can thus make use of it, we believe "bottom-up automation" based on the above-mentioned procedural approach is an indispensable element in any future "analog synthesis flow" (which we will explore further in Sect. 4.7).

4.6.3 Digital Design Flow

One of the main differences between analog and digital circuits is that while analog values and functions are processed in the former, digital values are processed in the latter. Since digital values are discrete with respect to time and magnitude, they are less susceptible to disturbances. Furthermore, fewer constraints are needed in the design to assure its proper functioning. Constraint handling can thus easily be automated in digital design. Hence, digital integrated circuits can be designed using well-understood synthesis algorithms; this can be done with *logic synthesis* (which replaces "Circuit Design" in Fig. 4.3), as discussed with respect to HDLs in Sect. 4.1.

Obviously, one can design much more complex circuits in an automated flow than manually. Hence, integrated digital circuits have become increasingly complex with each passing year, and today they can have billions of components and electrical interconnections (aka nets). At today's level of complexity, they can only be economically designed, verified and analyzed with the support of highly sophisticated design algorithms. The complexity continues to ramp up rapidly as a result of both miniaturization (Moore's Law, Chap. 1) and technological advances, which continue to permit ICs with larger and larger numbers of components to be fabricated.

Automated layout generation, often referred to as *layout synthesis,* is less challenging for digital circuits than with analog circuits, as there are fewer functional constraints. In addition, digital circuits are less sensitive to small voltage changes than their analog counterparts. The proper functioning of digital logic depends essentially on the reliable differentiation between a few different digital logic states. Also, these circuits comprise a relatively small number of different types of subcircuits (NAND, NOR, INV, etc.).

The scalability of the synthesis algorithms (for both logic synthesis and layout synthesis) is key due to the on-going verity of Moore's Law in digital integrated circuits. So-called heuristics are typically used in this context, as they enable the synthesis steps to yield practically "optimal" solutions very rapidly. The layout synthesis of digital circuits can thus be automated with relatively simple algorithms.

Verifications of layout synthesis steps are carried out to check for compliance with predefined constraints based on design rules. These verifications are an integral part of the automated design process, because the synthesis algorithms never take all constraints into account and are typically optimized for speed rather than quality—hence, the correctness of the output must be verified. The computational overhead in the verification as well as in the synthesis steps must scale with the complexity. The entire circuit cannot normally be fully analyzed, as this would require too much computation time. Instead, the verification process benefits from several filtering techniques, which narrow this partially complex task down to a few select sections of the complete circuit, that is, to specific critical nets or modules, for example.

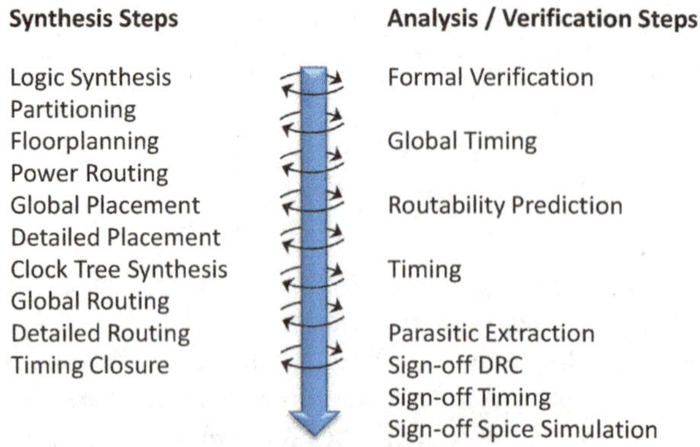

Fig. 4.25 Design flow for digital circuits with its typical synthesis-analysis loops [4]

Hence, today's design flow for digital circuits is characterized by a host of synthesis-analysis loops as shown in Fig. 4.25. On the one hand, the flow consists of a series of synthesis steps, which methodically concretize circuit geometry (Fig. 4.25, left). On the other hand, there is a set of verification steps alongside these synthesis steps. They check the correctness of each synthesis step and thus, ensure that the resulting circuit acquires the required electrical characteristics and functions, and meets the reliability and manufacturability criteria (Fig. 4.25, right).

4.6.4 Mixed-Signal Design Flow

The majority of today's ICs are mixed-signal designs. A *mixed-signal integrated circuit* combines both analog and digital circuits on a single semiconductor die. Mixed-signal ICs are often used to convert analog signals to digital signals so that digital circuits can process them. Mixed-signal ICs comprise a mix of digital signal processing and analog circuitry and they are usually designed for a very specific purpose. Designing these chips requires a high level of expertise (Fig. 4.26).

Both analog and digital designers claim their design tasks are "highly complex", and in fact both are correct, but in a different sense. As already pointed out, analog designs are characterized by a much richer and more complex set of design constraints that need to be considered simultaneously and which may span several domains (e.g., electrical, electro-thermal, electro-mechanical, technological, geometrical domains). Therefore, in typical mixed-signal ICs, the effort needed to design the analog part typically far exceeds the effort for the digital part. This is true despite the fact that analog modules typically contain only a small number of devices compared to digital ones. Therefore, when talking about complexity, we prefer to distinguish between (i) *quantitative* complexity, as observed in digital designs, referring mainly to the number of design elements (also called "More Moore" according to Moore's Law), and (ii) *qualitative* complexity. The latter is rooted in the diversity of the requirements to be considered (also called "More than Moore"), as found in analog designs.

Mixed-signal ICs are more difficult to design and manufacture than analog-only or digital-only integrated circuits. For example, fast-changing digital signals send noise to sensitive analog inputs, requiring extensive shielding techniques to be applied during the design stage. Automated testing of the finished chip can also be challenging.

Design methodologies must combine the advanced digital design methods and the rather primitive, mostly manual design methods for analog circuits. At the same time, special care must be taken with the interfaces and interactions of the two types of circuitry, requiring, for example, a chip-level verification methodology spanning analog and digital domains (see Fig. 4.26).

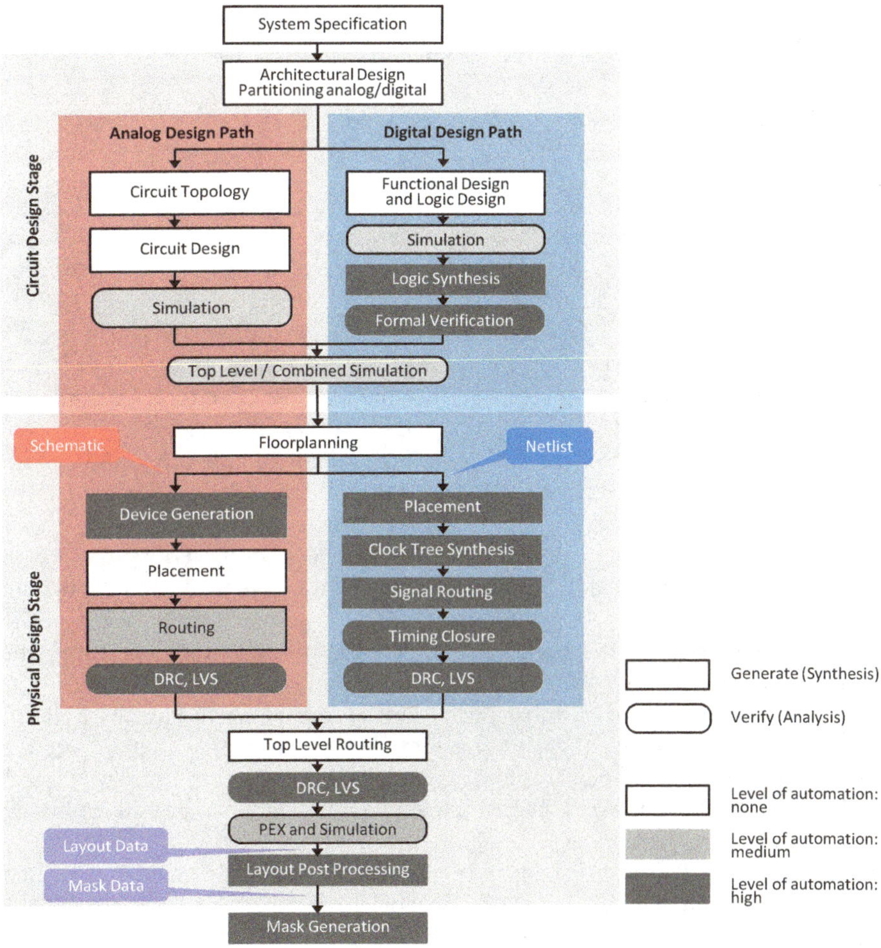

Fig. 4.26 Example of a design flow for mixed-signal circuits that is characterized by a combined simulation procedure and top-level routing, followed by a mutual verification method. Note the different levels of automation indicated by various shades of gray. Iterations (loops), which can be activated by analysis steps, are omitted to aid clarity

4.7 Visions for Analog Design Automation

As noted above, physical analog IC design has not been automated to the same degree as digital IC design. This gap is primarily rooted in the analog IC design problem itself, which is considerably more complex even for small problem sizes. Next, we will present two visions which could overcome this problem. While breaking with conventional design automation approaches, these two proposed paradigm changes could lead to a new class of (higher level) design techniques that brings us one step closer to the goal of full-scale analog design automation.

We want to emphasize that our proposals do not require a disruptive change in the way layout designers are used to working. Both proposals can be introduced gradually as an evolution of the common industrial design flows.

4.7.1 A "Continuous" Layout Design Flow

As discussed earlier and illustrated in Fig. 4.22, the degrees of design freedom are *gradually* reduced in the modern interactive analog layout design style. This is not an intrinsic characteristic of this style but is caused by the huge number of recursions required. These recursions result from repeating the same design steps, notably device generation, placement, and routing, again and again in order to make necessary modifications. Previously defined parameters, such as a transistor's folding characteristic or the width of a wiring segment, must be redefined due to constraints that emerge at a later stage in the design process and that therefore cannot be foreseen. The greatest amount of time and effort in analog layout work is spent on these modifications. Hence, the efficiency of the widely used interactive layout style can be greatly improved by reducing the number of these recursions.

Before outlining our proposal for a solution, we need to discuss the root cause of this problem, which is that the *edit commands* used in today's layout editors are simple implementations of *design steps* for the purpose of interactive usage. The problem arising from this similarity can be best explained by looking at how the editor commands affect the degrees of design freedom.

Each design parameter (i.e., the property of a design element such as a wire width) can be regarded as a degree of design freedom. When a design parameter is set to a value, the associated degree of freedom is eliminated. We will examine two typical layout tasks and their corresponding editor commands next to show their impact on the degrees of design freedom.

The task "routing a net" is usually performed by drawing paths. A path command simultaneously eliminates all degrees of freedom an electrical connection can have (i.e., layer assignment, x- and y-coordinates, Steiner nodes, wire width, and so on). This is an inevitable consequence of the command itself. The same thing happens when "placing" a device, where all related degrees of freedom (i.e., x-, y-coordinates of absolute position, orientation and, implicitly, all relations to other elements as well) are eliminated with one single mouse click.

The underlying reason for the recursions mentioned above is due to this "design-step-like" behavior of today's layout editors, which only allows the degrees of freedom to be handled in combination. The degrees of freedom are then *implicitly* eliminated. Thus, a designer is permanently forced to make implicit decisions concerning the degrees of freedom without having the appropriate information at the time of decision. The editing work is then done by trial and error and involves many recursions.

Despite its deficiency, this aspect of today's layout editors is accepted by the analog-designer community as this way of working seems "natural" and everyone has become used to it. We would like to raise awareness of this "blind spot" and outline a proposal for an innovative *continuous layout design flow* that addresses this issue.

We suggest that only those degrees of design freedom that are fully defined at the current design stage be removed. Then, functions such as place and route are decoupled from their respective *fixed* degrees of freedom such that these degrees can be accessed directly and thus managed independently. Hence, they are now eliminated *continuously* during the layout process, but each one is only eliminated when it is necessary and appropriate to do so according to its "definition status". This intrinsic flow operation is performed until we reach the physical mask layout which contains no further degree of design freedom. In such a *continuous layout design flow*, the layout would be generated first in an almost symbolic manner before getting more and more detailed with actual physical parameters until it finally "crystallizes" to a real physical design.

For example, using this continuous design flow, a net is laid out as follows: First the net routing region is assigned, afterwards the preferred routing layer is determined, and at a later stage, when the current flows are known, the appropriate wire widths are assigned to their associated net sections.

A continuous design flow would also support the re-use of previous layout solutions which is a well-known issue along with its current limitations. Some of these problems are that (i) the design is too application-specific, (ii) even small changes to the circuit may require large changes in layout, (iii) a new technology node is used, and (iv) the shape of a layout module does not fit. However, careful consideration reveals that the root cause is that the layout view does not encompass any remaining degrees of freedom.

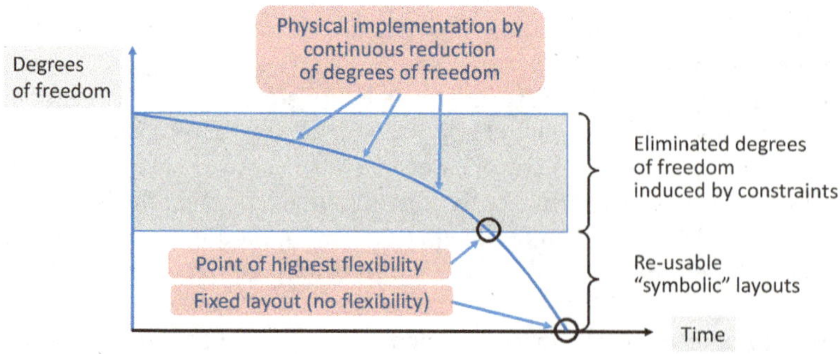

Fig. 4.27 In the proposed design flow, the degrees of design freedom are continuously reduced [8]. Re-using unfinished layout (i.e., symbolic level) supports the adjustment to new project-specific requirements because the symbolic level still contains degrees of design freedom needed for the adjustment

This issue can be addressed in a continuous design flow by re-using a layout at a *symbolic stage* defined as follows: A re-usable design may only contain design freedoms that do not impact constraints; hence, the remaining design freedoms are *unconstrained*. In turn, the absence of constrained degrees of freedom indicates that all constraints are met. In other words, all design decisions induced by fulfilling a constraint are maintained, which is in fact the (long-sought-after) re-use of the implemented expert design knowledge (Fig. 4.27).

The design can be modified for re-use to meet project-specific requirements with the help of the remaining degrees of freedom. The problems mentioned above can thus be overcome. A high number of remaining degrees of freedom means greater residual flexibility, and thus, greater "re-usability". And the more the design problem to be solved resembles the re-use candidate, the fewer the remaining degrees of freedom that are needed for modifications and the less work is needed. This is a major advantage over current re-use approaches which lack this ability to modify a design to suit a given project.

This "continuous" layout design flow can be implemented by enhancements of today's analog layout tools, which are based on graphic editors. Each new function allowing the elimination of single degrees of freedom could be instigated as new edit commands in a graphic editor. Special techniques for visualizing the status of a symbolic layout could be built using the existing capabilities of modern graphic editors.

4.7.2 A "Bottom-Up Meets Top-Down" Layout Design Flow

We concluded in Sects. 4.6.1 and 4.6.2 that top-down automation alone cannot solve the analog layout problem in its entirety. The best way to address the discussed problems in analog top-down optimization is to complement this strategy with appropriate bottom-up procedures (e.g., PCells). The rationale here is that bottom-up procedures can potentially provide the missing features in optimization-based approaches, as outlined next and visualized in Fig. 4.28.

A common issue with top-down automation is that algorithmically "invented" layout solutions are rejected by expert designers because they do not meet their expectations. Analog designers prefer to re-use existing, silicon-proofed design solutions which usually incorporate years of design knowledge, both from a human expert's personal experience and a company's design group portfolio.

In this regard, a strength of bottom-up automation is its intrinsic ability to augment the re-use of singular design solutions (i.e., "copy-paste") to a more sophisticated method of re-using design solution strategies. Therefore, novel techniques are needed that enable circuit and layout designers to efficiently translate their design strategies into novel automated procedures. Only then will we see real progress with bottom-up approaches. This is mainly a question of tool interfaces. The closer the interfaces to these techniques match the designer's way of thinking and the better they are adapted to his/her work style, the more easily the techniques will capture the valuable expert knowledge, skills and creativity that are mainly absent from mere top-down automatisms. We believe that these interfaces and techniques must be more than just novel description languages or tool wizards: They should be "schematic-like" for a circuit designer and "layout-editor-like" for a physical designer.

As mentioned in Sect. 4.6.1, analog design automation is severely handicapped by the qualitative complexity of analog constraints (aka "expert knowledge"). By restricting top-down optimization to "strategic constraints", such as high-level design requirements, and by delegating the remaining constraints to bottom-up procedures, this problem could be eliminated. The ability of bottom-up procedures to make use of implicitly integrated expert knowledge is an ideal supplement to optimization approaches. In this regard the optimizers can be regarded as "senior tools", which delegate special tasks to their "subordinate" procedural tools.

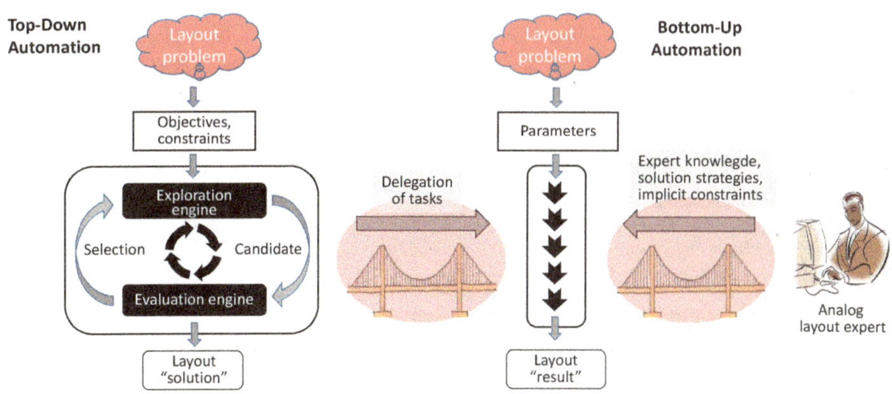

Fig. 4.28 The combination of top-down and bottom-up design approaches requires two bridges so that both design styles can be combined (middle) and human design experts can easily apply their design know-how (right)

Despite the widely held assumption that procedural automation is simply a matter of handcraft, we are convinced that developing the techniques mentioned above is an academically appealing and practically profitable challenge for future EDA research. Conflating the resulting bottom-up procedures with existing top-down automation may be the key to finally achieving the full analog synthesis flow that has been the holy grail in this field for many years now.

To make this vision a reality, we need at least two kinds of "bridges" (see Fig. 4.28). First, sophisticated techniques must be developed that enable human design experts to easily capture their design know-how in bottom-up automation procedures. Second, technical concepts are needed that intelligently combine the different automation paradigms of optimization-based (top-down) and procedural (bottom-up) approaches.

In summary, conflating top-down and bottom-up design styles to one *bottom-up meets top-down design flow* should enable us to incorporate the expert knowledge mentioned above while also addressing high-level design requirements.

References

1. D.D. Gajski, R.H. Kuhn, Guest editor's introduction: new VLSI tools. *IEEE Comput.* (1983). https://doi.org/10.1109/MC.1983.1654264
2. A.B. Kahng, J. Lienig, I.L. Markov, et al., *VLSI Physical Design: From Graph Partitioning to Timing Closure* (Springer, 2011), ISBN 978-90-481-9590-9. https://doi.org/10.1007/978-90-481-9591-6
3. A. Krinke, M. Mittag, G. Jerke, et al., Extended constraint management for analog and mixed-signal IC design, in *IEEE Proceedings of the 21th European Conference on Circuit Theory and Design (ECCTD)*, (2013), pp. 1–4. https://doi.org/10.1109/ECCTD.2013.6662319
4. J. Lienig, Electromigration and its impact on physical design in future technologies, in *Proceedings International Symposium on Physical Design (ISPD)*, (ACM, 2013), pp. 33–40. https://doi.org/10.1145/2451916.2451925
5. C. Maxfield, *FPGAs: Instant Access* (Newnes, 2008). ISBN 978-0750689748
6. A. Nassaj, J. Lienig, G. Jerke, A new methodology for constraint-driven layout design of analog circuits, in *Proceedings of the 16th IEEE International Conference on Electronics, Circuits and Systems (ICECS)*, (2009), pp. 996–999. https://doi.org/10.1109/ICECS.2009.5410838
7. R. Rutenbar, Design automation for analog: the next generation of tool changes, in *Proceedings International Symposium on Physical Design (ISPD)* and *1st IBM Academic Conf. on Analog Design, Technology, Modelling and Tools* (ACM, 2006), pp. 458–460. https://doi.org/10.1145/1233501.1233593
8. J. Scheible, J. Lienig, Automation of analog IC layout—challenges and solutions, in *Proceedings International Symposium on Physical Design (ISPD)*, (ACM, 2015), pp. 33–40. https://doi.org/10.1145/2717764.2717781

Chapter 5
Steps in Physical Design: From Netlist Generation to Layout Post Processing

Due to its complexity, the physical design process is divided into several primary steps. Having introduced in Chap. 4 the flow, constraints and methodologies of today's physical design process, we now investigate the various steps required to generate its output: a layout. These steps, which transform a netlist into optimized mask data, are dealt with one by one in this chapter.

A layout is generated from a netlist. We first describe how a netlist is created, that is, either by using hardware description languages (HDLs) in digital design (Sect. 5.1), or by deriving it from a schematic, as is common in analog design (Sect. 5.2). Then the physical design steps, comprising partitioning, floorplanning, placement, and routing, are presented in detail (Sect. 5.3). All of these steps are supported by highly sophisticated EDA tools in the case of digital designs, which is our focus here. We also discuss in this section the key aspects of symbolic compaction, standard-cell design and PCB design. (Physical design of analog circuits is covered in Chap. 6.)

When the physical design phase is completed, the resulting layout must be verified. This verification step confirms both functional correctness and design manufacturability. Methodologies and tools for comprehensive design verification, with a focus on physical verification, are covered in Sect. 5.4. Finally, we briefly touch on layout post-processing methodologies, such as resolution enhancement techniques (RET), that might impact physical design (Sect. 5.5).

5.1 Generating a Netlist Using Hardware Description Languages

5.1.1 Overview and History

The exploding complexity of digital electronic circuits since the mid 1960s (Moore's Law, Chap. 1, Sect. 1.2.3) has increasingly required circuit designers to use digital

J. Lienig and J. Scheible, *Fundamentals of Layout Design for Electronic Circuits*,
https://doi.org/10.1007/978-3-030-39284-0_5

logic descriptions that characterize the design at a high level, without being tied to a specific electronic technology. As part of this shift came a requirement for standard text-based descriptions of the structure of electronic systems and their behavior. Consequently, the first *hardware description languages* (*HDLs*) appeared in the late 1960s [2]. They provided a precise, formal description of electronic circuits and enabled them to be automatically analyzed and simulated. These languages also introduced the concept of *register transfer level* (*RTL*), which is a design abstraction that models a synchronous digital circuit in terms of the flow of digital signals between hardware registers, and the logical operations performed on these signals.

As designs became even more complex in the 1980s, the first modern hardware description language, *Verilog*, was introduced by Gateway Design Automation in 1985. Cadence® Design Systems later acquired the rights to Verilog-XL, the HDL simulator that would become the de facto standard of Verilog simulators for the next decade.

The United States Department of Defense launched the Very High Speed Integrated Circuit (VHSIC) program in 1980, which supported significant advancements in IC materials, lithography, packaging, testing, and algorithms. This program also led to the development of *VHDL* (VHSIC Hardware Description Language) in 1987 which, alongside Verilog, is the second major HDL in use today. VHDL borrows heavily from the Ada programming language in both concepts and syntax.

Initially, Verilog and VHDL were used to document and simulate circuit designs that had already been captured and described in another form, such as in a schematic file. This HDL methodology led to HDL simulation, which enabled engineers to work at a higher level of abstraction than simulation at the schematic level, further increasing design capacity from hundreds of transistors to thousands.

Yet another advancement that was driven by HDLs was the introduction of logic synthesis, which pushed HDLs from the background into the forefront of digital design. Synthesis tools compile HDL source files, e.g., written in a constrained format using the RTL concept, into a manufacturable netlist description in terms of gates and transistors. As a result, VHDL and Verilog have since emerged as the dominant HDLs in the electronics industry and are still in use today.

During the 1990s, an extended effort began to integrate analog functions into an HDL, to support the concurrent design of analog and mixed analog/digital circuits. One result of this effort was *VHDL-AMS*, which has become an industry standard modeling language for mixed-signal circuits. It includes analog and mixed-signal (AMS) extensions that define the behavior of these systems [3]. VHDL-AMS provides both continuous-time and event-driven modeling semantics. It is thus well suited for verifying complex analog and mixed-signal circuits.

During this same period, Verilog was also extended—*Verilog-AMS* is a derivative of the Verilog hardware description language that includes analog and mixed-signal extensions.

And today, there are several on-going projects for defining printed circuit board connectivity using language based, textual-entry methods.

5.1.2 Elements and Example

Figure 5.1 illustrates the VHDL syntax for describing an entity, exemplified by a half-adder gate. The half adder adds two binary digits and produces two outputs as sum and carry; XOR is applied to both inputs to produce "sum" and an AND gate is applied to both inputs to produce "carry". By using a half adder, one can implement simple addition with the help of logic gates. As shown in Fig. 5.1, the VHDL source code for this half adder consists of a description of its interface (left) and of its behavioral characteristics (right).

The process of writing an HDL description is highly dependent on the nature of the circuit and the designer's preference for coding style. The HDL is merely the "capture language", which often begins as a high-level algorithmic description such as a mathematical model written in C++. Designers often use scripting languages such as Python to automatically generate repetitive circuit structures in the HDL language. Special text editors offer features for automatic indentation, syntax-dependent coloration, and macro-based expansion of the entity/architecture/signal declaration.

A hardware description language looks much like a programming language such as C; it is a textual description consisting of expressions, statements and control structures. A fundamental difference between most programming languages and HDLs is that the latter support *concurrent* statements. Concurrent statements do not describe a step in a sequential control flow (as is the case in other programming languages), instead they describe a piece of hardware; these concurrent statements can be thought of as executing "in parallel" at the same time, rather than one after another. Concurrent statements can therefore appear in any order in HDL code. Another important difference is that HDLs explicitly include the notion of time.

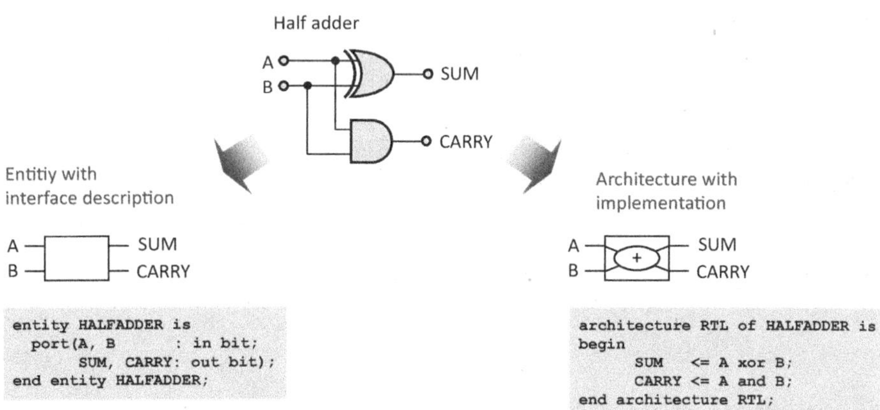

Fig. 5.1 The interface description of a VHDL entity—in this case a half adder—(left) and its related architecture implementation (right). Note the use of the keywords "entity" and "architecture"; the first is used to describe the interface, the second to describe the implementation, behavior and function of a VHDL object

5.1.3 Flow

We shall now describe how an HDL is applied in the design flow. Most designs begin as a set of requirements or a high-level architectural diagram (Chap. 4, Sect. 4.1). Control and decision structures are often prototyped in flowchart applications, or entered in a state diagram editor.

Essential to HDL design is the ability to simulate HDL programs. Simulation allows an HDL description of a design (called a model) to pass design verification. (Design verification is often the most time-consuming portion of the design process.) As depicted in Fig. 5.2, simulation also permits architectural exploration on system level. Here, one can experiment with design choices by writing multiple variations of a base design, then comparing their behavior in simulation (Sect. 5.4.3). At this early stage in the design process it is relatively easy (and, far less costly) to explore such design variations and architectural tradeoffs, as compared to later stages.

At the system level, the functionality and connectivity of each module must be defined, which is also called *functional design*. Each module has a set of inputs, outputs and description of timing behavior, i.e., only the high-level behavior is determined at this stage, not the detailed implementation within modules.

The resulting HDL code at the RTL level undergoes a code review, or auditing, in preparation for subsequent logic synthesis. During code review, the HDL description is subject to an array of automated checkers. This process aids in resolving errors before the code is synthesized into a netlist.

During *logic synthesis*, HDL descriptions are automatically converted into a netlist, i.e., a list of signal nets and specific circuit elements, such as standard cells. This netlist is then used in physical design of the circuit (Sect. 5.3). HDLs do not

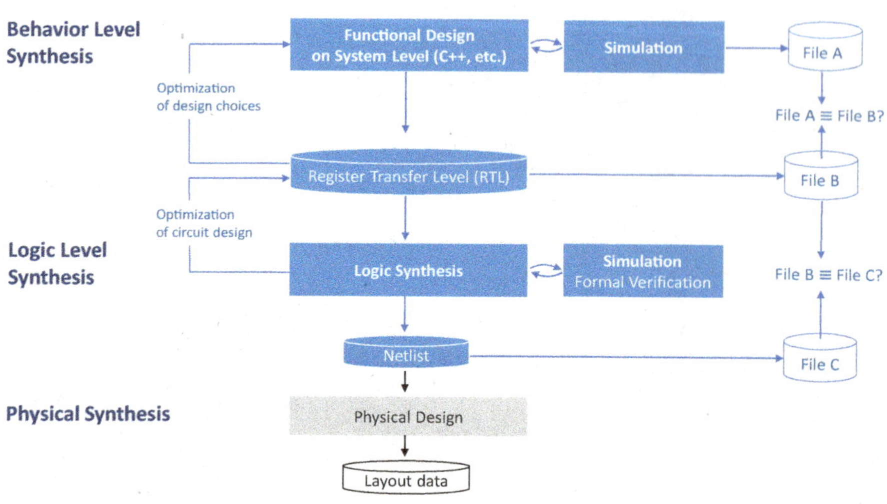

Fig. 5.2 The major steps in a design flow using a hardware description language (HDL)

play a significant role in physical design itself because the design steps become progressively more laden with technology-specific information, which (intentionally) cannot be stored in a generic HDL description.

As the routing (a step in physical design) has a major influence on the internal timing delays of a high-speed IC, modern synthesis tools are able to include place and route characteristics by providing a "pre-placed netlist". Here, the synthesis tool prepares an optimized layout, for example, by using buffer sizing or placement restrictions, in order to fulfil timing requirements. The term *physical synthesis* has been introduced for this, as it blurs the dividing line between logic synthesis and physical design.

5.2 Generating a Netlist Using Symbolic Design Entry

5.2.1 Overview

A netlist can also be generated using a circuit schematic (Chap. 3, Sect. 3.1.3). This approach, common for analog and PCB designs, is called *symbolic design entry* or *schematic entry*. It is done with a special graphics program, the schematic editor. Using such an editor, the designer positions symbols representing the devices and connects their pins by lines, representing the implementation's electrical connections. It is important to note that this topological arrangement bears no relation to the geometrical properties of the final circuit implementation—its primary purpose is to graphically document the devices in the circuit and their connections. The resulting graphics, called a *schematic* or a *circuit diagram*, will then be converted to a netlist which contains the same information, i.e., a list of the electronic components and a list of the *nodes* (aka *nets*) they are connected to.

Symbolic design entry or schematic entry is still the preferred method for small and medium-size designs, especially analog circuits and PCBs. However, analog circuits and PCBs are also marked by increasing complexity, and in order to cope with this complexity, the concept of hierarchical design is often applied. Here, higher design levels contain *block symbols* which represent design modules on lower levels, in addition to *elementary symbols* for basic electronic components, such as transistors. These block symbols may be instantiated multiple times on higher hierarchy levels. The schematic representation of each level is usually maintained as a separate file.

The data for a single symbol are stored in a symbol library. When a symbol is placed into a schematic, only a pointer to this library element is stored in the design. Since the same symbol might be instantiated (placed) multiple times in a design, individual devices must be differentiated by their *instance names*. Most schematic editors use an automatic numbering mechanism for this *device instantiation*.

The schematic entry is usually the first step in a design process. Besides graphically documenting the circuit, it also enables simulation and verification to be performed.

The circuit diagram is then converted into a netlist for use in further design steps, i.e., physical design.

5.2.2 Elements and Examples

Circuit diagrams contain the following elements:

- Symbols,
- Device labels (identification letter with consecutive numbers),
- Device type or rating,
- Electrical connections (interconnects, bus systems),
- Back-annotation data (optional),
- Frame and title block.

Device *symbols* are used to denote different functional units in a circuit. In this context, a device (which is associated with a symbol) may range in complexity from a basic electronic component, such as a transistor, to a combined sub-design or module, such as a complex logic cell. As mentioned earlier, the latter is usually assigned a block symbol. Both types of symbols are handled identically during schematic entry.

Symbols in circuit diagrams are typically designated with an *identification letter* depending on the component type, followed by a consecutive number (e.g., C4—capacitor no. 4, D1—digital gate no. 1, R12—resistor no. 12). The identification letters used are mostly the same as in the SPICE netlist format (Chap. 3, Sect. 3.1.3). The *value* of the component is also quoted with generic components, such as resistors, capacitors and coils (e.g., 2.2 μ[F]), whereby the units—"F" in this case—are omitted. One writes 2.2 μ (or 2.2u) instead of 2.2 μF for a capacitor, for example. The *type* should be specified for other electronic components, such as transistors or gates, as it appears in the components library (e.g., NAND gate 74ACT00).

There are two sets of symbols for elementary logic gates that are commonly used, one being defined in ANSI/IEEE Std 91-1984[1] (supplement ANSI/IEEE Std 91a-1991) and a second defined in IEC 60617,[2] respectively, and a "distinctive shape" set, based on traditional schematics (Figs. 5.3 and 5.4). The latter is used for simple drawings and derives from military standards from the 1950s and 1960s.

Digital gates are drawn individually despite typically being located inside a common IC package (circuit enclosure). These gates are represented in the schematic view by a number of symbols with the same identification letter and the same number that refer to a common circuit enclosure. The example in Fig. 5.3, lower left, depicts four 2-input NAND gates in a package D1. The individual gates are designated with D1A, D1B, D1C and D1D (or alternatively, D1.A, D1.B, etc.) in the schematic diagram.

[1]IEEE Standard 91-1984, IEEE Standard Graphic Symbols for Logic Functions, and IEEE Standard 91a-1991, Supplement to IEEE Standard 91-1984.

[2]IEC 60617 Graphical symbols for diagrams.

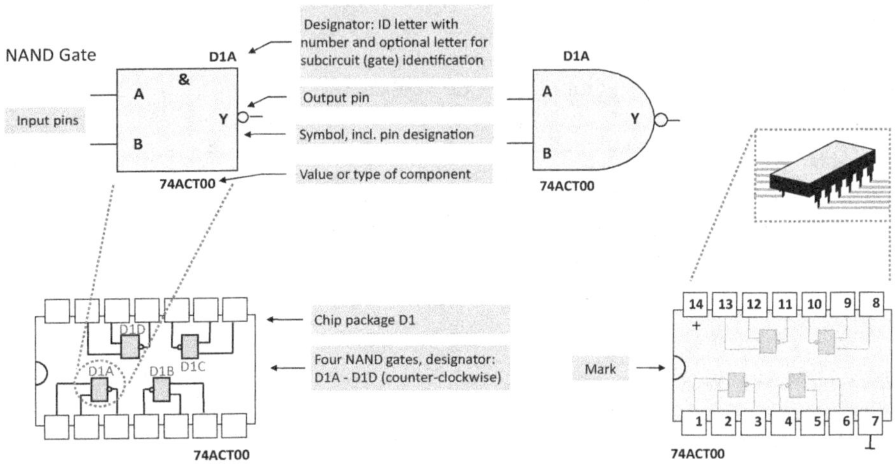

Fig. 5.3 The schematic of a logic NAND gate (above, left IEEE format, right traditional schematic) and its implantation inside an integrated circuit (below, left) [12]. As shown on the lower right, pins on a logic IC are typically numbered counter-clockwise, starting at the mark (pin assignment as per spec sheet, + supply voltage, ⊥ ground connection, IC as seen from above)

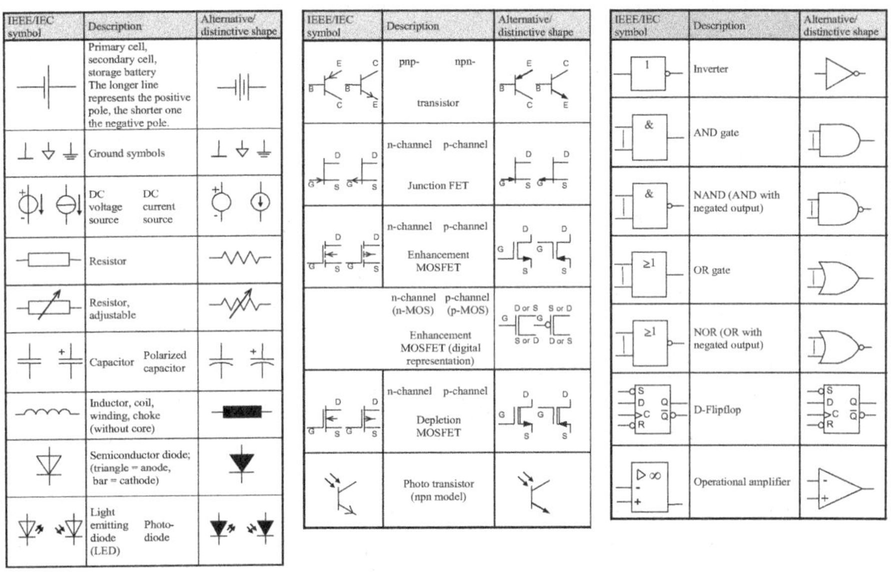

Fig. 5.4 Schematic symbols of electronic components of analog circuits (left), transistors (middle) and logic symbols used in digital designs (right) [12]. The transistor designations B, C, E and G, S, D mark the base, collector and emitter as well as gate, source and drain, respectively; they are not part of the actual symbol and added here to ease clarity

Figure 5.4 shows various types of symbols for different devices, such as those on analog and digital circuits as well as components on printed circuit boards. The decision as to which kind of symbol to use depends primarily on the target application.

In order to connect devices in a circuit diagram, wires are drawn from pin to pin. The following types of *connections* are deployed (Figs. 5.5 and 5.6):

- Wires: pin-to-pin interconnects of signal paths; signal name is optional,
- Bus systems: bundling together many signal paths; signal name is obligatory; every signal has the same name and a different index, and
- Lines: of no electrical significance; for decorative purposes only, e.g., borders.

Power supply pins, often labeled with VDD (PWR) and VSS (GND), are in most cases not visible in the schematic. These so-called *hidden pins* are connected automatically during netlist generation, using a *global node* mechanism. Exceptions are analog designs and PCB schematics where symbols for voltage supply and ground are typically used.

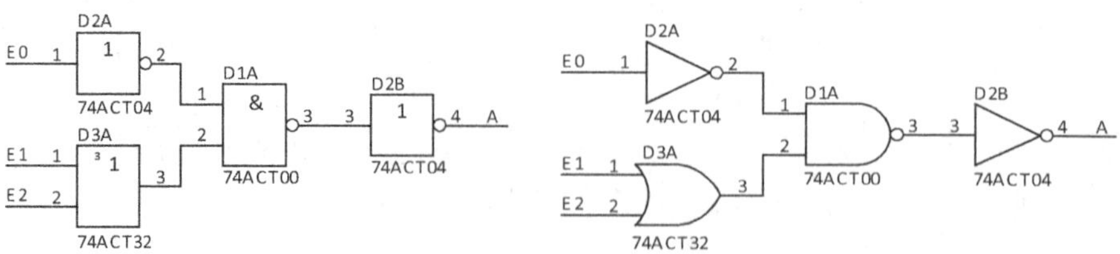

Fig. 5.5 Section of a circuit diagram with *digital* components (NAND gate D1A of Fig. 5.3, two inverters D2A, D2B, and OR gate D3A) in IEEE/ANSI/IEC standard format (left) and traditional schematics format (on the right). D1, D2 and D3 are copies (instances) of a library element and reference specific integrated circuits. These ICs may contain one or more gates that are identified by adding letters A, B, etc. to these labels (cf. Fig. 5.3). In this example, elements D2A and D2B are inverters in the same chip package D2 of IC 74ACT04

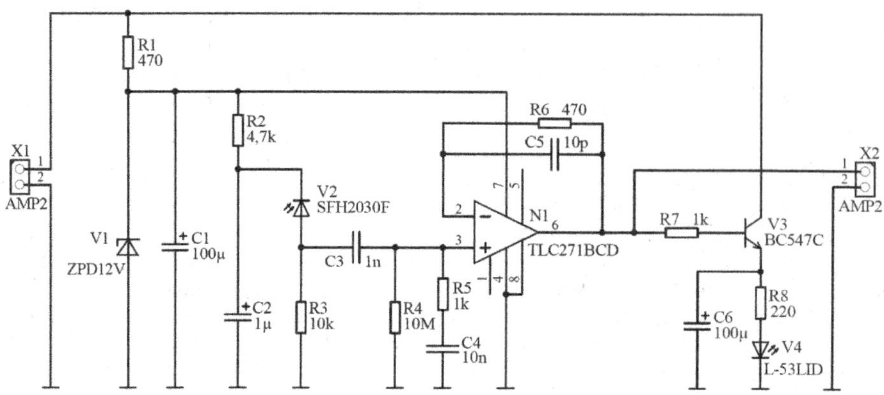

Fig. 5.6 Exemplary circuit diagram with an operational amplifier and different *analog* components (resistors, capacitors, connectors, photodiode, Zener diode, LED, transistor) [12]. Each symbol is followed by an identification letter with a serial no. and the type and/or value of the component. The engineering unit (e.g., ohm or Ω) is not cited in a circuit diagram

Back-annotation data contains details gathered during the design steps that follow circuit design, e.g., physical design, which is "written back" in the circuit diagram for future consideration. Current values for specific interconnects calculated in the layout simulation could be inscribed in the circuit diagram, for instance. Other examples are capacitance values on output connections or values from a DC analysis. Special symbol attributes are prepared for back annotation; the respective entries are assigned to these attributes and are then visible in the schematic.

5.2.3 Netlist Generation

Netlists establish the connection between the schematic design entry and the subsequent physical design of the circuit (Fig. 5.7). Prior to generating a netlist, the schematic is checked for errors and inconsistencies. One such analyzing program is the electrical rule check (ERC), which checks for unconnected inputs, multiple identical instance names, etc. The correctness of a schematic entry can also be verified by circuit simulation tools such as SPICE.

Netlists can be generated as *flat* or *hierarchical*, with the latter requiring a hierarchical schematic. The hierarchical information is preserved by using path structures in the device and node (net) names.

The information contained in a netlist might differ depending on its subsequent usage. A netlist generated for subsequent simulation contains not only node and device information, but also control information, such as model information for

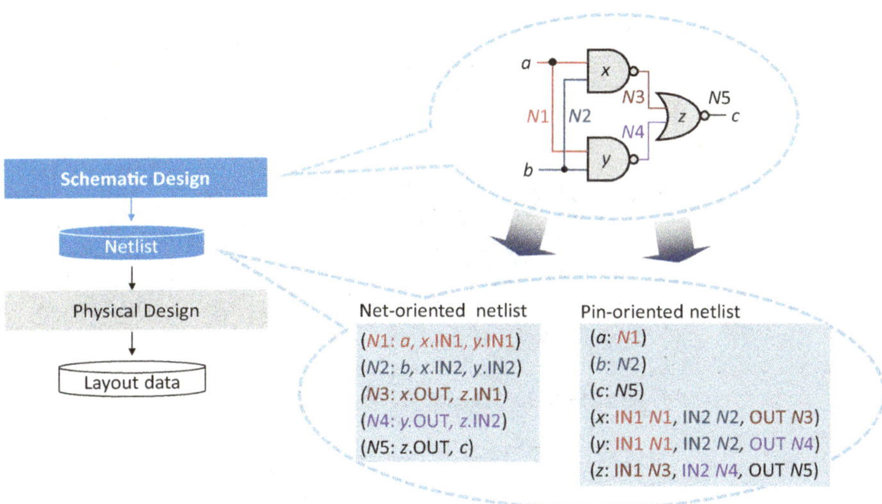

Fig. 5.7 Netlists generated out of a schematic. Netlists can be differentiated into net oriented (left), where each net has a list of device pins assigned, and pin oriented (right), with each device having a list of associated nets

simulation. A typical example is a SPICE netlist that contains control information in addition to circuit information. Obviously, a netlist that is transferred to physical design contains circuit information only, i.e., a collection of all signal nets (nodes) and the device pins they connect to. Netlists can be differentiated between net-oriented and pin-oriented lists, as introduced in Chap. 3 (Sect. 3.1) and illustrated in Fig. 5.7.

5.3 Primary Steps in Physical Design

Once a netlist is available, the circuit layout can be generated. We next investigate the various steps in physical design, which transform this netlist into optimized mask data, i.e., a detailed geometric layout on a (digital) die or a PCB. The inputs to physical design are (i) a netlist, (ii) library information on the basic devices in the design, and (iii) a technology file containing the manufacturing constraints (Fig. 5.8).

In the past, physical design was a relatively simple process. Starting with a netlist, a technology file, and a device library, a circuit designer would use a floorplanning step to determine where large blocks should be placed, and then a placement step would be used to arrange the remaining cells. Clock tree synthesis would follow, then signal-wire routing—and any (timing) problems would be resolved by iteratively improving the layout locally.

In contrast, today's massive circuits and multi-layer PCBs require a far more complex design flow. Large circuits are first *partitioned* in order to break down complexity and permit a parallel design process. *Floorplanning* of the resulting partitions has become quite complicated, and despite multiple floorplanning tools,

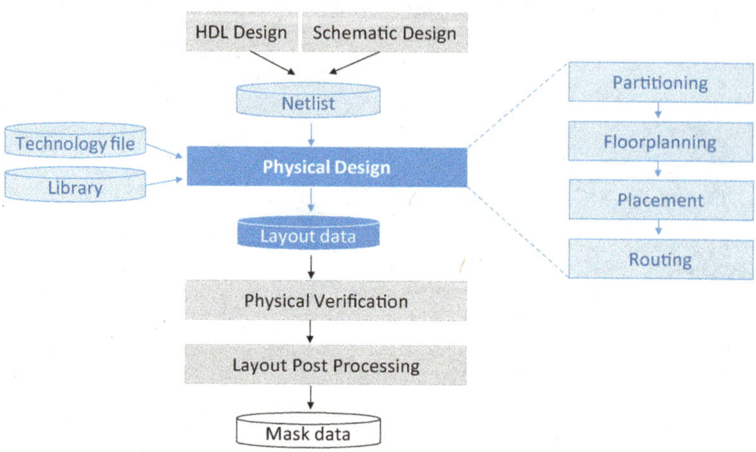

Fig. 5.8 The physical design step generates a circuit layout (aka layout data); physical design requires a netlist, and library and technology information as input

it is still largely a manual process. During floorplanning, soft blocks are assigned a specific shape and size, all blocks are arranged, and their external connections are designated to pin locations. Defining the power and ground connections on the top level as well as establishing a clock network is generally considered part of floorplanning too.

Once we have partitioned the circuit and arranged the blocks inside the floorplanning area, i.e., the top cell, these blocks can now be handled independently. *Placement* is the first step, which consists of an automatic global placement, followed by detailed placement (including legalization) to obtain local improvements. Buffer and wire sizing are then applied to meet timing constraints, as placement usually results in increased delays in some long interconnections.

Once all functional units have been placed satisfactorily, their pins must be connected by wires. Because of the complexity of this *routing* process, it is typically solved in two fully-automated steps: global routing, followed by detailed routing. The first assigns the nets to a coarse-grid structure, while the second finds the exact routes for all the nets.

These aforementioned steps are now presented in more detail (Sects. 5.3.1–5.3.3), followed by a discussion of three special applications: symbolic compaction and standard-cell and PCB design flows (Sects. 5.3.4–5.3.6). The physical design of another special application, analog integrated circuits, is covered in Chap. 6.

5.3.1 Partitioning and Floorplanning

A popular approach to decrease the design complexity of large circuits is partitioning them into smaller modules. As depicted in Fig. 5.9, these modules can range from a small set of electrical components to fully functional integrated circuits (ICs) or printed circuit boards (PCBs). Hence, the partitioner divides the circuit into several subcircuits (called *partitions* or *modules*). The primary goal of *partitioning* is to split (i.e., partition) the circuit such that the number of connections between subcircuits is minimized. Each partition must also meet all design constraints.

As noted previously, the results of partitioning are often labeled "partition" or "module"; however, as soon as the shape of this partition is considered, the term "block" is commonly used.

Blocks can be either "hard" or "soft". If a partition is defined only by its contents (cells and connections), then its dimensions are variable; this is called a *soft block*. In contrast, the dimensions and areas of *hard blocks* are fixed. Typical examples of hard blocks are (pre-existing) design modules that are being re-used, i.e., previously verified and optimized circuitry that has been used multiple times, such as (physically existing) intellectual property (so-called IP blocks).

The entire arrangement of blocks, including their locations, is called a *floorplan* (Fig. 5.10). In floorplanning, every partition is assigned a shape and a location (thus becoming a block), so as to facilitate subsequent "internal" placement (within

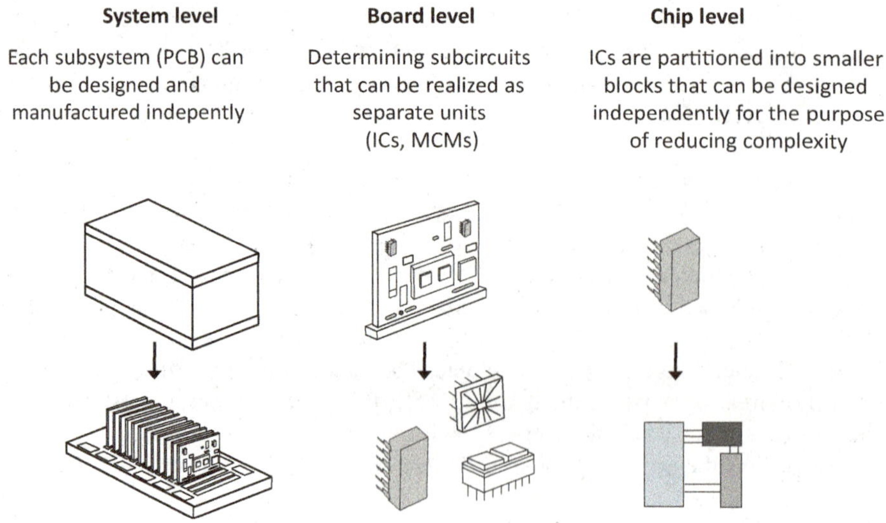

Fig. 5.9 Partitioning can be performed in the context of system-level design by breaking down a complex system into various boards (left), at the board level by differentiating between modules and ICs (middle), and at the chip level by splitting the circuit into different circuit blocks (right)

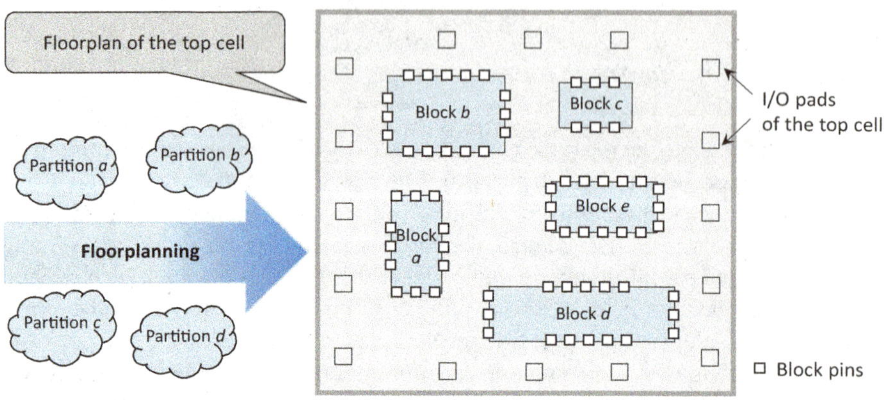

Fig. 5.10 Floorplanning defines the dimensions and shapes of the partitions and determines their external pin assignment; "soft partitions" become sized and placed blocks during the floorplanning process

the block); and every pin that has an external connection is assigned a net (pin assignment), so that internal and external nets can be routed.

To see this from a wider angle, the floorplanning stage determines the *external* physical characteristics—fixed dimensions and external pin assignment—of each partition. These characteristics are necessary for the subsequent placement

(Sect. 5.3.2) and routing steps (Sect. 5.3.3), which determine the *internal* physical characteristics of these partitions.

Floorplan optimization, which is still mostly performed manually, involves multiple degrees of freedom; while it includes some aspects of placement (finding locations) and routing (pin assignment), block-shape optimization is unique to floorplanning. (As mentioned earlier, we assign the term "block" to partitions once they have acquired, or are in the process of acquiring, a fixed shape.)

The major challenge of floorplanning is its multi-objective optimization problem, considering, for example, the simultaneous sizing of soft blocks and block placement. A floorplanner must be able to handle widely different shapes and sizes, congestions, and timing constraints [9]. Hence, most automatic floorplanners are augmented with manual optimization that is performed by an experienced designer.

Floorplanning often includes *power* and *ground structures*. Here, the floorplanning area, i.e., the top cell, is enclosed by one or more sets of power and ground rings. In most cases, horizontal and vertical power and ground segments are connected through an appropriate via cut. If these stripes run both vertically and horizontally at regular intervals, then this style is called a *power mesh* (Fig. 5.11).

It is strongly recommended to check for design rule violations and both power and ground connectivity after the power and ground networks have been laid out. One problem that should be checked for is wide metal that exceeds manufacturing tolerances. These limits are due to difficulties achieving consistent metal density, as wide metal lines cannot be produced with uniform thickness (they tend to become thin in the middle and thick at the edges in the CMP process, as described in Chap. 3, Sect. 3.3.2).

Another important consideration during floorplanning is that when both analog and digital blocks are present, extra care must be taken to ensure there is no noise between the blocks. This requires the development of a dedicated *power supply concept* for every mixed-signal IC. Specifically, any noise injection from digital

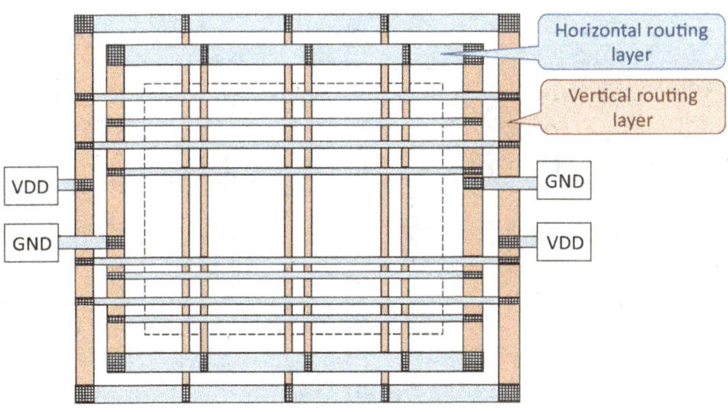

Fig. 5.11 Power planning during floorplanning with a ring and mesh structure of power (VDD) and ground (GND) connections [5]

blocks into sensitive analog blocks through power and ground connections must be avoided. This decoupling can be achieved by designing and planning the analog and digital power and ground connections separately.

Careful consideration must also be given to the placement of *macros* during floorplanning. Macros are large, pre-defined blocks that often consist of memories, individual subcircuits or analog circuitry. Proper placement of these macros has a large impact on the quality of the final (top-level) IC design. For example, special care must be taken to ensure there is enough space between large macros for interconnections.

Modern floorplanning tools usually perform the initial macro placement based on connectivity. Wire length optimization, i.e., length reduction, is the most important goal here, followed by equalized routing density, thermal requirements, and others.

One rule of thumb is that macros should be placed such that the remaining standard cell area is continuous (Fig. 5.12). A standard cell area with close to a 1:1 aspect ratio is recommended because it allows standard cell placers to utilize this area most efficiently and with minimum total wire length [5]. Extensive wire length is caused by routing connections that must cross the macro area(s) in order to connect the segmented standard cell regions (Fig. 5.12, left). Furthermore, macros should be placed such that their ports face the standard cells or the core area, and their orientation should be aligned with the respective routing layers (Fig. 5.12, right) [5].

One important aspect to be considered during floorplanning is *pin assignment*. This is usually performed after the relative placement of the blocks has been defined. During pin assignment, all nets (signals) are allocated to unique pin locations such that the overall design performance is optimized. Common optimization goals include maximizing routability and minimizing electrical parasitics both inside and outside the blocks [9].

Figure 5.13 illustrates this assignment using an example of a graphics processing unit (GPU). Here, each of the 90 external pins of this unit are connected to an I/O pad on the next hierarchy level (PCB). After pin assignment, each GPU pin has a

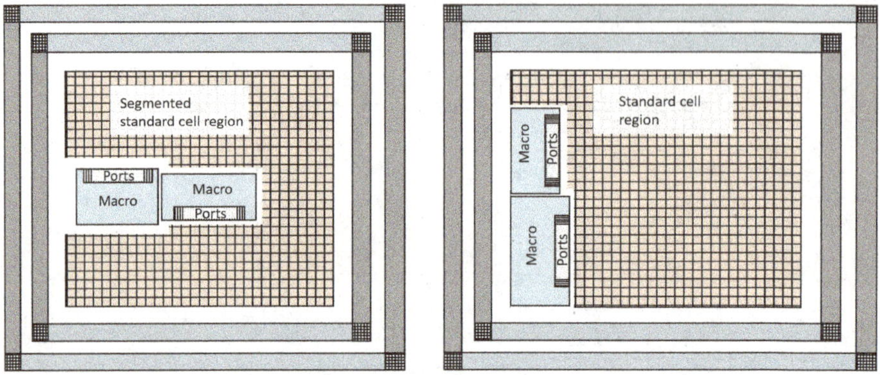

Fig. 5.12 Segmented floorplan (left) and floorplan with minimized wire length (right) [5]. Longer wires that are routed within the segmented standard cell region (left) may have to be routed "around" the macros, thereby increasing wire length, an issue that is avoided with the floorplan on the right

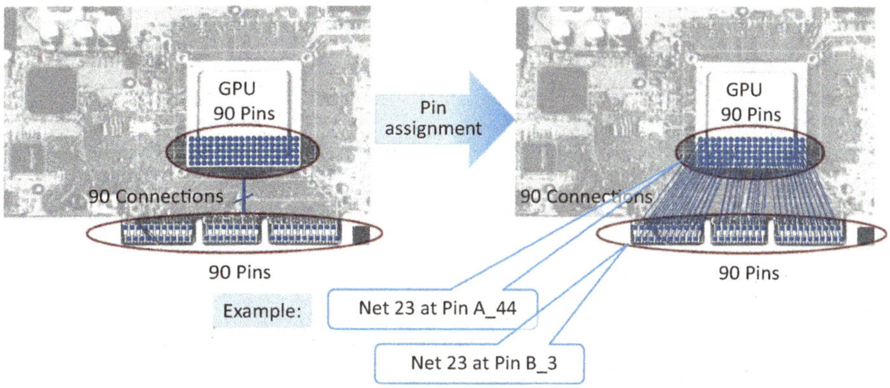

Fig. 5.13 Pin assignment allocates nets to external pins of the blocks. Here, each of the 90 pins on the GPU is assigned a specific I/O net so that the external route (between the GPU and the I/O pad) is optimized [9]

connection to a pin on the external device, preferably laid out in a planar manner, i.e., without crossings.

Clock planning is often part of floorplanning. An ideal implementation of clock distribution networks provides clock signals to all clocked objects (cells, macros, blocks) in the floorplan in a symmetrical-structured manner. Examples are mesh and tree structures (Fig. 5.14). Such a clock distribution network delivers the clock signals to all objects with minimized clock skew.

Despite sophisticated clock tree synthesis tools, high-performance and synchronized designs still depend on manually implemented clock networks. These networks must take line resistance and capacitance into account in order to minimize the skew between communicating objects. Hierarchical design is well suited for this because

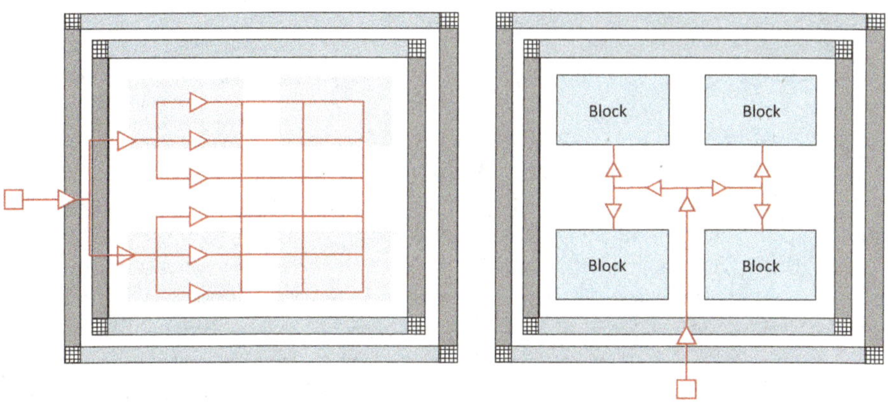

Fig. 5.14 Hierarchical clock planning with a clock distribution network using a mesh (left) and a tree structure (right) [5]

the main clock network can be manually drafted at the chip level such that it supplies the clock signal to each block. In order to minimize the clock skew among all leaf nodes, the clock delay for each block must be calculated and the design of the clock tree planned accordingly [5].

Finally, it is important to mention some special floorplanning rules for mixed-signal and smart power chips: Analog circuit blocks for sensor scanning as well as power stages (see Fig. 1.9 in Chap. 1) should be placed directly at the chip periphery, in close proximity to their respective bond pads.

Sensor-scanning circuits convert analog sensor inputs, that come into the chip, into digital signals inside the chip. Because these analog inputs are often very low-amplitude signals and are susceptible to interference, it is advantageous to keep their paths as short as possible, to avoid disturbance from other circuits or wires. Hence, sensor signals should be converted to digital signals as soon as possible, i.e., directly at their point of entry into the chip.

Power stages drive chip-external active devices, so-called actuators, such as electric motors, valves and alike by switching high currents. Because the conductivities of wires on a chip are limited, these power stages (i.e., large DMOS transistors) should be placed directly at the entry and exit points for these currents on the chip. Several bond wires operate sometimes in parallel to conduct very high currents (currents up to 10 A can be encountered). The bond pads in question are placed directly upon the sources and drains of the power DMOS transistors so that the currents do not flow laterally on the chip.

5.3.2 Placement

After partitioning the circuit into smaller partitions and floorplanning the layout to determine partition/block outlines, pin locations, and power/ground supply lines, *placement* seeks to determine the locations of basic devices, such as standard cells, within each block (Fig. 5.15). During this process, it considers optimization objectives, e.g., minimizing the estimated total length of connections between devices.

Applying placement to PCBs, this step is often divided into two phases: (i) placing all components that have fixed locations, such as connectors and pads, on the board, and (ii) placing all remaining devices, by iteratively improving their initial placement with regard to wire lengths and other objectives.

In contrast, placement techniques for large ICs encompass global and detailed placement. (Sometimes *legalization* is counted as a third step; however, we consider this as part of detailed placement.) Global placement assigns general locations to movable objects. As such, it often neglects specific shapes and sizes of placeable objects and does not attempt to align their locations with valid grid rows and columns. Some overlaps are allowed between placed objects, as the emphasis is on judicious global positioning and overall density distribution.

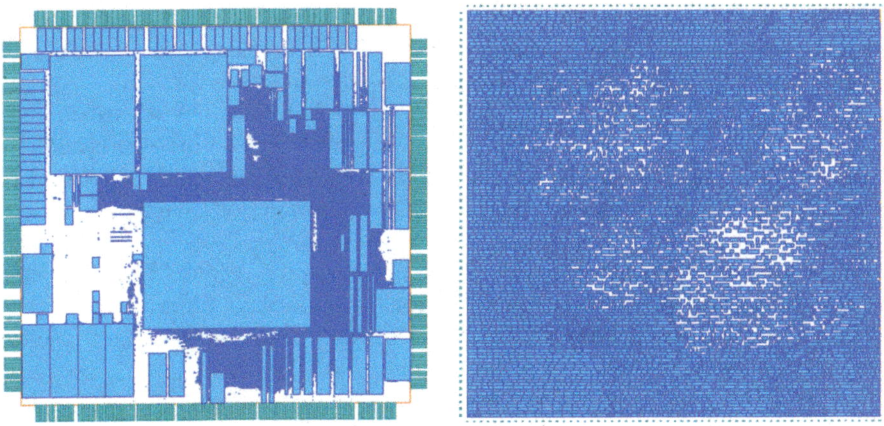

Fig. 5.15 Placement result of differently sized macros (light blue) and cells (dark blue) on the left; placement of standard cells (right) [15]

Legalization is performed during detailed IC placement. This step seeks to align placeable objects with rows and columns, and to remove overlap, while trying to minimize displacements from global placement locations as well as impacts on interconnect length and circuit delay. Detailed placement incrementally improves the location of each basic device (e.g., standard cell) by local operations (e.g., swapping two cells) or shifting several cells in a row to make space for another object.

Global and detailed placement typically have comparable runtimes, but global placement often requires much more memory and is more difficult to parallelize.

The term "cluster" is used to refer to a number of cells that are placed near each other (Fig. 5.16, left). The purpose of clustering is to control the closeness of timing-critical components during IC placement, similar to a module definition within the

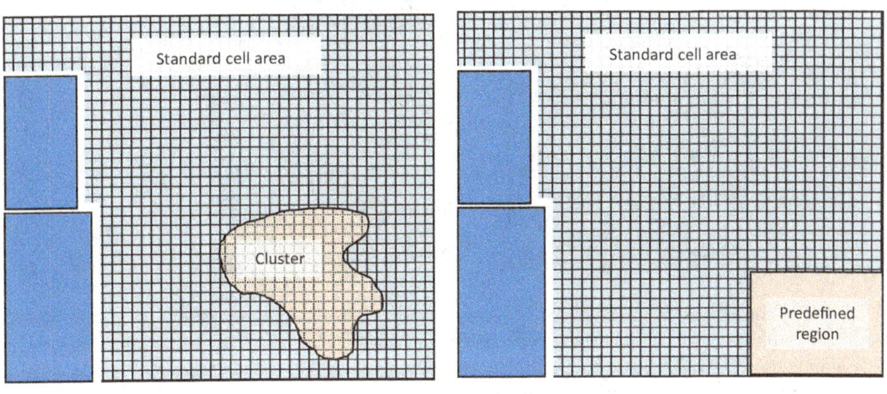

Fig. 5.16 Visualizing the concept of a cluster, which is defined by a number of cells that are placed near each other anywhere on the chip, and a region, which is a pre-defined placement area [5]

netlist [5]. In most cases, the cluster location remains undefined until all cells have been placed within it. This is in contrast to the "region" concept where a region's location is defined prior to cell placement (Fig. 5.16, right).

After clusters and regions have been defined, global placement distributes all (standard) cells uniformly across the available placement area, aiming for minimum global wire lengths and optimized congestion. A detailed placement algorithm is then executed to refine their placement based, for example, on congestion, power requirements, and timing constraints.

Congestion-driven placement takes the required routing resources into account and aims to equalize the expected routing congestion. Its critical impact is illustrated in Fig. 5.17. Almost all of today's automatic placers consider routing congestion due to its importance in ensuring a successful routing step afterwards.

Timing-driven placement can be classified as either path- or net-based. A path is a sequence of nets through which a signal passes when moving through a circuit. The path-based approach works on all or a subset of paths, i.e., constraints are applied to delay paths of entire (sub)circuits. The net-based approach deals with nets only, with the hope that if we handle the nets on the critical paths well, the entire critical path delay may be optimized implicitly [5]. This can be achieved by assigning a weight to these nets. Modern placers often use a hybrid approach that combines both methods. They interleave weighted connectivity-driven placements with timing analysis, which we will further discuss in Sect. 5.4.4.

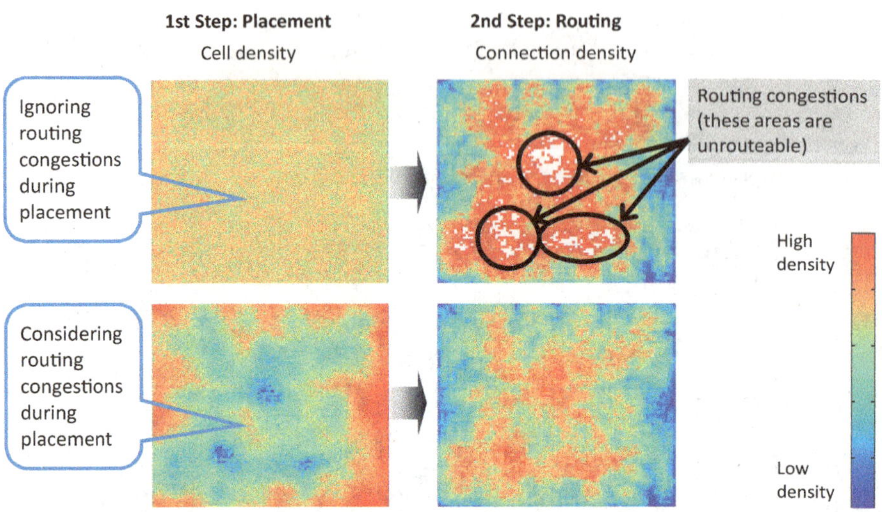

Fig. 5.17 Visualizing the routing results (distributions) if routing congestion is not considered during placement (top, all cells are equally distributed during placement) and if congestion-driven placement is applied (bottom, where cells are placed unevenly) [15]. This example illustrates that considering the routing congestion during placement is essential

5.3.3 Routing

Placement is followed by *routing*, where pins requiring the same electric potential are connected using wire segments. This is one of the most complicated and time-consuming physical design steps. Even after a (seemingly) successful placement, routing can fail; it can take an unacceptable amount of execution time; or, as it is often the case in PCB layouts, it may deliver only a partial routing solution (with some remaining open, un-routed nets).

Before considering the routing of signal nets, special nets such as clock or power and ground connections are embedded. As we have discussed this process earlier (Sect. 5.3.1), it is only briefly considered here. We visualize the routing of clock nets in Fig. 5.18 and power and ground networks in Fig. 5.19. Another special routing methodology, *differential-pair routing* (Chap. 7, Sect. 7.3.2), which uses differential signaling, is depicted in Fig. 5.20. Here, the signal is transmitted through a pair of tightly coupled wires, one of these carrying the signal, the other carrying an equal but inverted image of the signal. As illustrated in Fig. 5.20, differential

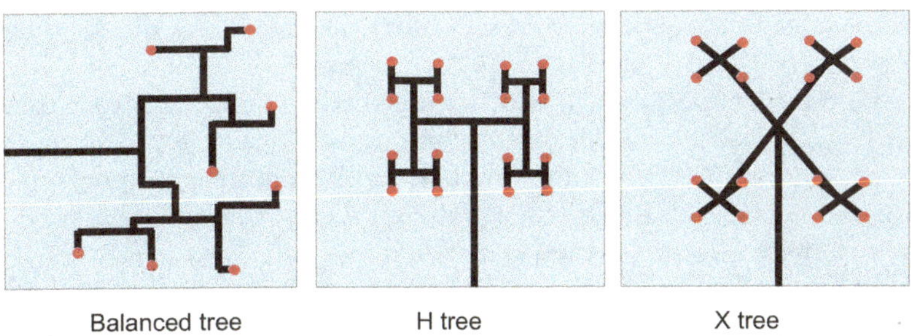

Fig. 5.18 Illustration of several methodologies for clock routing which aim for minimized skew, for example, by equalizing the routing lengths to all cells. The nets of the H and X tree are schematic delineations that illustrate the underlying principle

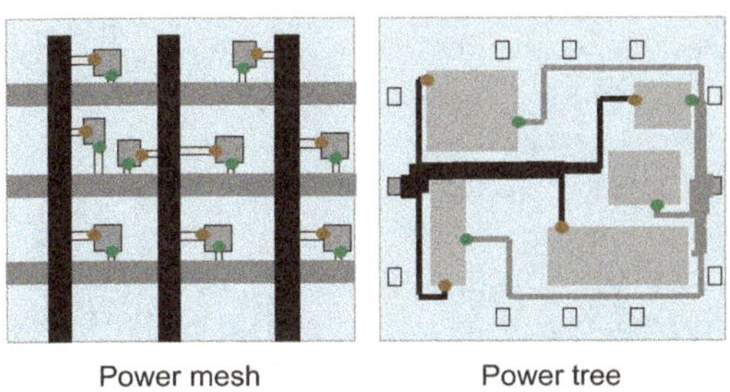

Fig. 5.19 Power and ground networks can be routed either by using a power/ground mesh on different layers (left) or applying a planar tree structure (right). The latter is chosen, for example, when a metal layer with extra thickness is required to drive the currents and only one such layer is available

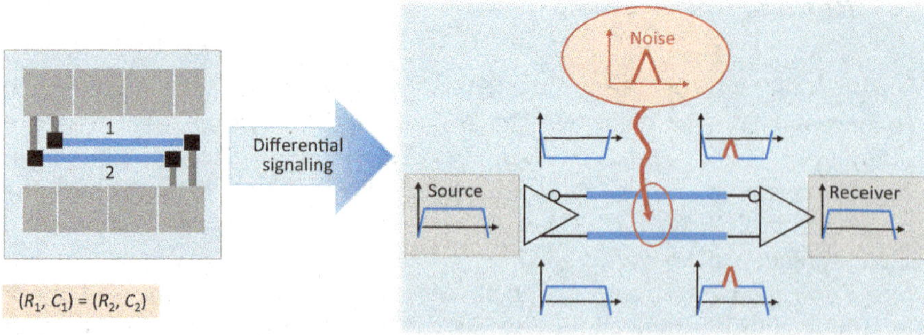

Fig. 5.20 Illustration of differential-pair routing, where two wires are routed "equally" in order to achieve identical electrical impedances (R, C) on both lines (cf. Fig. 7.13 in Chap. 7). Using two complementary signals ensures maximum noise robustness because (i) the noise tends to affect both conductors identically and (ii) the receiver detects only the difference between the signals (this effectively eliminates the noise-induced signal changes)

signaling is inherently immune to common mode electrical noise. Another advantage is minimized electromagnetic interference (EMI) generated from the tightly placed signal pair.

After introducing special nets, which are embedded into the given placement structure first, let us now consider the routing of the remaining "common" signal nets. Depending on the application, different types of routing methodologies are used. The routing task for digital circuits is split into *global* and *detailed routing*, while analog circuits and PCBs use *area routing* in most cases.

Let us first discuss the routing of digital circuits. Here, the huge complexity of routing millions of nets requires that this procedure be divided between global and detailed routing. During global routing, the net topologies' wire segments are tentatively assigned (embedded) within the chip layout. A global routing for a net is thus a "general" path that the connection should follow, but does not specify individual, detailed routing resources. In order to perform global routing, the chip area is represented by a coarse routing grid, and available routing resources are represented by edges with capacities in a grid graph. Nets are then assigned to these routing resources. Routing regions created by global routing are commonly known as global routing cells, called *gcells*. The capacity of these cells is defined by the number of routing layers, cell dimension, and the wires' minimum widths and spacings (layer-wise).

After global routing is performed, almost all routers report the over- or underflow of these gcells in a so-called congestions map (Fig. 5.21). This over- or underflow is the ratio of the gcell's capacity and the number of nets assigned to it. Detailed routing should only be attempted if this ratio does not exceed the value 1 in too many places.

Detailed routing seeks to refine global routes and typically does not alter the configuration of nets determined by global routing. Hence, if the global routing solution is poor, the quality of the detailed routing solution will likewise suffer.

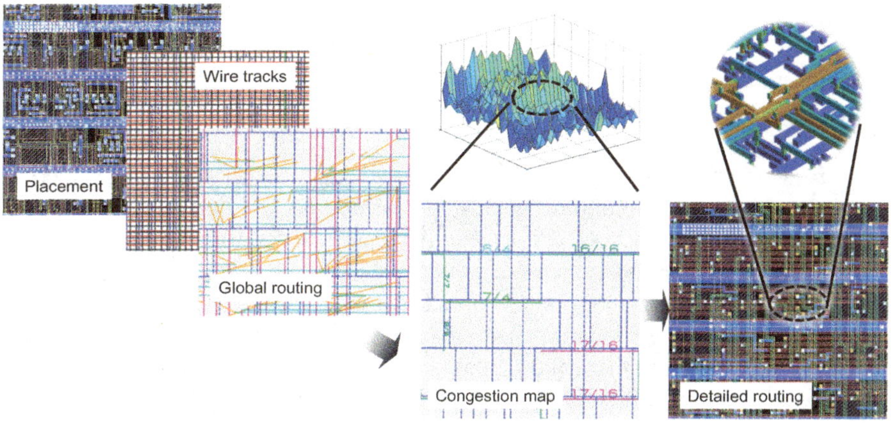

Fig. 5.21 Visualizing global routing where the number of wire tracks is defined by design rules. The global router takes these into consideration to determine the capacity of its gcells in which the wires are assigned. The resulting congestions map (middle) indicates the expected outcome of the detailed routing, which assigns the wire segments to detailed routing tracks (right)

During detailed routing, the wire segments are assigned to specific routing tracks (see Fig. 5.21). Detailed routing involves a number of intermediate tasks and decisions such as *net ordering*, i.e., which nets should be routed first, and *pin ordering*, i.e., in what order should the pins be connected within a net. These two issues are major challenges due to the unavoidable sequential nature of detailed routing, where nets are routed one at a time. For example, the routing resources used by a net that is routed first cannot be used by a net that is routed later. Obviously, the net and pin orderings can have dramatic impacts on final solution quality.

To determine net ordering, each net is given a numerical indicator of importance (priority), known as a net weight. High priority is given to nets that are timing-critical, connect to numerous pins, or carry out specific functions. High-priority nets should avoid unnecessary detours, even at the cost of detouring other, less important nets.

As mentioned earlier, dividing the routing step into global and detailed stages is common for digital circuits. The main reason for this splitting is the enormous complexity of the task of routing millions of nets; this can only be achieved by some pre-assignment prior to detailed routing. For analog circuits, multi-chip modules (MCMs), and printed circuit boards (PCBs), global routing is often unnecessary due to the smaller number of nets involved, in which case only detailed routing is performed; which is then called *area routing*.

When applying area routing, nets are embedded directly in a single step, without prior global routing. This routing methodology consists of the following primary steps: (i) determining the net order in which the nets are to be routed, (ii) assigning nets to routing layers, (iii) performing a path search [9] for each net within the assigned routing layers, and (iv) applying a rip-up and reroute strategy for nets that could not be routed due to the blockages caused by earlier routed nets.

Most area routers are *grid-based*. The router imposes a grid on the layout area that consists of evenly spaced routing tracks running both vertically and horizontally across the routing area. The router then follows these tracks, either vertically or horizontally. As these routing tracks are defined by design rules (width and spacing rules), a design-rule-violation-free route is guaranteed this way.

Grid-less routers can use various wire widths and spacing(s) without considering a routing grid. As their solution space can be considered infinite in size, they often deliver superior solutions, however, at the cost of significantly higher runtime. Hence, they should only be applied to small circuits or the routing of specific nets, such as the clock net or nets that conduct substantial currents.

Detailed (area) routers are often hampered by nets that have been routed early on but which later prevent other routes from being connected. In these cases, a *rip-up and reroute methodology* is applied which removes these early routes ("rip-up"), embeds the current connection, and then reroutes the removed route on a different path. An example of this strategy is presented in Fig. 5.22. Obviously, this procedure is very time consuming and error-prone; hence, it should only be applied to a few remaining open nets.

Detailed (area) routers are often supplemented by a *search-and-repair phase*. Here, the detailed router tries to resolve all types of physical design rule violations, such as wire shorts, fill notches and metal spacing [5].

As vias, i.e., connections between metal layers, impose performance and reliability losses, their number should be minimized. Hence, via *minimization* is often applied after all nets have been routed. This process removes as many vias as possible by reducing the number of jogs associated with a wire connection.

Another via-optimization methodology is via *doubling*. Here, single vias are doubled as long as there is no negative impact on the routing area. The benefits of this process are (i) increased via yield, (ii) reduced electrical resistance, and (iii) better immunity against electromigration (Chap. 7, Sect. 7.5.4).

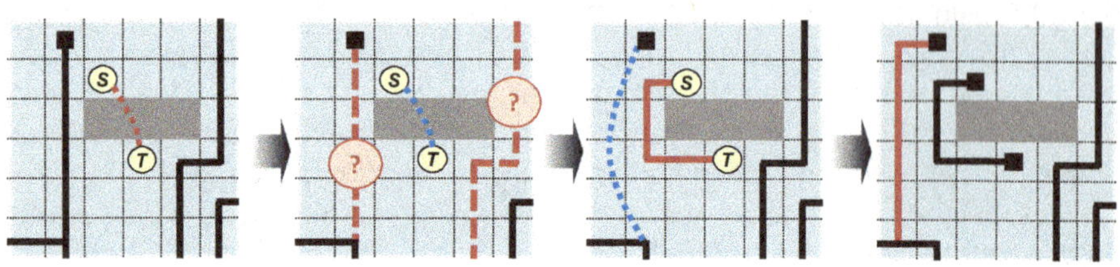

Fig. 5.22 Illustration of a rip-up and reroute strategy. In order to connect points *S* and *T*, a previously routed net has to be identified as an obstacle to be removed. The net in question is then routed on the (now) unoccupied grid cells, and the removed net is re-routed afterwards

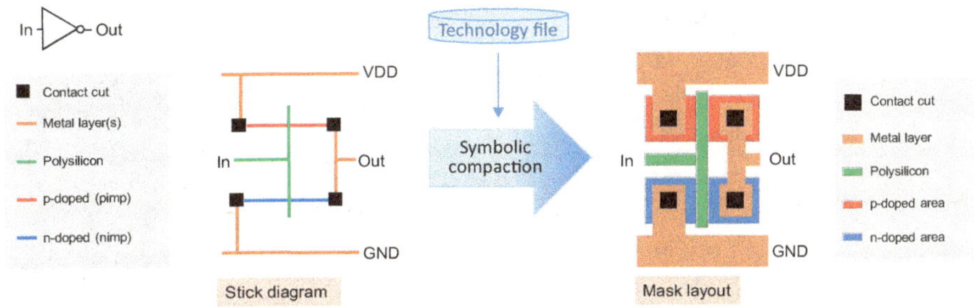

Fig. 5.23 Symbolic compaction of a CMOS inverter. Symbolic compaction is an efficient technique for adjusting a library of (symbolically) laid-out standard cells to various technology nodes

5.3.4 *Physical Design Using Symbolic Compaction*

In general, *compaction* rearranges a (placed and routed) topological group of objects in order to minimize the layout area, while maintaining minimum spacing conditions between the objects. Hence, compaction preserves both the circuit topology and design rule correctness while compressing the layout.

Symbolic compaction is an important design step when a *symbolic layout design* methodology is adopted [14]. Here, designers create layouts by using abstract objects, so-called *stick diagrams*, rather than actual layout objects, such as wires and contacts (Fig. 5.23) [1, 7]. Once the symbolic layout is generated using the aforementioned steps (Sects. 5.3.1–5.3.3), the compaction tool is applied when a specific technology is assigned. Here, symbolic compaction generates the actual mask layout, taking the minimum spacing and width rules and all other design rules into account.

The main advantage of the symbolic layout approach is its design-rule independence: As long as the basic construction principle of the devices does not change, the layout can be generated without a specific technology in mind, and then easily adjusted to any design-rule set. Hence, the compactor is used to generate different mask layouts for different technologies, derived from the same symbolic layout.

5.3.5 *Physical Design Using Standard Cells*

As mentioned in Chap. 4, standard cells are designed with fixed cell height and defined locations for power (VDD) and ground (GND) ports. Cell widths vary depending on the transistor network implemented. Because of this restricted layout style, all cells are placeable in rows (Fig. 5.24).

The final standard-cell layout consists of a number of rows with power and ground running at the bottom and top of each row. These power and ground lines connect to vertical power bars, which, again, connect to the outer power and ground pads (Fig. 5.25).

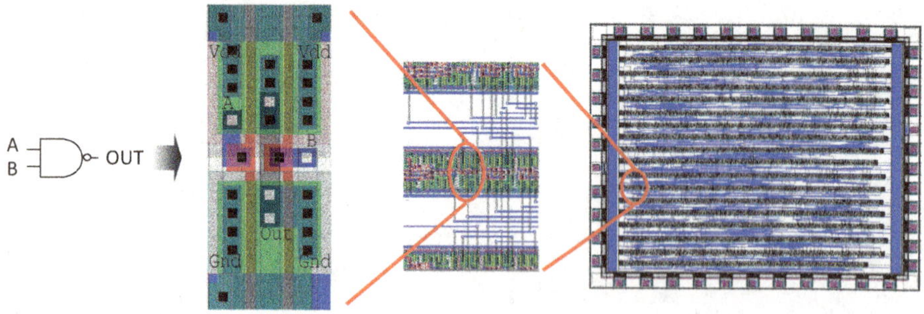

Fig. 5.24 A NAND standard-cell implementation in a standard-cell design. The signal ports of the NAND gate (A, B, OUT) are visible; they are connected vertically into the channel (middle) to be routed to other cells

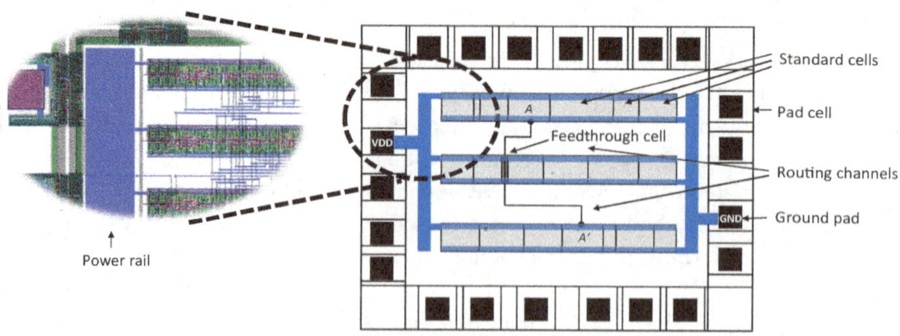

Fig. 5.25 Simplified layout of a standard-cell chip that illustrates how the power and ground pads connect to all standard cells

Prior to placement, the user can assign (i.e., request) a specific height-width ratio to the resulting standard-cell core. The automatic placer then tries to achieve this ratio. The automatic placement of standard cells usually minimizes wire lengths; fulfilling timing constraints can be another objective.

The space between the cells (in a two-layer structure) or above the cells (if more than two routing layers are available) is used for subsequent routing. In the first case, the area reserved for routing is called a *channel*. All (signal) cell ports are connected during the routing stage, either within the adjacent channel or by using multiple channels. Using multiple channels requires various crossings of the standard-cell rows, for example, by feedthrough cells (see Fig. 5.25). The advantage of this channel structure is its flexibility—the height of each channel is adjusted according to the number of nets to be routed inside.

If more than two routing layers are available, then the cells' ports (usually located in the first metal layer) are no longer a routing obstacle. Routing can then be performed "over the cell" (so-called OTC routing). In this case, channels can be omitted and the standard cell rows are placed side by side, ideally sharing the power and ground structures by flipping every other row (Fig. 4.13, right, in Chap. 4).

The finished standard-cell layout can either be used as a single chip or it may become part of a larger circuit. In the latter case, it can serve as a digital macro cell in a mixed-signal floorplan (Sect. 5.3.1).

5.3.6 Physical Design of Printed Circuit Boards

A printed circuit board (PCB) mechanically supports and electrically connects electronic components. These components are basic devices, such as resistors, capacitors, etc., and larger modules, for example, integrated circuits (ICs). Devices and modules are generally soldered onto the PCB to both electrically connect and mechanically fasten them to it. While physical design of a PCB generally follows the aforementioned steps, some aspects need special consideration—they are the topic of this section.

The design of modern PCBs with high package density requires powerful EDA tools to facilitate their layout generation. At the same time, manual intervention is still common. Figure 5.26 illustrates the main PCB design steps, which consist of (i) schematic entry, (ii) layout design, and (iii) post processing which includes generating the data files required for PCB manufacturing.

Let us now investigate these steps in more detail. As mentioned earlier, *schematic entry* involves taking symbolic representations of components (resistors, capacitors, ICs, etc.) and wiring them together into a visual diagram that can be easily viewed to understand circuit functionality. The resulting schematic representation is often partitioned into several pages and hierarchies to increase its readability. Schematics also support electrical checks, such as the use of electrical rule check (ERC) tools. These tools verify conformance to basic design and electrical rules, flagging, for

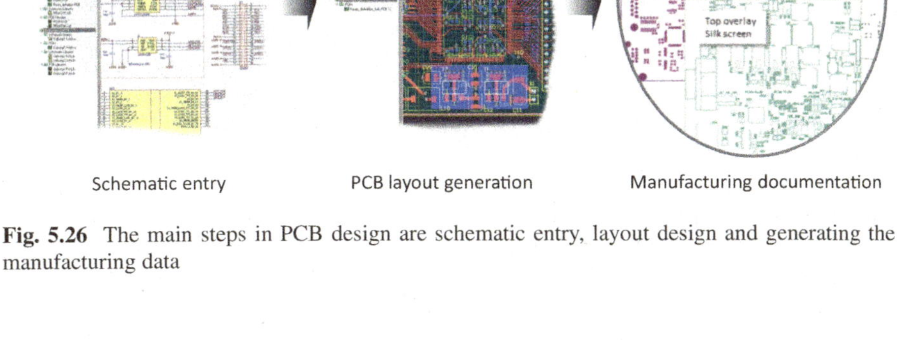

Fig. 5.26 The main steps in PCB design are schematic entry, layout design and generating the manufacturing data

example, unconnected pins and ports, identical device references or connections between different power nets. *Simulation* tools are also used to evaluate the behavior of real world components in a virtual environment, thereby allowing the designer to perform advanced analysis on the PCB schematic by simulating and visualizing the board's characteristics (Sect. 5.4.3).

If all required functions are performed and verified by simulation, the PCB netlist is automatically generated from the schematics. This netlist contains all listed symbols with their pins and their respective connections (signal and power nets). The netlist may also include additional properties, such as those needed for further layout post processing.

The next step, *generating the PCB layout*, defines the board outline, places all components and generates the connections between them. During the PCB layout stage, the design is defined as it will eventually appear when manufactured.

First, each symbol in a schematic is associated with a *landpattern*, also referred to as a *footprint* (Fig. 5.27). A landpattern visually represents the physical dimensions of an electronic component. It consists of all entities needed to connect a component to the PCB, such as pads, solder areas and holes. Essentially, the landpattern translates the symbol of a schematic component into a package that is placed on the PCB layout. This placement can be performed interactively, where the designer decides where to place specific components, or it can be done automatically.

The manual placement approach is usually accompanied by automatic rule checking in order to prevent invalid placements. The interconnections are displayed as rubber bands, illustrating which pins are connected by the same net. The designer picks

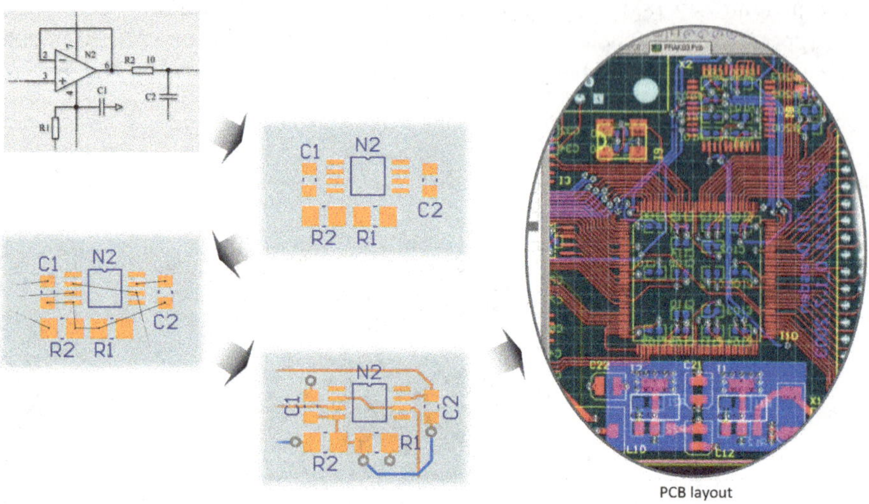

Fig. 5.27 Illustrating the placement and routing steps. While placement defines the landpattern of the components and places them on the board (orange areas), the subsequent routing connects their pins using traces in various routing layers

and places the components; he/she is guided, among others, by overflow information regarding the feasibility of routing.

Manual placement can be effectively combined with automatic placement. When this is the case, all critical or pre-assigned components are first placed manually; they are then protected ("fixed"), and subsequently all remaining components are placed automatically by the PCB layout tool.

After all components have been validly placed, routing is performed. This creates a set of traces that connect the component pins on the PCB using specified routing layers (see Fig. 5.27). This step is done interactively, fully automatically, or by a combination of both.

Routing layers can be differentiated into power layers and signal layers. The first ones contain power supply nets, i.e., power and ground, that connect components' power and ground pins. Signal layers primarily contain the traces that connect the signal pins on components. Some PCBs also allow for power and signal traces to be routed on the same layer(s). Connections between layers are provided by vias, which can be differentiated into through-hole vias, blind vias and buried vias (Fig. 1.2 in Chap. 1). The last two types are technologically more challenging and therefore more expensive.

Most PCB layout programs include a tool—the so-called "autorouter"—that automatically assigns the conductor tracks. With this tool, the routing can be automatically produced in compliance with the design rules. The tool is normally only used for digital circuit parts, as the number of different constraints that must be simultaneously observed on a PCB is very high. A typical application benefitting from the use of this tool is a bus system, where a large number of data lines are routed between components. In analog circuitry, such as sensor-signal conditioning circuits or power output stages or in mixed-signal circuit parts (both analog and digital subcircuits), manual design still outperforms the autorouter.

Keepout areas (aka *region keepouts*) are important PCB features; they are areas that should be free of components and traces. A keepout area is placed as a layer-specific keepout object or an all-layer keepout to act as a placement and routing barrier. As an example, Fig. 5.28 shows keepout areas around holes that are needed for screws in order to fasten the PCB inside a housing; the keepouts ensure that the components and traces do not overlap with the screw heads, which are often much larger than the holes themselves.

Fig. 5.28 Example of keepout areas (indicated by red circles) around screw holes for mounting a PCB

PCB layout is very susceptible to electromagnetic compatibility (EMC)-related problems, such as interference emissions as well as inductive and capacitive coupling. *EMC-compliant layout design* addresses these issues by preventing signal coupling and defining appropriate reference grounds, etc. There are a multitude of design rules available for enforcing EMC-compliant PCB layout design. As the theoretical background to these design rules is quite complex, they are not covered in this book. The reader is kindly requested to refer to [12, Chap. 6] for a detailed introduction to this topic. Having said that, we would still like to summarize the basic rules of EMC-compliant layout design; they are illustrated in Fig. 5.29, as follows:

- Avoid (current) loops in both signal and power lines. Circuits require signal and return conductors; run signal and return conductors close to each other on a PCB (Fig. 5.29, top left). Interference and interference coupling are approximately proportional to the loop area.
- Implement defined return paths for currents. Current always takes the path of least impedance. Ground planes are better than separate return lines. Run the return line near the signal line on PCBs with no ground plane, ideally taking an identical route on a different layer. Implement the return line in the same way as the (forward) signal line if both lines cross (Fig. 5.29, top right).
- If there is a break in the ground plane, route the (forward) current trace around the break (Fig. 5.29, bottom left).

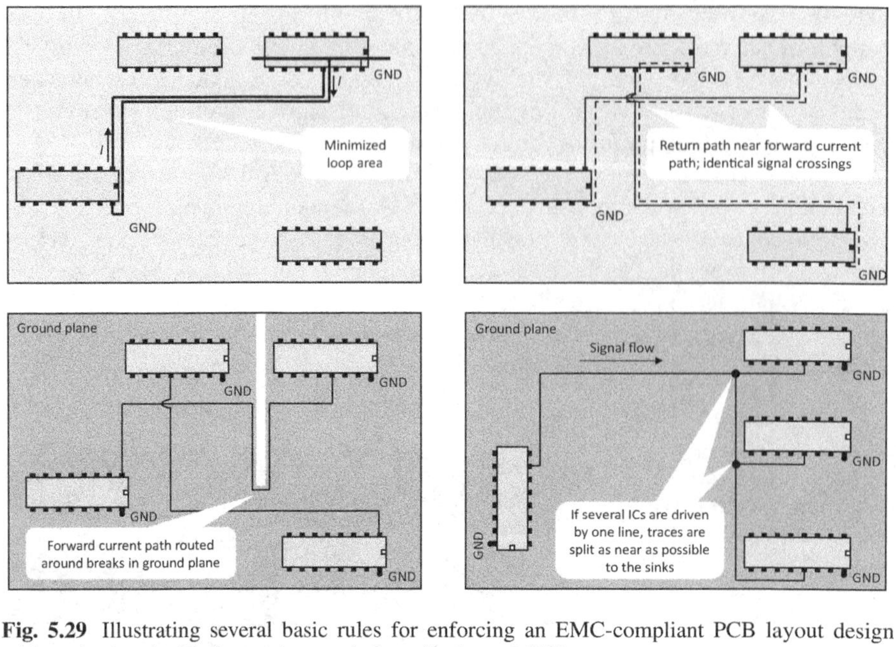

Fig. 5.29 Illustrating several basic rules for enforcing an EMC-compliant PCB layout design without (top) and with (bottom) ground plane (dark gray) [12]

- If several modules are to be driven from one logic output, for example, routing a clock signal to a number of devices, split the trace as near as possible to the destination modules (Fig. 5.29, bottom right).
- If you want to provide capacitive decoupling between two signal traces on a board, introduce a grounded trace between the two traces. Route lines with rapidly switching signals, that is, with high current spikes or high voltage spikes, away from "sensitive lines", such as analog inputs.
- Analog and digital modules should be grounded separately.

Enforcing these and other rules for EMC-compliant layout design requires experienced PCB designers, as knowledge of coupling and grounding mechanisms is needed to select suitable counter measures. Debugging EMC problems at a later stage by shielding and filtering, for example, involves extensive effort and should be avoided by proper place and route design of the PCB [12].

After the place and route procedures have been successfully completed, the PCB layout is checked by a *design rule check* (DRC) tool. The DRC ensures that all traces and other layout patterns are placed according to the design rules; passing the DRC run without errors certifies that the PCB layout can be produced in the intended manufacturing technology.

In a final step, the *manufacturing documentation* is generated. It includes files for all layers as well as additional manufacturing information (Fig. 5.30). In general, this output data can be differentiated into *documentation data*, which is independent of the technology, and *manufacturing data*, which is manufacturer (and thus technology) specific. Documentation data encompasses the schematic, the netlist and the list of components. Manufacturing data includes mask descriptions for each layer, drilling tables and further PCB assembly data.

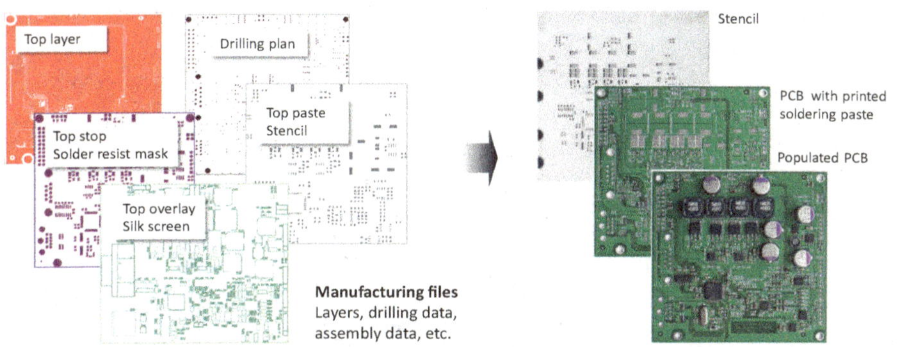

Fig. 5.30 Generating the output files (left) that are needed for manufacturing a PCB (right)

5.4 Verification

When physical design is complete, the layout must be fully verified. This verification step validates functional correctness and design manufacturability. Effectively, the main objective of verification is to ensure the correct functionality of the design and to minimize the risk of problems occurring during manufacturing.

Designing an electronic system is both challenging and time-consuming. Issues can be encountered during the process that jeopardize or completely scupper the design. Research and practical experience have shown that if a fault is left undiscovered it becomes far more costly to correct at a later stage. In fact, the cost of correcting a fault increases by an order of magnitude for every layout-processing step in which it remains undiscovered. The immediate goal is therefore to discover faults as early as possible in the flow. This requires multiple verification steps.

The product cannot be tested or validated during the design process as it does not yet exist at this stage. However, during design reliable and informative criteria need to be specified for checks that will occur in subsequent stages. These criteria can be derived from the technological and functional constraints; we describe this operation in Sect. 5.4.1, and then elaborate on the individual verification techniques in subsequent sections.

As visualized in Fig. 5.31, any comprehensive design verification process includes the following checks: *formal verification* (Sect. 5.4.2), *functional verification* (Sect. 5.4.3), *timing verification* (Sect. 5.4.4), and *physical verification*. Physical verification can be further split into *geometric verification* (Sect. 5.4.5) and verification based on *extraction* and *netlist comparison* (LVS, Sect. 5.4.6). As a physical

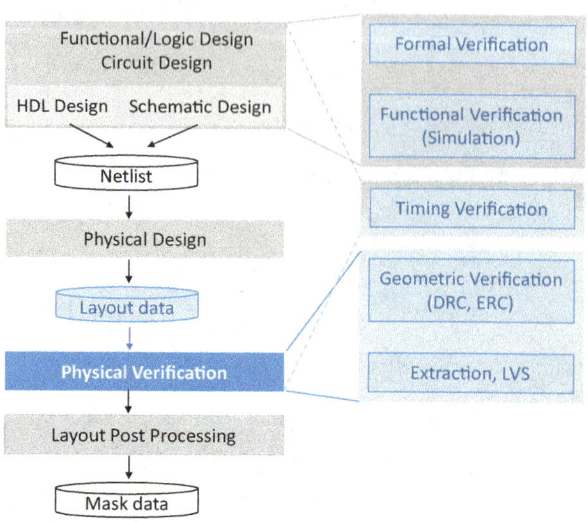

Fig. 5.31 Verification steps (right) that are discussed in Sects. 5.4.2–5.4.6

designer needs to know which verification steps have been applied prior to layout generation, we cover these "early-on" verifications (formal, functional and timing) first, before presenting in detail physical verification procedures (DRC, ERC, Extraction, LVS).

5.4.1 Fundamentals

As we know, electronic systems are designed in many steps due to their high complexity. Each design step produces a new intermediate result that brings us ever closer to the final design. These intermediate results must be checked for violations of the specified technological and functional constraints. The functional constraints are derived from the project, including some that are defined at project kick-off, one example being the signal-to-noise ratio of a signal defined in the specification. Others could be defined during the design, an example here being the maximum permissible IR drop in a line. The technological constraints are specified by the fab (aka the fabricator); they are based on the manufacturing limits of the technology being used.

We noted in Chap. 2 (Sect. 2.3.4) that it is critical that the result of an IC design, i.e., the finished layout, is correct, because if an IC with inherent design flaws is fabricated, serious financial losses would be incurred by the defective chip. Violations are therefore checked for with advanced computer-based tools. To enable this to happen, the relevant constraints must be converted to a format that can be used by these verification tools.

The constraints are converted in a two-stage mapping process. Figure 5.32 illustrates this for all constraints. Constraints based on physics/reality are converted in the first mapping to formal rules or standards, which are formulated in a readable format as text, for example. This formal description of the constraints is a meta format (middle column in Fig. 5.32). These meta descriptions are then converted to the data formats needed for the verification tools (right column in Fig. 5.32).

Following the second mapping, the constraints are available for the verification tools as a *technology file*, for example (right-hand column in Fig. 5.32). A verification tool, such as a DRC tool, can then check the intermediate result of a design step automatically for compliance with preset criteria. If a violation is found during the check, it is labeled as an *error*.

Figure 5.32 shows clearly the significance of the different categories of constraints. (We introduced these categories in Sect. 4.5.2.) The technological and functional constraints are converted into the physical design domain (right column in Fig. 5.32) to enable an automatic check. Compliance with these checks determines whether a chip or a PCB can be manufactured, and whether or not it is fit for purpose. The purpose of the remaining constraint category, the design-methodology constraints, is to enable this mapping and thus automatic processing.

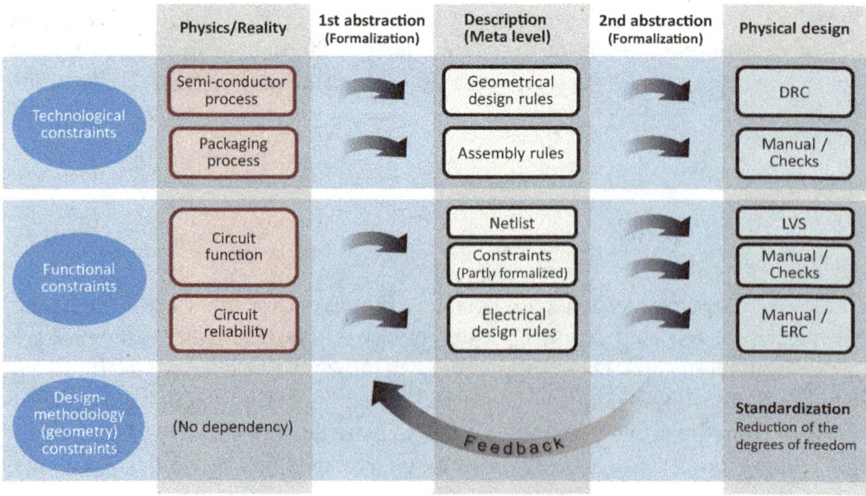

Fig. 5.32 Making technological and functional constraints accessible to verification tools requires two conversions. The first conversion results in formal rules (middle column), which are then converted into the data formats of the respective verification tool (right column). Design-methodology constraints provide a feedback from the second to the first abstraction; they also limit the degrees of freedom within physical design

Let us consider an example: DRC tools need the design rules written in a specific format. Complex design rules can be described in this format by routines comprising numerous commands (examples are given in Chap. 3, Sect. 3.4.3). Although the syntax and semantics of these languages are very powerful, they are still limited. Hence, the rules need to be described in such a manner in the meta level that they enable the second mapping in the physical design without any information loss.

Design-methodology constraints are restrictions that must be adhered to when the design rules of the meta level are drawn up so that these rules can be programmed subsequently. Hence, the design-methodology constraints in Fig. 5.32 represent a feedback from the second mapping to the first. Within physical design (i.e., the result of the second mapping in Fig. 5.32, that is, the column on the right), design-methodology constraints promote standardization by restricting the degrees of freedom.

Design engineers should always remember that both mappings are abstractions, as the real constraints (shown on the left in Fig. 5.32) are formalized by these mappings. This means that a formal constraint, such as a physical design constraint, is not equivalent to the real requirement in the technology or in the physical design solution. We will illustrate this with an example of a technological constraint.

It may not always be possible to exactly model a technological constraint by a design rule because of the complexity of modern technologies. In these cases, a small margin of safety is introduced into the design rule w.r.t. the real requirement. This causes layout results to fail the DRC even though they have not violated the real technological constraint. This type of error is called a *dummy error* or *false error*.

The impact of dummy errors can be waived when interpreting DRC results, and the layout structure is left unchanged. However, the design engineer must be very experienced and understand exactly the underlying technology to properly handle these cases.

Geometric rules within the physical design domain (right column in Fig. 5.32) are typically conservatively formulated to ensure that the real requirements (i.e., the original technological constraints) are met with certainty. Technological constraints therefore tend to be met with a safety factor. In fact, manufacturability is typically one hundred percent assured with a perfect DRC result. Contrast this with the situation for functional constraints, where they can only be modelled in part in modern design environments. As such, they cannot be completely checked.

This is an important observation as it is the reason for further checks, such as simulations, and also one reason why a valid *verification* ("Has the circuit been correctly designed?") can nevertheless result in an invalid *validation* ("Has the correct circuit been designed?") (Chap. 4, Sect. 4.4, cf. Fig. 4.18).

Before we discuss different verification methods in physical design in the following subsections in detail, we first classify them. Table 5.1 presents an overview of

Table 5.1 Different options for verifying an electronic circuit as presented in the following subsections. For the sake of completeness, we also include testing, i.e., to validate a circuit design from a customer perspective

Check	What is checked?	How is it checked?	Method
Model check	Logical characteristic (Assumption is true?)	Mathematical models	Formal verification
Equivalence check	Logical equivalence of two descriptions	Mathematical models	Formal verification
Simulation	Circuit behavior versus specification	Virtual experiment (stimuli and output)	Functional verification
DRC (OPC, RET)	Layout versus technological constraints (manufacturability)	Geometrical design rules	Geometric verification
LVS	Layout versus schematic	Netlist extraction from layout, rule based	Geometric verification
PEX (plus simulation)	Impact of parasitics on circuit behavior	Parameter extraction from layout, rule based; followed by simulation	Geometric and functional verification
ERC	Layout versus electric process boundaries (reliability)	Connectivity extraction from layout, rule based	Geometric verification
Testing	Compliance for practical usage	Real experiment, customer checking	Validation

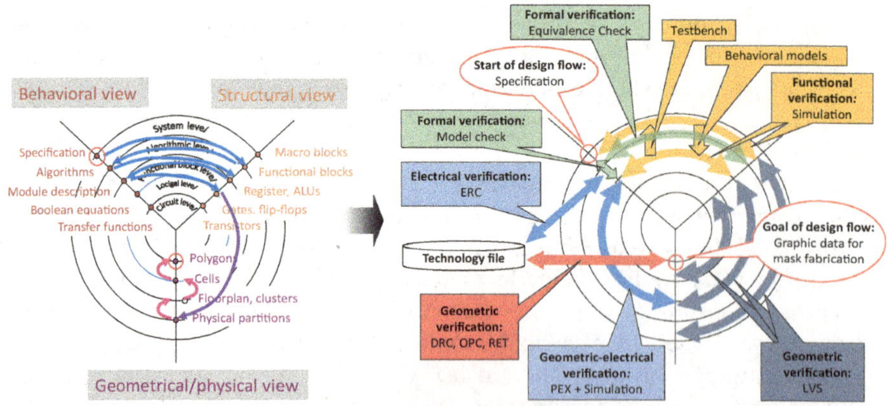

Fig. 5.33 Illustration of the various verification methodologies (cf. Table 5.1) using the Y-chart (right), where the top-down design style is visualized (left)

the various options available for verifying a circuit. Table 5.1 also includes a method that is beyond the scope of this book, testing, which *validates* a circuit design with regard to a customer's requests (see Fig. 4.18 in Chap. 4). Please also note that we omit *assertion-based verification* (ABV) where designers use assertions to capture specific design intent. This verification methodology can be addressed nowadays by formal verification techniques (model checking) as well as traditional simulation strategies.

Figure 5.33 relates the verification methodologies to the Y-chart (Chap. 4, Sect. 4.2.2).

5.4.2 Formal Verification

The goal of *formal verification*, also called *formal functional verification*, is to prove the correctness of a circuit implementation with respect to its specification. More specifically, it shows the correctness of an intended circuit regarding a specific formal specification or property, using formal mathematical methods. The best known formal verification methods are "model checking"—often called "property checking" in commercial tools—and "equivalence checking".

Model checking verifies a certain property of a design or an implementation. It proves (or disproves) that a design under verification, often described in HDL code, satisfies its specifications, i.e., that it behaves as expected in every way (and only as expected). Both the design model under verification and the specification are formulated using precise mathematical language. Essentially, a given structure must satisfy a given logical formula in this check.

Equivalence checking, on the other hand, compares two circuit descriptions. It exhaustively checks that two design representations, such as HDL code and a derived gate level netlist, provide the same functional behavior. There are different approaches to executing this type of proof. For example, both circuit descriptions can be represented by a normalized notation, such as a netlist syntax, to simplify the comparison. Equivalence checking is the primary methodology for synthesis verification.

Formal verification delivers either a successful verification result or demonstrates (i) that the circuit description does not meet a desired property (model checking), or (ii) that two circuit descriptions are not the same (equivalence checking).

Formal verification is part of the early design steps, such as HDL-based netlist generation, that we covered in Sect. 5.1. As the layout designer does not deal directly with this pre-layout verification method, we will not explore it further here; more information on formal verification can be found in the literature: for example, [13] is a well-written and easy-to-grasp introduction to the topic.

5.4.3 Functional Verification: Simulation

A circuit's functional correctness can be verified by simulation. Here, typical input patterns, so-called *stimuli*, are used to check whether the simulated outputs are identical to the intended outputs. Alternatively, the stimuli can be applied to the design behavioral description and to the final gate description. In this case, their responses are compared and evaluated.

Any differences in simulation results can be caused by (i) design errors or (ii) simulation errors or inaccuracies. In both cases, further investigations are required. If, however, the simulation results are identical with the design values, confidence in the correctness of the design is boosted. Unfortunately, simulation can never guarantee that a design is correct in its entirety.

Simulating a circuit's behavior before actually building it can greatly improve design efficiency by flagging design faults early in the flow and providing insight into the circuit's behavior. Almost all IC design relies heavily on simulation. The best-known analog simulators are based on the principle of (or directly derived from) SPICE (Simulation Program with Integrated Circuit Emphasis); digital simulators are often using Verilog or VHDL syntax (Sect. 5.1).

Popular simulators frequently include both analog and event-driven digital simulation capabilities, known as mixed-mode simulators. This means that any simulation may contain components that are analog, event-driven (digital or sampled-data), or a combination of both. Mixed-mode simulation is performed on three levels; (i) with primitive digital elements that use timing models and the built-in digital logic simulator, (ii) with subcircuit models that use the actual transistor topology of the IC, and (iii) with in-line Boolean logic expressions. An entire mixed-signal analysis can be driven from one integrated schematic.

Simulation tools usually interface to a schematic editor, a simulation engine, and on-screen waveform display (Fig. 5.34). These tools allow a designer to quickly

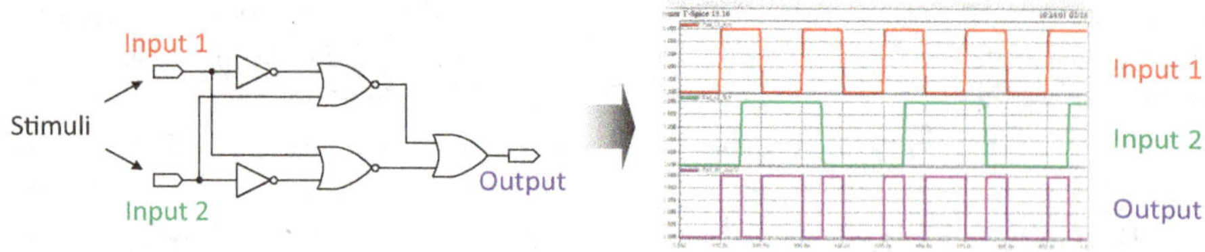

Fig. 5.34 Example of a five-gate circuit with XOR functionality and a (correct) waveform display

modify a simulated circuit and see what effect the changes have on the output. Simulators also typically contain extensive model and device libraries.

When verifying a circuit by simulation, we should always keep in mind that simulation-based verification requires a long execution time, especially for large designs. Worse still, there is usually a lack of a comprehensive set of stimuli to validate the entire design. It is impossible to consider all possible input patterns and circuit states even for fast simulators and small circuits. (For example, the exhaustive simulation of a multiplier for two 32-bit binary numbers would require 2^{64} input patterns, which would take 5,849 years of execution time even with a simulation rate of 100 million multiplications per second [8].) This often forces the designer to rely on some method of random stimuli generation, despite the requirement of full design coverage [5]. Hence, some design errors may remain undetected due to "wrong stimuli".[3]

5.4.4 *Timing Verification*

The term *timing verification* is used to describe the process of checking whether a digital circuit's timing is still valid after its layout has been produced.

Critical paths in a circuit are calculated during logic synthesis; all paths are checked for worst-case delay times caused by changing signals. The resulting critical path defines the fastest possible clock rate at which the circuit can produce correct output signals. The circuit's layout must meet timing constraints, as well. Essentially, the circuit must pass two timing checks: maximum delay, which is related to setup (long-path) constraints, and minimum delay, which relates to hold (short-path) constraints (Fig. 5.35). Setup checks characterize the performance, whereas a non-passing hold check indicates a faulty circuit.

One approach for timing verification is *dynamic timing analysis*. Here, all wire capacities and resistances are extracted from the layout and the circuit is simulated considering these values. This method is very time consuming as many stimuli must be considered. Restricting dynamic timing analysis to the critical path (defined during

[3]The so-called "Pentium FDIV bug" is an infamous example, where a well-simulated Intel processor returned incorrect binary floating-point results when dividing a number, causing a $475 million loss for Intel [6].

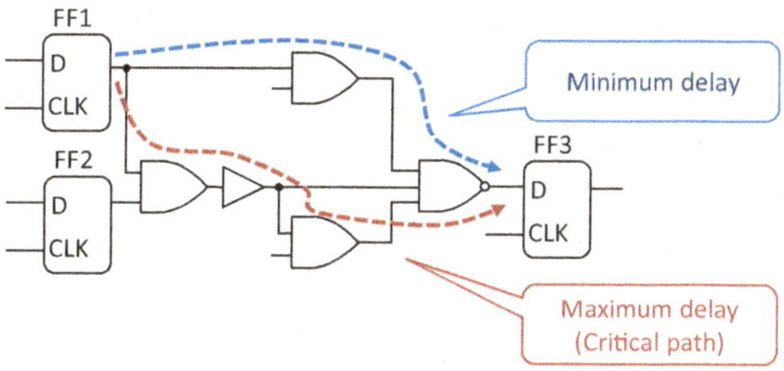

Fig. 5.35 Illustrating minimum and maximum delays resulting from the shortest and longest paths

logic synthesis) does not help because this path is undefined at this later stage—place and route could have easily generated a different critical path than was calculated during logic synthesis.

Static timing analysis (STA) has been developed as a more efficient timing verification method. It is based on the netlist extracted from the layout considering wire capacities and resistances too. The signal delays calculated on all paths are compared to the timing constraints defined by the designer. Specifically, STA propagates actual arrival times (AATs) and required arrival times (RATs) to the pins for every gate or cell. STA quickly locates timing violations, and diagnoses them by tracing out critical paths in the circuit that are responsible for these timing failures [8]. In the past, logic synthesis was then repeated using more restrictive timing constraints on these critical paths; nowadays, STA produces as output an optimized netlist.

Logic gates and wires along with their respective delays are inputs for dynamic and static timing analysis. While gate delays are specified in the timing models in a library, wire delays are calculated using a variety of techniques. Among these techniques, the moment-based technique is widely applied today where impulse responses from the RLC network are analyzed by means of time-frequency transformation methods [5]. Another moment-based interconnect delay calculation uses the first moment of the impulse response; this is known as the Elmore delay model [4].

Any timing-related circuit simulation *after* layout generation requires layout-dependent timing information for the current operating condition to be simulated. The Standard Delay Format (SDF) is used for this timing information. The SDF file contains interconnect delays, gate delays and timing checks that are exported from the physical design tools into an abstracted format. Most important here are the delays associated with the interconnections between devices and ports, i.e., the wire-segment delays as laid out during physical design.

Timing verification also requires checking for resistive and capacitive coupling. For example, crosstalk-induced noise occurs when signals in adjacent wires transition between logic values—and capacitive coupling between these wires causes a charge transfer [5]. This capacitance also has a serious impact on the adjacent wire delays. It is therefore crucial that an accurate timing engine is available to calculate the delay of a coupled system in post-layout timing verification.

Verbal expressions such as “the design has closed timing” are commonly used when the design satisfies all timing constraints. More precisely, the term *timing closure* denotes the process of satisfying timing constraints through layout optimizations and netlist modifications [9]. These layout optimizations include timing-driven placement and timing-driven routing. As they are of importance for a layout designer, let us elaborate on these two procedures with additional details.

Timing-driven placement optimizes circuit delay, either to satisfy all timing constraints or to achieve the highest possible clock frequency. It uses the results of STA to identify critical nets and attempts to improve signal propagation delays through these nets. As we introduced in Sect. 5.3.2, timing-driven placement can be categorized as net-based or path-based. There are two types of net-based techniques—(i) delay budgeting assigns upper bounds to the timing or length of individual nets, and (ii) net weighting assigns higher priorities to critical nets during placement [9]. Path-based placement seeks to shorten or speed up entire timing-critical paths rather than individual nets. Although it is more accurate than net-based placement, path-based placement does not scale to large, modern designs because the number of paths in some circuits, such as multipliers, can grow exponentially with the number of gates [9].

After detailed placement, clock network synthesis and post-clock network optimization, the *timing-driven routing* phase aims to correct the remaining timing violations. It seeks to minimize one or both of (i) maximum sink delay, which is the maximum interconnect delay from the source node to any sink in a given net, and (ii) total wirelength, which affects the load-dependent delay of the net's driving gate [9]. Specific methods of timing-driven routing include generating minimum-cost, minimum-radius trees for critical nets, and minimizing the source-to-sink delay of critical sinks [9].

If outstanding timing violations still remain, further optimizations, such as re-buffering, are applied.

5.4.5 Geometric Verification: DRC, ERC

The term *geometric verification* summarizes all checks that are executed at the finished (geometric) layout or during layout design. Most notable here are the design rule check (DRC) and the electrical rule check (ERC).

Every chip manufacturer provides geometrical design rules (Chap. 3, Sect. 3.4) for their technology to the chip-designing organization (cf. Fig. 4.19 in Chap. 4). They are stored in *technology file* which is part of the design suite for a given technology (process design kit, PDK). These rules are a prescription for preparing photomasks that can be applied during IC design and which deliver a manufacturable layout. More precisely, a design rule set specifies certain geometric and connectivity restrictions to ensure sufficient margins to account for variability in the applied semiconductor manufacturing process.

As discussed earlier (Chap. 3, Sect. 3.4.2 and Chap. 4, Sect. 4.5.2), geometrical design rules can be separated into width, spacing, extension, intrusion, and enclosure rules (Fig. 5.36, left; cf. Fig. 3.20 in Chap. 3). Another category of rules that can be

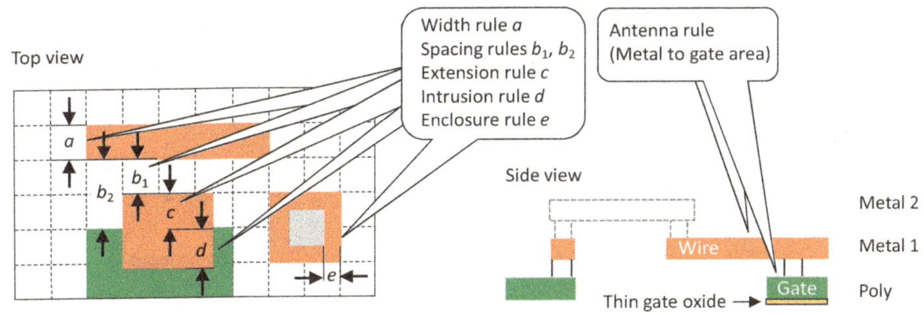

Fig. 5.36 Visualization of the basic DRC checks (width, spacing, extension, intrusion, and enclosure rules, cf. Fig. 3.20 in Chap. 3) and of the antenna rule, which is the allowable ratio of poly or metal area to gate area (right)

checked during geometric verification are antenna rules (Fig. 5.36, right), which we will elaborate on below.

DRC software uses the aforementioned technology file, sometimes called a *DRC deck*, during the verification process; the layout data is usually provided in the GDSII/OASIS standard format. DRC has evolved from simple measurements and Boolean checks to much more sophisticated rules that modify existing features, insert new features, and check the entire design for process limitations such as layer density. Modern design rule checkers perform complete verification checks on geometrical design rules (Chap. 3, Sect. 3.4). The DRC tool either flags any violations directly in the layout (Fig. 5.37) or it produces a report of design rule violations.

In some special design cases, the designer may not choose to correct any DRC violations. Here, carefully "stretching" or waiving certain design rules is a tactic

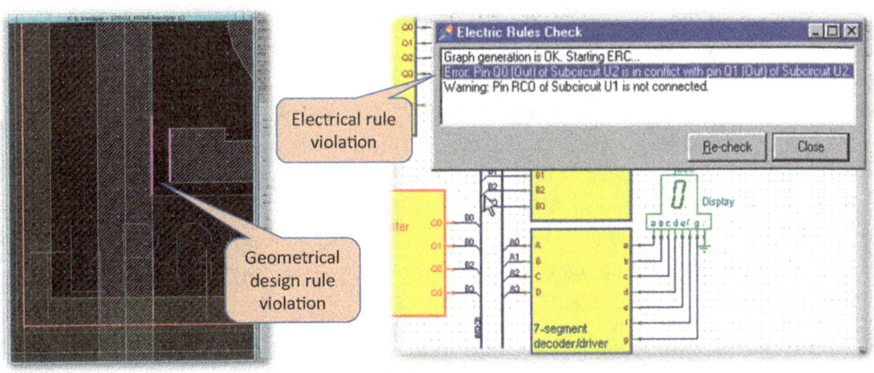

Fig. 5.37 The design rule check (DRC) verifies that geometrical design rules are met, exposing a minimum distance violation (left). In contrast, the electrical rule check (ERC, right) checks for inconsistencies in the electrical network that can be determined from the geometry and connectivity in the circuit schematic or layout. Simply speaking, DRC does syntax analysis on the layout and ERC performs syntax analysis on the network

used to increase performance and component density at the expense of yield. Obviously, the more conservative the design rules are, the more likely the design will be manufactured correctly; however, performance and other objectives could suffer.

As already mentioned, *antenna rules* can be included in the DRC. A so-called *antenna* is an interconnect, i.e., a conductor such as polysilicon or metal, that is only partially complete during the manufacturing process. During this time, as the layers above are not processed yet, this interconnect is temporarily not electrically connected to silicon or grounded during the wafer processing steps (see Fig. 5.36, right). Charge can accumulate on these (temporary dead-end) connections during the manufacturing process to the point at which leakage currents are generated and permanent physical damage can be caused to thin transistor gate oxide that lead to immediate or delayed failure. This destructive phenomenon is known as the *antenna effect*. Fabs normally supply antenna rules that are often expressed as an allowable ratio of polysilicon and metal area to gate area. There is one such ratio for each interconnect layer, which is then verified during the DRC. Sometimes a specific ratio of the polysilicon and metal shapes' circumference to the gate area is additionally required because the charges are preferably collected at the antenna edges.

As design for manufacturability (DfM) has gained importance, DRC tools increasingly include checks for manufacturability which go beyond the basic geometrical design rules. This encompasses the Boolean operations and sizing functions that we covered in Chap. 3; relations between different layers can also be included in an automatic verification. Again, these rules are directly provided by the IC manufacturer.

Finally, we must point out that DRC can be extremely runtime intensive as the checks usually run on each sub-section of the circuit to minimize the number of errors that are detected at the top level. Modern designs can have DRC runtimes of up to a week. Most design companies require DRC to run in less than a day in order to achieve reasonable cycle times since the DRC will likely be executed several times prior to design completion.

So far we have hopefully conveyed to the reader that the DRC ensures that the circuit will be *manufactured* correctly. It should also be clear that correct *functionality* cannot be checked this way, this is left to the simulators and verifiers that manipulate circuit behavior and that we covered earlier in Sects. 5.4.2–5.4.4.

Now let us investigate the "middle ground" between simple layout checking and complex behavioral analysis, which is the domain of *electrical rule checkers* (*ERC*). Electrical (design) rules make a circuit more robust, for example, by guarding it against damage from electro-static discharge; they also improve its reliability by reducing aging due to electrical overstress. These rules are highly dependent on (i) the applied semiconductor technology, (ii) the circuit type, and (iii) the circuit's future use as a component in a large system environment. In addition, electrical rules are often complemented by design-house specific rules and experience-based rules.

Hence, electrical rule checking is a methodology used to validate the robustness and reliability of a design both at the schematic and layout levels against various "electrical design rules". It verifies the correctness of power and ground connections and checks for floating nets or pins and open and short circuits. For example, by propagating the power, ground, input, and clock signals through the circuit's schematic and/or layout, it is possible to check for incorrect output drives, inconsistencies in signal specifications, unconnected circuit elements, and much more. The results are either visualized inside the schematic/layout editor or presented in a table (see Fig. 5.37, right).

Electrical rules are often specified as topological structures rather than single device/pin checks. Geometrical rules from the layout are also associated with these topologies to ensure proper design function, performance, and yield. Some rules, such as voltage-dependent metal spacing rules, combine both geometrical and electrical checks.

5.4.6 Extraction and LVS

The *layout versus schematic tool*, often abbreviated as the *LVS check*, compares the original netlist, that was used to generate the layout, with a netlist that has been extracted from the layout produced. This proves finally that the generated layout corresponds exactly with the original netlist. More precisely, the LVS check ensures that circuit design and layout design match by checking for (i) the electrical connectivity between device instances, (ii) the correct device instances in the netlist and the layout, and (iii) function-critical device instance parameters. This tool and the DRC are the most important verification tools in any IC design flow.

In order to compare both netlists, the LVS tool must first *extract* a netlist from the layout data. This is performed in an extraction step. It requires a technology-dependent extraction file containing three definitions:

- How are the layers connected, i.e., what forms a *net*?
- What combination of polygons and layers form a *device*?
- Which device polygon properties determine the electrical *parameters*?

Figure 5.38 visualizes the contents of such an extraction file. The contents of a netlist can only be derived from a given layout with these three pieces of information (layer connections, devices, device parameters), as the layout, after all, consists only of polygons.[4]

The extraction algorithm is able to generate a netlist from the graphics data of the layout based on this description. The procedure is as follows:

(1) Defining the basic devices
    (a) Determine all geometrical structures that represent the basic devices.
    (b) Separate the basic devices from the other layout structures.

---

[4]It is important to note why we do not take any other layout information into account, such as library information. This would, of course, greatly simplify the task and speed up netlist recognition. However, any error in the library would then be considered as well. The final netlist check would then check identical netlists as both lists would be affected by the same library-based error(s). This would render the LVS useless.

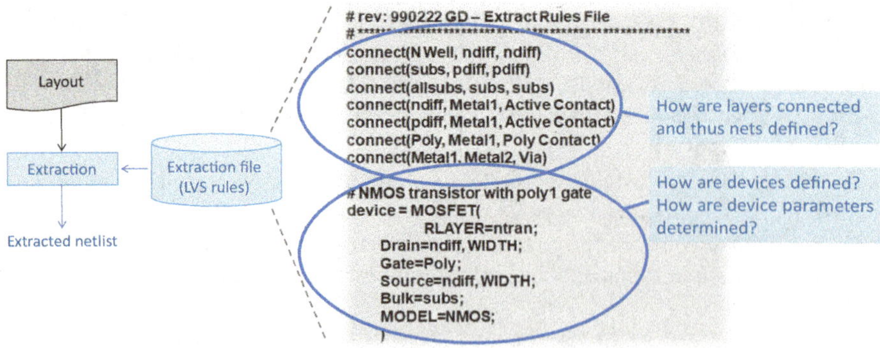

Fig. 5.38 An extraction file is needed to extract the netlist from the layout polygons as this can only be achieved by knowing which polygon configuration forms a via or a device

(2) Determine electrical nodes
Determine all geometrical structures that form electrically linked units. This is an intra-mask operation.

(3) Generating the netlist
   (a) Determine the nodes to which geometrical structures adjacent to basic devices belong.
   (b) Assign the connection types (e.g., gate and source in the case of transistors).

The contents of this netlist are then compared with a netlist derived from the circuit schematic. The entire LVS procedure is depicted in Fig. 5.39.

The LVS compares the output data (layout) with the input data (circuit schematic) w.r.t. the following three circuit-diagram attributes:

- Nets: Are all electrical connections in the circuit schematic—and only these connections—in the layout as well?
- Type of devices: Are all devices from the circuit schematic—and only these devices—present in the layout?
- Parameters of devices: Do all devices in the layout have the electrical parameters specified in the schematic?

The result of the LVS is a report file (see Fig. 5.39) that contains the number and types of devices as well as nodes in the original netlist (from the schematic) and the netlist that was extracted from the layout. This file also lists all non-matching components in both netlists. It is up to the designer to investigate these issues further, as these comparison errors or warnings can be serious faults or simply unrecognizable features flagged by the extraction tool.

One of the major issues with LVS verification is the repeated iterations of design checking required to find and remove these non-matching components between both netlists [5]. As this can be very time consuming, hierarchical verification features (rather than a flat comparison) should be used. Here, memory blocks and other

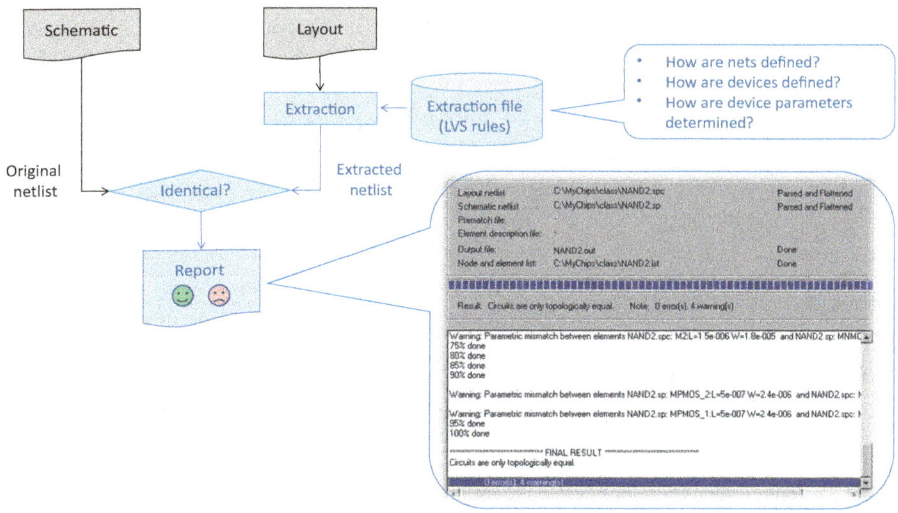

Fig. 5.39 The LVS methodology is based on a netlist extraction from the layout. This netlist is compared with the original netlist that was used to generate the layout

intellectual-property (IP) elements are compared in a hierarchical manner, while other design elements, such as analog blocks and macro cells, maintain a flat representation [5]. Consequently, verification (debugging) time can be drastically reduced.

So far we have seen how the extraction tool generates a netlist from a layout. Extraction tools also feature *parasitic extraction* (*PEX*). Here, the parasitic effects in the interconnects are calculated. The parasitics in question are: (i) parasitic capacitances, (ii) parasitic resistances, and (iii) parasitic inductances.[5]

Parasitic extraction is required in order to create a more accurate analog circuit model. Based on device models and PEX results, detailed simulations can emulate actual digital and analog circuit responses. Another factor in the rise of interest in parasitics is the importance of wiring capacity in advanced technology nodes: interconnect resistances and capacitances started making a significant impact on circuit performance below the 0.5-μm technology node. Interconnect parasitics cause signal delays, signal noise, and IR drops—all important issues affecting circuit timing and performance especially of analog circuits. In summary, timing analysis, power analysis, circuit simulation, and signal integrity analysis rely on parasitic extraction.

Parasitic extraction methodologies can be broadly divided into (i) field solvers, which provide physically accurate solutions, and (ii) approximate solutions with pattern matching techniques. Since field solvers can only be applied to small problem instances, pattern matching techniques are the only feasible approach to extract parasitics for complete modern IC designs.

[5]Additional parasitic coupling effects are caused by the chip substrate, which is common to all devices. However, these effects are not considered in all simulation tools.

The extraction tool can also be used for antenna checks (Sect. 5.4.5). Here, the gate area and the area of the conductor(s) are extracted, and their ratio is calculated and compared with a reference value.

Finally, the extraction tool is also required for specific ERC functions (Sect. 5.4.5). An example is pin-to-pin checks within the ERC where a specific resistance value should not be exceeded in order to meet ESD requirements.

5.5 Layout Post Processing

Traditionally, after an IC specification had been converted into a physical layout, the timing verified, and the polygons certified to be DRC-clean, the physical design was ready for fabrication [10]. The data files for the various layers were given to a mask shop, which used mask-writing equipment to convert each data layer into a mask. Afterwards, the masks were shipped to the fab where they were used to manufacture the designs in silicon. Thus, layout creation and verification terminated the actual design process.

Nowadays, the physical-design data of ICs require extensive post processing, which we have covered in detail in Chap. 3 (Sect. 3.3). There we introduced and defined the *layout post-processing* step where amendments and additions to the chip layout data are performed in order to convert a physical layout into data for mask production (Fig. 5.40).

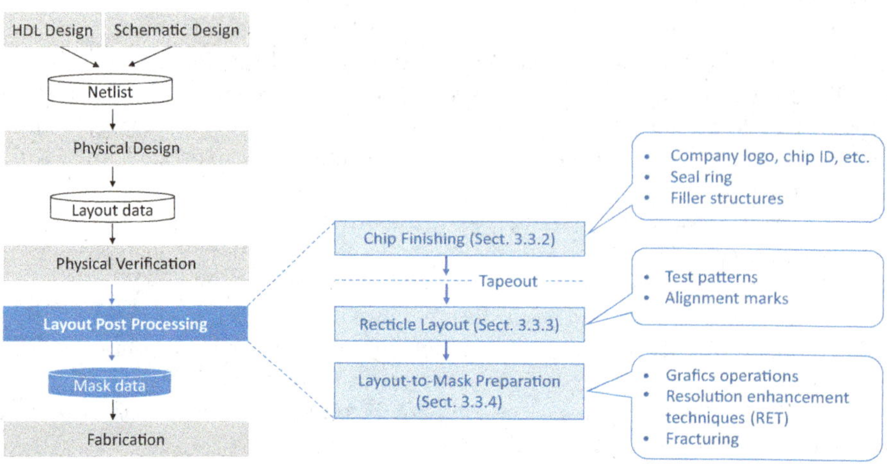

Fig. 5.40 The major steps in layout post processing where amendments and additions to the layout data are performed in order to convert it into mask data (cf. Fig. 3.14 in Chap. 3)

As visualized in Fig. 5.40, layout post processing can be divided into three steps:

- *Chip finishing* which includes custom designations and structures to improve manufacturability of the layout (Chap. 3, Sect. 3.3.2),
- Producing a *reticle layout* with test patterns and alignment marks (Chap. 3, Sect. 3.3.3), and
- *Layout-to-mask preparation* that enhances layout data with graphics operations and adjusts the data to mask production devices (Chap. 3, Sect. 3.3.4).

While the first two steps are not directly related to the actual physical design process (and, hence, we refer the reader to Chap. 3 for their discussion), layout-to-mask preparation might impact physical design directly. Here, the layout is amended with graphics operations, then subjected to operations to enhance the optical resolution, and finally adapted to the mask production devices.

Resolution enhancement techniques (*RET*) play a key role in layout-to-mask preparation (Chap. 3, Sect. 3.3.4). They must be applied to cutting-edge ICs to counter manufacturing and optical effects caused by their extremely small feature sizes. Due to the possible impact of RET on physical design, such as layout restrictions [11], we briefly cover them in this section.

Even though the final layout may be what is required in silicon, it must still be further processed before the IC masks are fabricated. These layout alterations are carried out by RET as illustrated in Fig. 5.41. They can be broadly classified as (i) distortion corrections and (ii) reticle enhancement [10, 11].

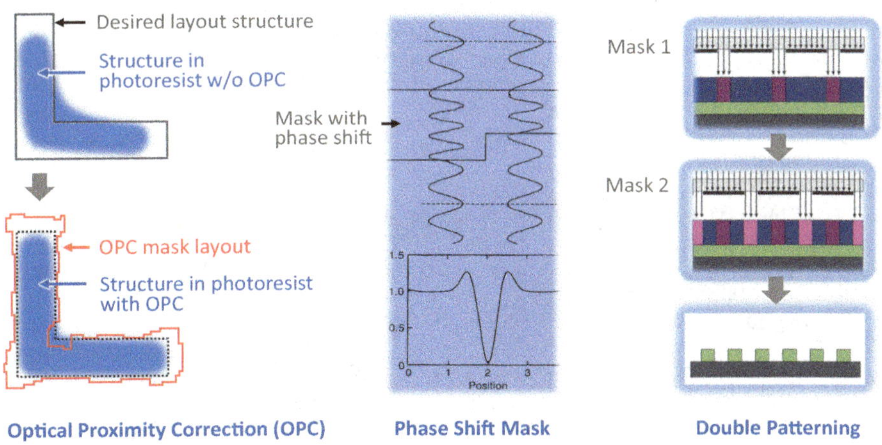

Fig. 5.41 Illustration of resolution enhancement techniques (RET), such as optical proximity correction (OPC), phase shift masks, and double patterning. OPC (left) distorts the patterns on a mask to counter diffraction effects. Phase shift masks (middle) alter the phase of the light passing through some areas of the mask, thus reducing the defocusing effect of mask dimensions that are less than the wavelength of the illuminating light. Double patterning (right) splits dense patterns into two interleaved ones on two masks

Distortion corrections compensate for distortions inherent in the manufacturing process. An example is optical proximity correction (OPC), which addresses image errors due to diffraction effects. OPC counteracts these effects by slightly enlarging the mask opening at underexposed places and slightly shrinking it at overexposed places, as shown in Fig. 5.41 (left). (We discuss diffraction effects in photolithography and possible corrective measures in Chap. 2, cf. Fig. 2.9.)

Reticle enhancement improves the manufacturability or resolution of the lithography process. Examples are (i) phase shift masks, which are photomasks that take advantage of the interference generated by phase differences to improve image resolution (Fig. 5.41, middle) and (ii) double or multiple patterning. Here, multiple lithographic exposure enhances the feature density by splitting dense patterns into two interleaved patterns of less-dense features, using two (or more) masks (illustrated in Fig. 5.41, right).

Double/multiple patterning imposes new layout constraints in order to avoid subsequent decomposition violations that a designer would have to resolve [16]. For example, the mask layers are assigned colors, based on spacing requirements, which are then used to split the original drawn layout into (two or more) new layers.

As both layout post processing in general, and RET in particular, are constantly improving and adjusting to new technologies, we will not investigate these any further but refer the reader to up-to-date literature on these topics.

References

1. R.J. Baker, CMOS circuit design, layout, and simulation, in *IEEE Press Series on Microelectronic Systems*, 3rd edn. (Wiley-IEEE Press, 2010). ISBN 978-0470881323
2. M.R. Barbacci, A comparison of register transfer languages for describing computers and digital systems. *Technical Report* (Carnegie Mellon University Research Showcase @ CMU, Department of Computer Science, 1973)
3. E. Christen, K. Bakalar, VHDL-AMS-a hardware description language for analog and mixed-signal applications. *IEEE Trans. Circuits Syst. II Analog. Digit. Signal Process.* **46**(10), 1263–1272 (1999)
4. W.C. Elmore, The transient response of damped linear networks with particular regard to wideband amplifiers. *J. Appl. Phys.* **19**, 55–63 (1948). https://doi.org/10.1063/1.1697872
5. K. Golshan, *Physical Design Essentials* (Springer, 2007). ISBN 978-0-387-36642-5. https://doi.org/10.1007/978-0-387-46115-1
6. T.R. Halfhill, An error in a lookup table created the infamous bug in Intel's latest processor. *BYTE* (20), 163–164 (1995)
7. M.Y. Hsueh, Symbolic layout compaction, in *Computer Design Aids for VLSI Circuits*, ed. by P. Antognetti, D.O. Pederson, H. de Man. NATO ASI Series (Series E: Applied Sciences), vol. 48 (Springer, 1984). ISBN 978-94-011-8008-5. https://doi.org/10.1007/978-94-011-8006-1_11
8. D. Jansen et al., *The Electronic Design Automation Handbook* (Springer, 2003). ISBN 978-14-020-7502-5. https://doi.org/10.1007/978-0-387-73543-6
9. A. Kahng, J. Lienig, I. Markov et al., *VLSI Physical Design: From Graph Partitioning to Timing Closure* (Springer, 2011). ISBN 978-90-481-9590-9. https://doi.org/10.1007/978-90-481-9591-6

10. L. Lavagno, G. Martin, L. Scheffer, *Electronic Design Automation for Integrated Circuits Handbook* (CRC Press, 2006). ISBN 978-0849330964
11. L. Liebmann, Layout impact of resolution enhancement techniques: impediment or opportunity?, in *International Symposium on Physical Design (ISPD)* (2003), pp. 110–117. https://doi.org/10.1145/640000.640026
12. J. Lienig, H. Bruemmer, *Fundamentals of Electronic Systems Design* (Springer, 2017). ISBN 978-3-319-55839-4. https://doi.org/10.1007/978-3-319-55840-0
13. B. Murphy, M. Pandey, S. Safarpour, *Finding Your Way Through Formal Verification* (CreateSpace Independent Publishing Platform, 2018). ISBN 978-1986274111
14. S.M. Sait, H. Youssef, *VLSI Physical Design Automation, Theory and Practice* (World Scientific, 1999)
15. P. Spindler, Personal communication (TU Munich, 2008)
16. B. Yu, D.Z. Pan, *Design for Manufacturability with Advanced Lithography* (Springer, 2016). ISBN 978-3-319-20384-3. https://doi.org/10.1007/978-3-319-20385-0

Chapter 6
Special Layout Techniques for Analog IC Design

While the physical design steps introduced in Chaps. 4 and 5 are universal, analog integrated circuits present further challenges that require additional layout techniques. There are many differences between analog and digital, and as such, the design flows and tools vary in both cases. Analog circuits are generally less complex in terms of transistor count and are designed in a manual fashion. The distinct lack of design automation means that manual design is still in widespread use today, requiring specialist knowledge that is unique to analog design. This specialist knowledge is covered in this chapter.

We discussed analog design flows in Chap. 4 (Sects. 4.6 and 4.7), previously. Now we present layout techniques that accompany these analog flows, which an analog layout designer must be fully aware of. We start with an introduction to sheet resistances and wells (Sects. 6.1 and 6.2) as this knowledge is needed for the sizing and understanding of analog devices, which we then cover in Sect. 6.3. The methodology for cell generators, which produce such analog devices, is presented in Sect. 6.4. An explanation of the fundamental importance of symmetry and a treatise of resulting matching concepts (Sects. 6.5 and 6.6) conclude this chapter on special layout techniques for analog design.

6.1 Sheet Resistance: Calculating with Squares

When a current I flows through an electrically conducting material with a magnitude that is proportional to the applied voltage V, this is called ohmic behavior. The ratio of V over $I (V/I)$ is known as the ohmic resistance R of the wire/conductor. In other words, $R = V/I$. This ratio is also known as *Ohm's law*. If the conductor is made of a homogenous material, we can express the ohmic resistance as

$$R = \rho \frac{l}{A}. \tag{6.1}$$

J. Lienig and J. Scheible, *Fundamentals of Layout Design for Electronic Circuits*,
https://doi.org/10.1007/978-3-030-39284-0_6

The specific electrical resistance of the material is denoted by ρ, the length of the wire along which the current flows is l, and A is the cross-sectional area through which the current flows. We encounter primarily planar (flat) features in semiconductors. If a current flows laterally in such a sheet, the resistance can be expressed as

$$R = \rho \frac{l}{t \cdot w}. \tag{6.2}$$

Here, the current flows through the rectangular cross-section of the sheet of width w and thickness t. Hence, the cross-sectional area is $A = t \cdot w$ (Fig. 6.1).

The thicknesses of single layers on an IC are specified by the process technology. This applies to the doped areas as well as the metal layers. For a given process technology, thickness t of a layer is considered a constant, and only the lateral dimensions w and l of a wire are variable during the layout process.

Thus, in addition to the material constant ρ, a second constant term associated with every layer in Eq. (6.2) is the layer thickness t, which is specific to a given process technology. We can now define a new value based on the quotient in Eq. (6.2) formed with these two terms

$$R_{\mathrm{Sh}} = R_{\square} = \frac{\rho}{t}, \tag{6.3}$$

where R_{Sh} represents the *sheet resistance* or *sheet resistivity*. If we substitute Eq. (6.3) into Eq. (6.2), we get the resistance of a sheet where the current flows laterally

$$R = R_{\mathrm{Sh}} \frac{l}{w} = R_{\square} \frac{l}{w}. \tag{6.4}$$

If we set $l = w$, we find that the sheet resistance R_{Sh} equals the resistance of a square portion of the sheet, *where the size of the square does not matter*. In other words, the resistance of a square conductive sheet is the same no matter what size it is so long as it remains a square. Hence, the factor R_{Sh} is often named $R_{\square}$. The physical unit of $R_{\square}$ is the same as for a standard resistance, that is, Ω. It is also sometimes labeled Ω/□ (ohm divided by a "square sign") to indicate that it is a sheet resistance.

The sheet resistivities $R_{\square}$ of doped layers and metal layers are defined in every process design kit (PDK). Subsequently, Eq. (6.4) allows us to calculate resistances of all layout features in which a current flows laterally. One can visualize this equation as follows: Simply count the number of square sections (squares) along the wire where the current flows and multiply this number by $R_{\square}$ to obtain the wire's resistance.

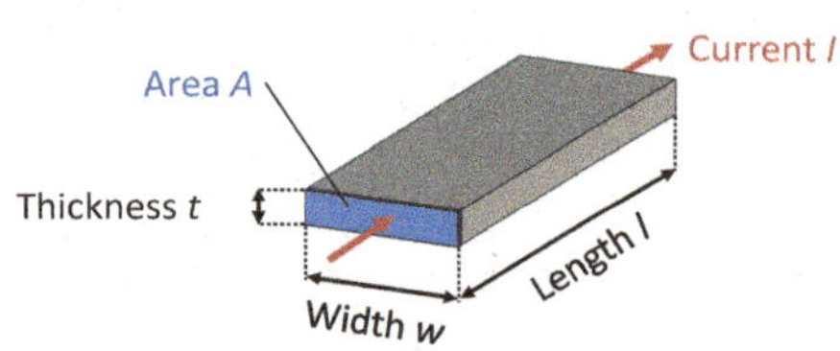

Fig. 6.1 Illustration of a wire segment with width (w), length (l) and thickness (t)

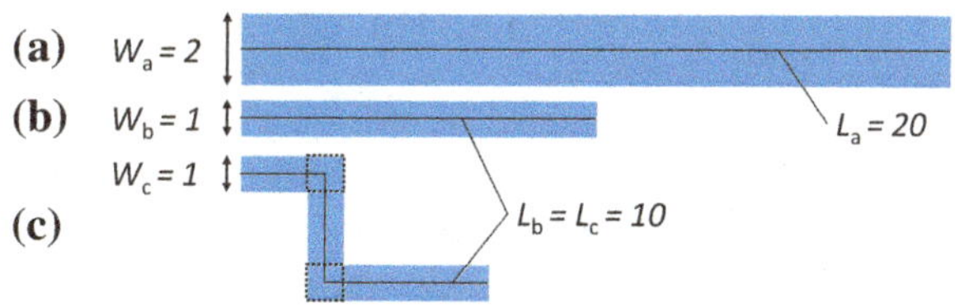

Fig. 6.2 Examples of calculating the resistance values of interconnects. Both wires (**a**) and (**b**) enclose the same number of squares (i.e., have the same length-to-width ratios) and, hence, both have the same sheet resistivity $R_a = R_b = 10R_\square$ despite being differently sized. Wire (**c**) has the same length and width as wire (**b**), however, due to the two corners in (**c**), this wire's resistance is reduced by (approximately) one square to $R_c = 9R_\square$

This method is commonly used with interconnects to estimate their parasitic resistances. The two upper interconnects (a) and (b) in Fig. 6.2 have the same length-to-width ratios $l_a/w_a = l_b/w_b = 10$. Both interconnects are made up of 10 squares each; remember, the respective *size* of a square does not matter, and here interconnect (a) is comprised of 10 squares of size 2×2, while interconnect (b) is comprised of 10 squares of size 1×1. Therefore, they both have the same resistance values $R_a = R_b = 10R_\square$.

The use of squares to calculate resistance is easy to follow. Some words of caution, though: First, only nominal values can be calculated with this method. Resistance values in the real world often deviate greatly from nominal values because of process tolerances.

Furthermore, calculating the resistance by counting the squares only leads to valid results if the current flow is homogenously distributed across the wire cross-section. Calculations will be incorrect in places where the direction of current flow changes, e.g., at corners, and in places where the cross-section changes.

Changes in direction occur in interconnects, as in Fig. 6.2c, for example, where there are right-angle bends in the wire. In this case, the resistance is estimated by counting only half of the values of the squares at the corners (drawn with dashed lines). This rough estimate is sufficient, given the manufacturing tolerances. Wires (b) and (c) in the example in Fig. 6.2 have the same length and width. However, the total resistance of wire (c) is reduced by the resistance of two half squares when the two corners are considered. We therefore estimate the resistance of wire (c) to be $R_c = 10R_\square - 2R_\square/2 = 9R_\square$.

Current usually flows vertically through contact holes and vias, into or out of a layer. The current flow is inhomogeneous at these locations, as well. The contribution of these current entry and exit points to the total resistance must be estimated with other methodologies. These effects are normally considered in the resistance values quoted in PDKs for contacts and vias. We shall review these issues when we come to talk about resistors in Sect. 6.3.2.

6.2 Wells

For ICs to function properly, components must be electrically isolated from one another. Some devices are automatically isolated, e.g., NMOS-FETs in a p-substrate. Specific measures are needed for the electrical isolation of other devices. Elements called *wells* come to the rescue in these cases.

6.2.1 Implementation

Wells are doped regions in an IC that are electrically isolated from their surroundings. They are used to hold one or more devices. Figure 6.3 shows different types of wells based on a lightly p-doped base material.

The main methodologies for creating isolated wells are:

(1) The existing conductivity type is doped as the complementary conductivity type at the desired places. Figure 6.3a–c shows three examples.
(2) A region with the desired conductivity type is enclosed with a barrier of the complementary conductivity type. The barrier is formed by redoping the existing

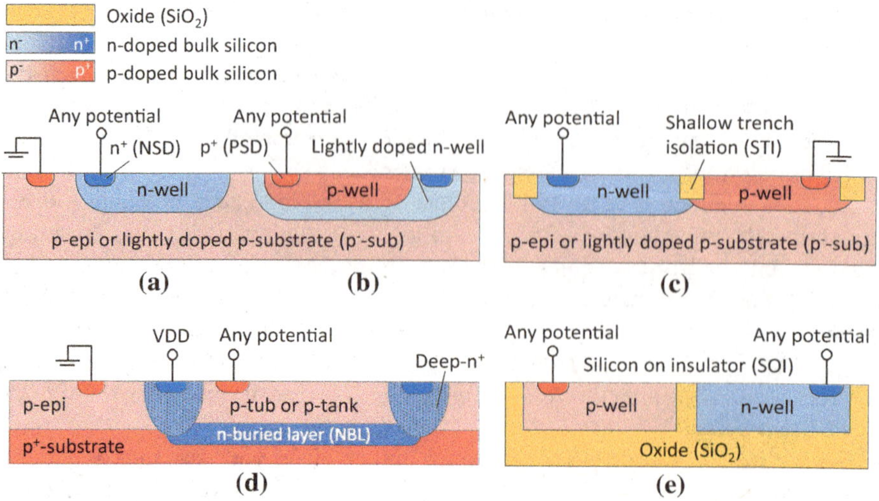

Fig. 6.3 Patterning wells in CMOS processes with different isolation techniques: junction isolation (JI) (**a**, **b**, **d**), shallow trench isolation (STI) (**c**) and silicon-on-insulator (SOI) (**e**). Heavily doped contact points for making electrical connections are also shown. The p-epitaxy (p-epi) is needed in (**d**) to create the buried layer. The other options can be implemented as well in lightly doped p-substrate

area. Thus, a p-well is produced in the p-epi with a buried n-doping ("NBL", n-buried layer) and a deeply embedded n-doping ("Deep-n^+", sometimes called a *sinker*) in the example shown in Fig. 6.3d. This well is also called a *tank* or a *tub*.

(3) A region with the desired conductivity type is encapsulated with a dielectrical oxide barrier (see Fig. 6.3e). These types of wells are created with SOI technology (silicon-on-insulator) [5, 9], where buried oxide layers can be created.

The resulting p–n junctions must be polarized in the reverse direction when using methodologies (1) and (2) to maintain the electrical isolation. This method is known as *junction isolation* (*JI*).

If the substrate is p-doped, as in our examples, it is tied down to the lowest potential in the circuit, which is generally called *ground* (GND) and is by definition at potential 0 V (Fig. 6.4). The procedure is as follows: A bond pad is connected to an external ground. This ground is distributed throughout the entire chip by metal interconnects and connected by (heavily p-doped) *substrate contacts* with low resistance to the p-substrate (or the p-epi).

A dedicated net, separate from the current-carrying GND net, is recommended to ground the substrate. The "SUB" net in Fig. 6.4 fulfils this function. Its topology is called *star routing*. As the "star point" connecting SUB and GND lies directly on the bond pad, there should be no current, or almost no current, flowing through this net. This prevents the substrate potential from rising locally (due to the IR drop) over the interconnects.

When the p-substrate is at 0 V, every n-well can be at any required potential. If an n-well is used to hold a p-well, the highest potential in the circuit (normally defined by the supply voltage VDD) is generally selected for the n-well to allow the p-well to have any potential. The wells are electrically connected to this potential (like the substrate contacts) by metal interconnects and correspondingly heavily doped areas. These areas are marked in Fig. 6.3 with n^+ and p^+. (The NSD and PSD doped regions,

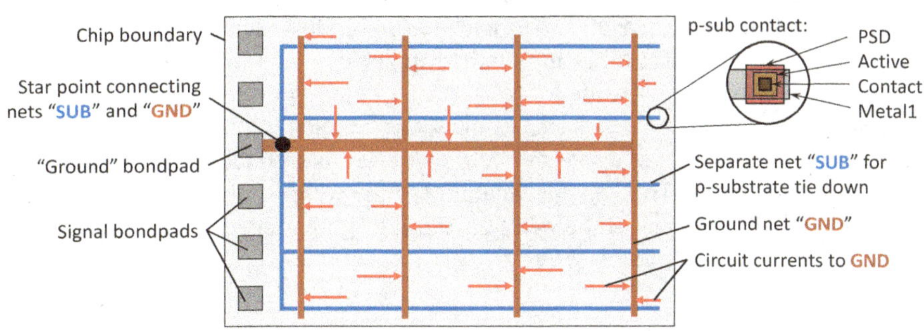

Fig. 6.4 Tying down the p-substrate in a chip to ground by a separate ground net ("SUB" in this example). The "SUB" net is connected directly to the bond pad by a star point so that (almost) no current flows through it

which serve as source and drain regions, are typically used for this purpose in the standard CMOS process).

Isolated wells can also be created by method (3) where the dielectric exclusively isolates the well from its surroundings. This technique is called *galvanic isolation*.

The above techniques are combined these days in processes with trench isolation (see Fig. 6.3c). Here, devices are electrically isolated laterally from neighboring features by STI (shallow trench isolation), while electrical isolation to the semiconductor base (bottom) is realized by junction isolation. This process is described fully in Chap. 2 (Sect. 2.9.3).

6.2.2 Breakdown Voltage

The dielectric strength of an insulating material is given by the maximum electric potential difference that the material can withstand without breaking down, also known as *breakdown voltage*. The breakdown voltage of dielectrically-isolated wells can be set by adjusting the oxide thickness. Vertical layer thicknesses are always specified by the process, and as such, can only be changed by selecting a different process.

The situation is more critical if the wells are electrically isolated by means of junction isolation (JI, Sect. 6.2.1). In addition to biasing and dimensions, the doping strengths of the silicon forming the junction impact the isolation in this case. We shall explain this effect only briefly here. The reader is kindly referred to [9] as a valued resource for further information on this topic.

Given the large differences in charge carrier concentrations at p–n junctions, the majority carriers (holes in the p-type region, electrons in the n-type region) diffuse to the other side of the junction, where they become minority carriers and predominantly recombine. Thus, the region around the p–n junction suffers a depletion of free charge carriers. A *space-charge region*, also called *depletion region*, forms because of the (stationary) ionized dopant atoms.[1] These space charges generate an electrical field. The currents caused by the diffusion and this field work against each other and cancel each other out. The field inhibits the majorities, thus causing the isolating effects between the n- and p-sides.

A lower space-charge density causes the space-charge region to expand and thus increases the breakdown voltage of the well. This effect is produced by weaker doping and is explained in [9]. The voltage capability of the well benefits greatly if *at least one* of the two doped areas of the p–n transition is lightly doped. The field can then spread in the lowly doped area(s).

[1] A space-charge region is also called a *depletion region* because it is formed by the removal of all free charge carriers, leaving almost no free charge carriers to "transport" a current. We prefer to use the synonymous term *space-charge region* throughout this chapter as the space charges generate the electric field whose effects are essential for our discussion.

Consequently, a layout designer should keep the following two rules in mind when using junction isolation:

- The breakdown voltage and, hence, the blocking capability of a p–n junction decreases with increasing dopant concentration.
- A high blocking capability requires that at least one side of the junction is lightly doped.

6.2.3 Voltage-Dependent Spacing Rules

Following the considerations in Sect. 6.2.2, we can now explain why voltage-dependent spacing rules often apply to wells in processes that use lateral junction isolation.

We should recall that, during a doping operation through the wafer surface, the maximum dopant concentrations occur near the silicon surface and that they decrease dramatically deeper into the silicon (Chap. 2, cf. Fig. 2.16, 2.18, and 2.19). In the case of lateral junction isolation (cf. Fig. 6.3a, b, d), the space-charge region spreads at the surface mainly in the lightly p-doped well environs. This spreading effect is depicted in Fig. 6.5.

Hence, the higher the voltages applied to the wells, the greater the required distance between two neighboring wells. This is why wells are classified according to voltage in automotive electronics where processes for higher voltages are common. Spacings are derived from these classifications. Figure 6.5 shows how a spacing rule for n-wells is produced. The rule is composed of twice the outdiffusion plus the expansions of the two space-charge regions in the p-substrate plus an electrical spacing between the two space-charge regions. The space-charge regional expansions are defined by the voltage classes.

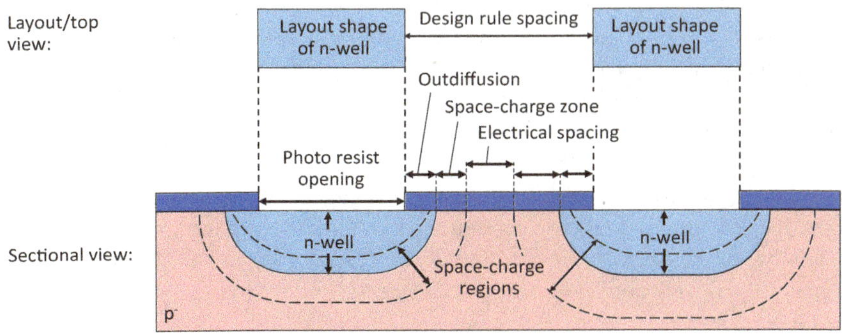

Fig. 6.5 Voltage-dependent spacing rules for wells. The voltage of the well on the left is lower than for the one on the right, as evidenced by the latter's larger space-charge region

Voltage-dependent spacing rules are normally not required when the wells are galvanically isolated. Hence, wells with trench isolation can be considerably more densely packed. The deep vertical p–n junctions (cf. Fig. 6.3c) do not negatively impact the packing density, as the space-charge regions can expand vertically in these regions. The dopant concentrations in these wells are generally much lower here, as well.

6.3 Devices: Layout, Connection, and Sizing

We shall discuss the most important devices in a standard CMOS process in this Sect. 6.3. We show the layouts with associated cross-sections and schematic symbols.

The layouts were generated with cell generators from the process design kit "GPDK180" [1]. This is a generic PDK for a typical 180 nm CMOS process. The cross-sections are based on these layout results.[2] The layer names are the same as in Chap. 2, where we introduced the CMOS process. Although contacts and metal pins are also created in the layouts, we leave them out of the cross-sectional views, where they are not needed to understand the subject matter, and to simplify the presentation.

We examine the following questions for each device: (i) How are the devices constructed, (ii) how are they contacted electrically, and (iii) how are they sized?

6.3.1 Field-Effect Transistors (MOS-FETs)

We have already gotten to know the MOS-FET, which is the most commonly used device, in Chap. 2 (Sect. 2.9.1). There are two MOS-FET types: the NMOS-FET and the PMOS-FET; they are depicted in Figs. 6.6 and 6.7, respectively.

When a suitable voltage is applied between the control electrode (gate G) and the bulk (backgate B), a conductive channel is created between the source and drain doped areas (S and D). Current can then flow through this channel.

The width w and length l of this channel are the critical parameters for the electrical behavior of the MOS-FET. These two parameters are defined in the layout, when a MOS-FET is "sized".

The width w is defined by the field-oxide opening. In the layout, this is the dimension of the structure in the layer "Active" (often called the "active region") perpendicular to the current flow. Please note that in Fig. 6.6 the NSD regions and in Fig. 6.7 the PSD regions (each forming the sources and drains) are not visible in the cross-sectional views (ii) because the respective cutting lines (ii) intersect the transistors in the channel region.

---

[2]These layouts are slightly different to those in Chap. 2, where, for example, we did not separate bulk and source/drain gates by oxide.

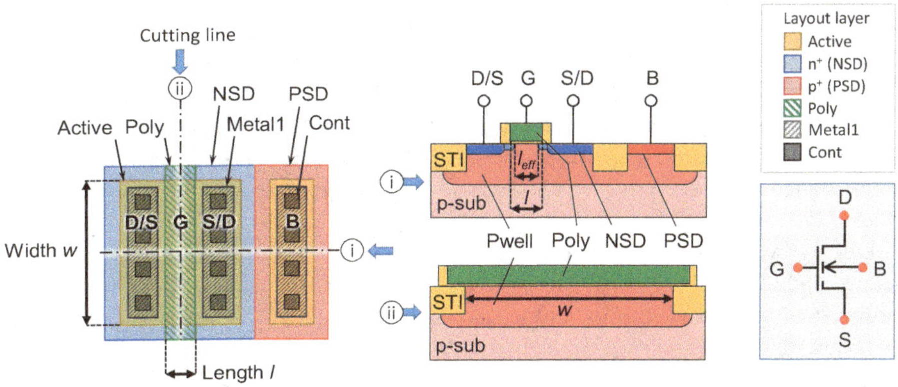

Fig. 6.6 NMOS-FET: Layout (left); sectional views (middle), drain and source contacts are interchangeable; schematic symbol (right). Drain and source pins of the schematic symbol are arrranged in such a way that the voltages are positive from top to bottom

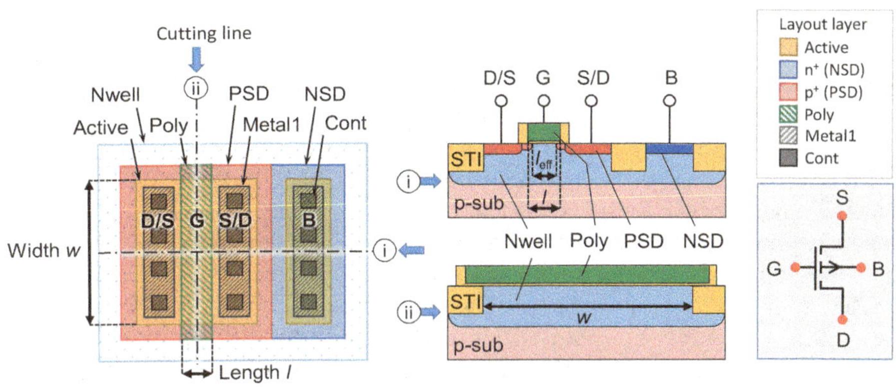

Fig. 6.7 PMOS-FET: Layout (left); sectional views (middle), drain and source contacts are interchangeable; schematic symbol (right). Drain and source pins of the schematic symbol are arrranged in such a way that the voltages are positive from top to bottom

The gate defines the spacing between the source and drain doped areas and, thus the channel length, by means of a self-alignment process. Hence, the length l is defined by the structural dimension in the direction of current flow in the layer "Poly". These variables are annotated in the layout structures and in the cross-sectional views in Figs. 6.6 and 6.7. The associated circuit schematic symbols are also shown (see Figs. 6.6, right and 6.7, right).

It is clear from the cross-sectional views (see Figs. 6.6 and 6.7, top and cutting line i) that the electrically effective channel length l_{eff} is shorter than the nominal channel length l (defined by the poly size). The correction factor dl is considered in the MOS-FET simulation models. The simulation models also include a similar correction factor dw for the channel width w, such that

$$l_{\text{eff}} = l - dl \tag{6.5}$$

$$w_{\mathrm{eff}} = w - dw. \tag{6.6}$$

The value *dw* is not significant for processes with STI (shallow trench isolation). The situation is very different in processes with LOCOS field oxide, where the *bird's beak effect*[3] considerably shortens the electrically effective channel width w_{eff} by twice the length of the bird's beak. The *dw* value is approximately the same as the field-oxide thickness.

Due to the symmetrical layout, which side is the source and which is the drain is determined by the wiring in the circuit. Given that source and backgate are normally on the same potential, it is logical to use the contacts next to each other as source and backgate contacts.

Folding Field-Effect Transistors

FETs with very large w/l ratios are often used for large currents or small resistances across the channel. Here, the channel width can exceed the channel length by orders of magnitude. In these cases, the transistors are *folded* to avoid an unfavorable aspect ratio in the layout. What does this mean?

When the FET is folded, it is split into n equal subtransistors, with the same length l, but a smaller width w/n. These widths are added together by shunting the subtransistors to achieve the required value w again.

In Fig. 6.8 we show the folding principle using the example of an NMOS-FET of width $w = 20$ and length $l = 2$, which is folded with $n = 4$. First, the device is split into four imaginary subtransistors each of width $w/4 = 5$ (step 1) and each containing four vias, whose number and spacing are determined by the design rules. Then, two of these imaginary subtransistors are flipped such that the respective sources and drains are facing each other (step 2). Since these regions are always at the same potential, they can be shared by the subtransistors. This is done by virtually pushing these regions on top of one another (step 3). This results in a very compact configuration whose contacts are now routed as a parallel circuit (step 4).

The layout is not actually generated like this in the real world. We are simply illustrating the internal composition of a folded FET with this example. It is electrically the same as the circuit schematic drawn on the left, which we would not encounter in the real world either. The layout in Fig. 6.8 at the bottom was produced by a *cell generator* (Sect. 6.4), where the number of required folds is defined by a parameter.

The poly gates produced by folding are often called *fingers* because of their shape. Care is needed with the terminology associated with fingers as the fingers are perpendicular to the current flow. When someone talks about the "width" (or "thickness") of a poly finger, he or she means the "length" of a channel! Accordingly, the "finger length" stretches along the "width" of the channel.

[3]When performing LOCOS steps, a bird's beak effect is commonplace. The oxide grows laterally under the nitride mask, which is meant to block the oxide from growing on the silicon surface. The resulting geometry of the oxide represents a bird's beak (Chap. 2, Sect. 2.5.4 and Fig. 2.13).

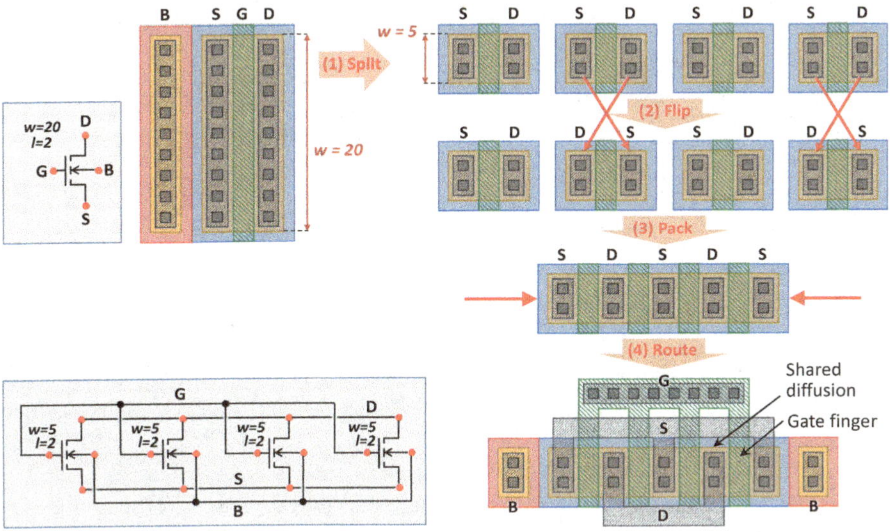

Fig. 6.8 Illustration of how a field-effect transistor is folded. The transistor of width $w = 20$ on the top left is folded with $n = 4$. Four transistors, each with a width $w = 5$, are connected in parallel on the bottom right

Folding Attributes and Layout Notes

As we have seen, the aspect ratio of a MOS-FET in the layout can be set within wide bounds by folding to generate a compact device. One mostly aims for a near-quadratic layout. Generally, space can be saved, as well, by using shared diffusions.

When sizing the finger length it is important to remember that the parasitic gate capacitance C_{GB} between gate and backgate has to be charged and discharged when a MOS-FET is switched on and off, respectively. To operate the device, the charging current must overcome the parasitic gate resistance R_G. The response times depend on the time constant $R_G \cdot C_{GB}$. Specific response times can only be assured by limiting R_G. PDKs therefore usually contain constraints for the maximum length of gate fingers or for the number of allowed squares. Using Eq. (6.4), we can estimate the gate resistance as

$$R_G = R_{\square,\mathrm{poly}} \frac{l_G}{l} \Big/ 2. \tag{6.7}$$

The sheet resistance of polysilicon $R_{\square,\mathrm{poly}}$ in this equation typically has double-digit values of Ω.[4] The gate finger length is l_G and its width is the channel length l. This value is divided by 2 in Eq. (6.7) because the charging current I_G does not flow completely through the poly finger in contrast to a standard resistor.

[4]Poly is doped by "silicidation" in some processes, thereby achieving very high doping concentrations. This process option enables single-digit values of Ω for the sheet resistance of poly.

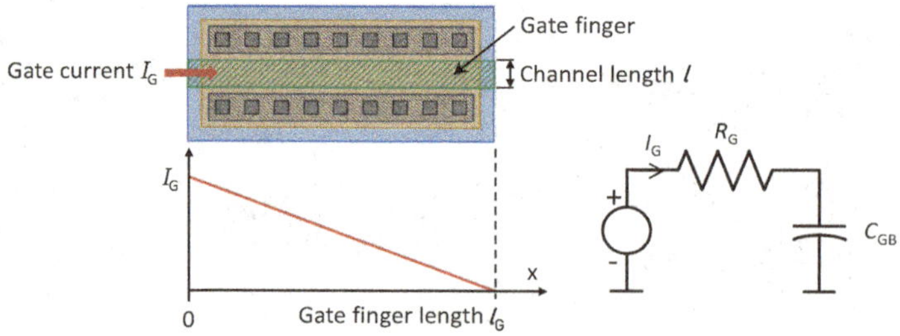

Fig. 6.9 Gate current I_G distribution along a gate finger

The current distribution along the finger is plotted in Fig. 6.9. The total resistance is halved due to the linear reduction in current. (The gate resistance R_G can be further halved by energizing the gate from both ends.) If, for example, $R_{\square,poly} = 50\ \Omega$, the gate resistance R_G, as per Eq. (6.7), reaches 0.5 kΩ with only 20 squares.

Hence, folding minimizes the gate resistance. Folding the finger n times reduces the finger length to l_G/n, i.e., the finger resistance is reduced as per Eq. (6.7) by a factor n. The resulting n subtransistors are connected in parallel, so that the total resistance shrinks by a further factor n. Hence, when a transistor is folded into n subtransistors, the gate resistance is reduced by $1/n^2$. This, in addition to the smaller footprint, is a huge advantage, and a reason that folding is used so often.

Source and drain terminals should always be connected by metal with as many contacts as possible. This ensures that the source-drain current is uniformly routed from a metallic wire to the channel, thereby minimizing the resulting (parasitic) resistance $R_{DS,on}$.

6.3.2 *Resistors*

Diffusion Resistors

All conductive layers can be used to make passive resistors by sizing their layout areas appropriately. We next examine some typical examples. Resistors from the doped layers NSD, PSD, and Nwell are depicted in Fig. 6.10. This type of resistor is commonly called a *diffusion resistor*.[5] The current enters and leaves at the contacts marked "R1" and "R2", respectively.

The *NSD resistor* is located in the p-well, which is tied to ground along with the p-substrate. Hence, this device does not require any further electrical isolation. The schematic symbol therefore normally has no further contact of its own. We have included a substrate contact in the layout and in the sectional view in Fig. 6.10 to visualize the biasing.

[5]This naming is still common despite the fact that the doping is performed by implantation rather than by diffusion in modern processes.

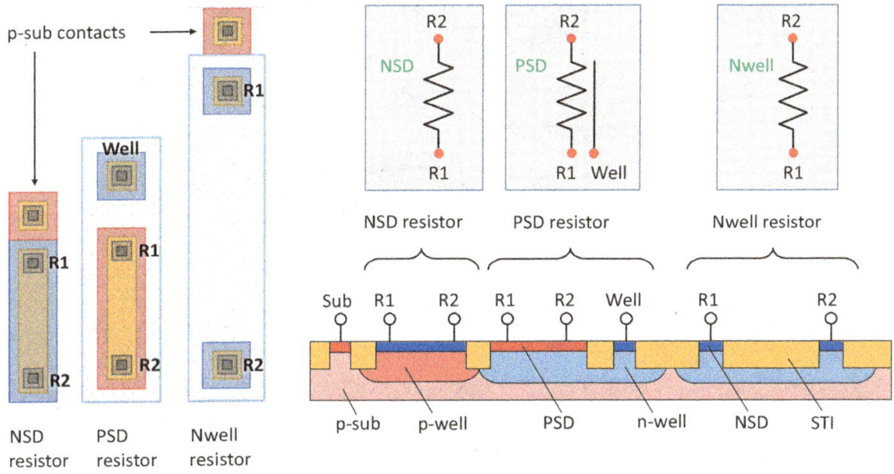

Fig. 6.10 Layout structures for NSD, PSD, and Nwell resistors (left), as well as sectional views and schematic symbols for these devices (right). The abbreviations NSD and PSD denote the n- and p-dopings used for source and drain regions of MOS-FETs

The *Nwell resistor* is encapsulated with p-substrate; and its electrical contacts are the same as for the NSD resistor. The sheet resistance of the Nwell resistor is significantly higher than for the PSD and NSD resistors. This is not only because it is more lightly doped, but also because the conductive layer is depleted by the field oxide (realized here by STI) in the upper portion of the resistor, where the dopant concentration is much higher. The current therefore can only flow through the lower section of the resistor, where the resistance is significantly higher.

The *PSD resistor* is situated in an n-well which is electrically connected with a third pin labeled "Well". This pin's potential should be selected so that the Nwell-PSD diode is polarized in the reverse direction. This is the case when the well potential is at least as high as the higher potential at R1 and R2. This is done in practice either (i) by connecting Nwell to the highest potential in the circuit or (ii) by connecting it with the resistor contact that is at the higher potential.

The latter approach (ii) works only if this potential is present throughout the entire operating period, i.e., if the resistor is not connected to AC. This method has two advantages. To explain the first advantage, let us assume that the potential at R1 is higher than at R2. By short-circuiting the well with R1, routing is easier as no other potential must be supplied through a wire. Secondly, the sheet resistance is more precisely defined. To understand this scenario, it is necessary to be aware that the effective resistor cross-section is bounded from the bottom by the depletion region between Nwell and PSD. If Nwell and R1 are tied to the same potential, this depletion region adapts itself to the resistor voltage level, i.e., it becomes independent of it.

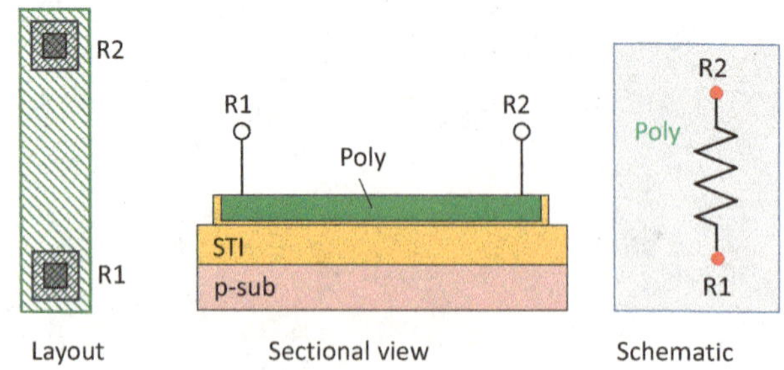

Fig. 6.11 Polysilicon resistor; layout, sectional view and schematic symbol

Poly Resistors

Although polysilicon is quite heavily doped for making low-resistance poly gates, it can be a good base material for resistors. As it can be patterned quite thinly, many "squares" can be placed on a small area to achieve appreciable resistance values. A simple poly resistor with associated sectional and schematic views is depicted in Fig. 6.11.

Poly resistors are the most commonly used type of resistor because they offer several benefits. Their primary advantage is that they are galvanically isolated from other devices. Hence, there is no occurrence of parasitics associated with p–n transitions and they have a higher breakdown voltage (defined by the strength of the dielectric surrounding them) in relation to their surroundings. There are also the savings on space as no wells or space-charge regions are required with poly resistors. In addition, the temperature coefficient of poly resistors is lower than that of diffusion resistors.

Sizing Resistors

Resistors are sized by setting the length and width of the resistor body, whose resistance value is calculated with Eq. (6.4). Please note that the current at both contact points R1 and R2 enters and exits vertically through contact holes, which forces it to alter its flow direction in the resistor body. This enforced change of direction must be considered in the calculations. These current-density-related considerations are shown in Fig. 6.12 using an NSD resistor as an example. The same issues need to be considered for other types of resistors, as well.

Both the effective cross-section through which the current flows and the flow direction change simultaneously in the regions where the current enters and exits a resistor, leading to an inhomogeneous current flow. This section of a resistor is called the *resistor head* (Fig. 6.12). The additional resistance value R_{H} of the resistor head must be considered twice for every resistor.

Two differently sized NSD resistors are illustrated in Fig. 6.13. The resistor heads R1 and R2 scale with the width during sizing, since as many contacts as possible are used, as shown in Fig. 6.13 (bottom). Changing the resistor length does not affect the resistor head. With this in mind and applying Eq. (6.4), we can express the resistance of a real resistor R_i as

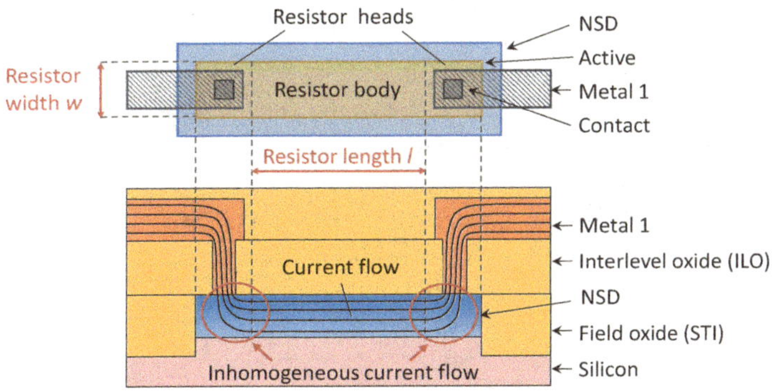

Fig. 6.12 An NSD resistor with special emphasis on the current-density distribution; layout (top), sectional view (bottom)

Fig. 6.13 Two differently sized NSD resistors

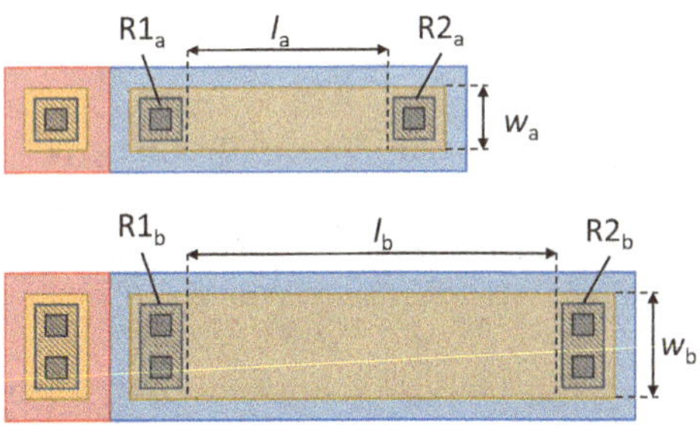

$$R_i = R_\square \frac{l_i}{w_i} + 2R_{\mathrm{H}}(w_i), \tag{6.8}$$

where $R_\square$ is the sheet resistance, l_i is the length of the resistor body where the current flow is homogenously distributed,[6] w_i is the resistor width and $R_{\mathrm{H}}(w_i)$ the resistance value of a resistor head (which is a function of the width). A PDK typically includes *cell generators* (Sect. 6.4) that automatically create resistor layouts according to Eq. (6.8).

6.3.3 Capacitors

Let us first recap on the well-known formula for the capacitance C of an ideal parallel-plate capacitor:

[6]The current flow in diffusion resistors is also inhomogeneously distributed in the z-direction due to the decreasing dopant concentration downwards, as indicated in Fig. 6.12. This effect can be ignored, however. The corresponding sheet resistance $R_\square$ defined in a PDK is an RMS value.

$$C = \varepsilon_0 \varepsilon_r \frac{A}{d}. \tag{6.9}$$

In this formula, the term $\varepsilon_0\varepsilon_r$ is the permittivity of the dielectric between the plates, d the distance between the plates and A the surface area of the plates. Given that there is very little surface area available on a chip and that the surface area that is available is valuable, we need to use stratifications that provide a very small clearance d between electrodes in order to achieve useful capacitances.[7]

The gate oxide is a very thin and high-quality oxide layer. This is why MOS-FETs lend themselves very well for use as capacitors in the standard CMOS process. The capacitance between gate and backgate is used in this case. Here, a number of configurations are available. A popular approach is to generate the backgate electrode by short-circuiting the source, drain and backgate contacts. An example of this configuration is shown in Fig. 6.14 with an NMOS-FET used as a capacitor (also known as an "NMOS-Cap"). A PMOS-FET can be turned into a "PMOS-Cap" with the same circuit.

Although such NMOS- and PMOS-Caps can be easily constructed, they have serious electrical issues. In particular, it is not advisable to operate them near the transistor threshold voltages V_{th} (i.e., not in depletion mode), as the capacitance would drop considerably. The circuits can be operated with the transistors in reverse mode (channel formation by accumulation of minority carriers) as well as in accumulation mode (further accumulation of majority carriers, i.e., no channel formation). However, this is only feasible by means of negative voltages with the NMOS-FET, given the p-substrate is always tied to 0 V. Hence, this operating mode is not viable for the NMOS-Cap. The problem with these capacitors in reverse mode is that the capacitance collapses at high frequencies. They perform best when the bulk electrode is connected to AC ground.

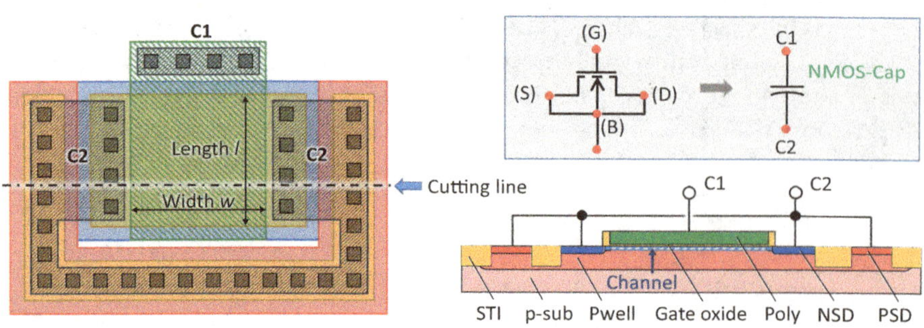

Fig. 6.14 Using an NMOS-FET configuration as a capacitor; layout (left), sectional view and schematic symbol (right)

[7] Dielectrics with higher permittivities than SiO_2 are utilized in modern processes, as well. However, they are generally not used to increase the capacitance. The most common reason these "high-k" dielectrics are applied is to enable greater layer thicknesses d as this reduces leakage currents while maintaining the capacitance levels.

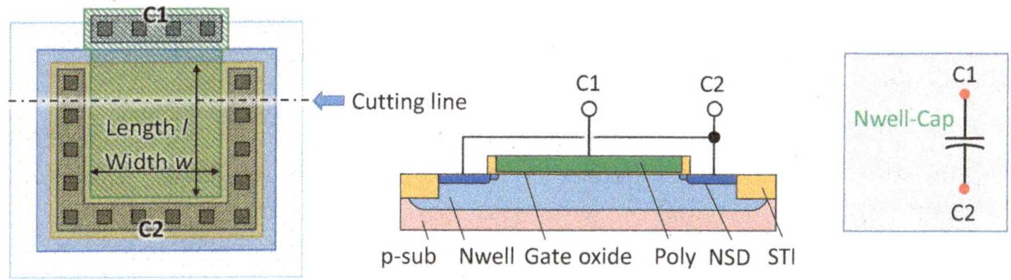

Fig. 6.15 Layout (left), sectional view (middle) and schematic symbol (right) of a Poly-Nwell capacitor

Using only the Nwell layer as a counter electrode to the polysilicon is another way of using the gate oxide as a dielectric in the standard CMOS process. This is effectively a PMOS-FET with no source or drain (Fig. 6.15). The main issues with this constellation are the high parasitic body resistance of the Nwell electrode and the space-charge region of the Nwell substrate diode, which forms a series parasitic capacitance to ground.

Several process extensions are available for fabricating even better capacitors—for example, additional very thin oxide layers are placed above the gate oxide for use as dielectrics along with additional conductive layers made of poly or metal acting as electrodes. These extensions enable so-called PIP caps (poly-insulator-poly) and MIM caps (metal-insulator-metal). MIM caps can only be deployed in processes with very good planarization (such as CMP, chemical-mechanical polishing) as they are located between the metallization layers that are used for routing. An example of a MIM capacitor between "Metal2" and "Metal3" is visualized in Fig. 6.16. These structures benefit from minor parasitic effects through the use of metal and the distance from the silicon surface.

Sizing. Capacitors are sized by defining the electrode surface areas A used to calculate the capacitance as per Eq. (6.9). Rectangular capacitor plates are generally

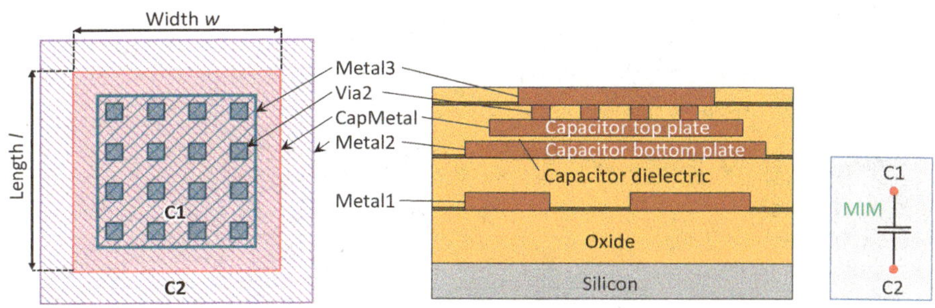

Fig. 6.16 Layout (left), sectional view (middle) and schematic symbol (right) for a MIM (metal-insulator-metal) capacitor

preferred. The devices are stretched horizontally and vertically in the layout such that the product of the width w and length l, drawn in Figs. 6.14, 6.15, and 6.16, produces the required total area A.

There are other designs as well with stripe metal electrodes that use lateral fields; these are so-called *fringe* or *flux* capacitors. We kindly refer the reader to the literature, such as [9], for more information on these designs.

6.3.4 Bipolar Transistors

The first chips in the 1960s had only bipolar transistors and no field-effect transistors. The design of bipolar transistors is based on phenomena occurring in two consecutive p–n transitions separated by very small distances. There are two possible stratifications: n–p–n (NPN) and p–n–p (PNP).

The so-called *bipolar processes* were tailored for the NPN transistor. These had a heavily doped p-substrate on which a lightly doped n-epitaxy was deposited. The n-epi was divided into n-wells with deep p-doping (typically called "iso"), which formed the collector surroundings (see [3], for example).

NPN Transistors

We can build very good NPN transistors nowadays by extending state-of-the-art CMOS processes by a few layers. These processes are called *BICMOS* processes (bipolar and CMOS). Henceforth, we refer to a widespread version with the additional layers "Deep-n^{+}", "NBL" (n-type buried layer) and "Base". NBL is a heavily n-doped buried layer, which can only be produced by allowing an epitaxial layer to grow. We use this configuration in our example, a configuration that is also common in standard CMOS processes.

We first explain how the transistor operates using the cross-section in Fig. 6.17 (middle), where the current flow is also indicated. If the base-emitter diode, which

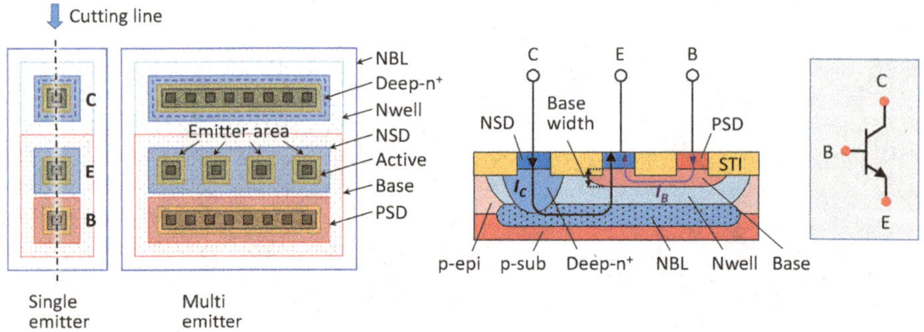

Fig. 6.17 NPN transistor; single-emitter and multi-emitter layout (left), sectional view (middle) and schematic symbol (right)

is comprised of NSD (the n-doping for source and drain of NMOS) and “Base”, is forward biased (“allows current”), the emitter injects electrons into the base. These electrons are minority carriers there and they diffuse further until they are attracted by the electrical field created by the reverse-biased (“blocked”) base-collector diode. They drift through this field until they reach the collector where they find a low impedance path through the heavily doped NBL and Deep-n$^+$ to the collector pin. This is the collector current I_C.[8]

Some of the electrons recombine with holes in the base. This happens in the base *and* in the emitter, as the base injects holes into the emitter too. The current produced by recombinations is the base current I_B. We can therefore express the emitter current as $I_E = I_B + I_C$.

The ratio $B = I_C{:}I_B$ characterizes the bipolar transistor and is called the *current gain*; its value can be higher than 100. The (large) collector current can be controlled by the (small) base current. Thus, the fewer the number of electrons that recombine from I_E, the better the transistor. The recombination risk is minimized by two factors: (i) the base doping should be as low as possible to ensure the number of holes available is low, and (ii) the path in the base, namely, the *base width*, should be as short as possible (see Fig. 6.17, middle).

Armed with this knowledge, we should be able to understand the construction of the NPN transistor. The base must be produced in a very lightly n-doped environment so that it does not have to be heavily doped itself. The parasitic collector resistance would be very large if no further measures were taken. This body resistance can be effectively minimized with the heavily doped Deep-n$^+$ (often called “sinker”) and NBL. The short base width is defined by the difference between two diffusion depths (“Base” and NSD). This value can be very accurately checked in the process. Importantly, note that it is not affected at all by the layout.

Sizing. The behavior of the NPN transistor is a function of the process parameters, i.e., dopant concentrations and layer thicknesses. The amount of current it can carry is influenced by the layout. The current is a function of the emitter area, which is defined by the opening in the field oxide (see Fig. 6.17, left). The transistor can be stretched when sizing such that this surface area has the required size.

Generally, the current is not equally distributed over the emitter surface. One of the reasons for this is the “current crowding” effect described in [3]. This effect causes more current to be injected in the proximity of the base contact, as the base-emitter voltage is slightly higher near this contact. Hence, we should stretch the emitter only in the direction perpendicular to the cutting line shown in Fig. 6.17 (left).

Altering the emitter can cause further nonlinearities. To avoid this, the entire emitter surface is often not continuously stretched, but is reproduced by replicating a single emitter, as depicted in Fig. 6.17. (There are further, matching-related reasons for this “integer sizing”, which we shall discuss in Sect. 6.6.1.)

[8]The definition of the “technical” current is based on the perception of “positive” charge carriers. This is the reason why the current depicted in Fig. 6.17 (middle) flows in the opposite direction, i.e., from the collector to the emitter.

PNP Transistors
A PNP transistor works in the same way as an NPN transistor, except that the roles of electrons and holes are swapped and the collector is made up of p-substrate. This so-called "substrate-PNP" is seldom used because of the substrate currents, which change the ground by means of the IR drop.

It is not really worthwhile using even more layers to create better vertical PNP transistors, given that PNP transistors are hardly ever used due to the availability of MOSFETs and NPN transistors. Nevertheless, we can build a horizontal PNP transistor without additional layers by creating concentric circles in the Nwell with a p-doping (e.g., "Base"), as depicted in Fig. 6.18 (left).

As illustrated in Fig. 6.18, the inner ring is the emitter, and the outer ring the collector. The base is the lightly doped n-well. The base width is defined by the difference in the circle radiuses. The main current (holes) flows radially from the emitter to the collector. Hence, these doped areas must not be separated by STI (shallow trench isolation). The base width is not impacted by alignment tolerances, given that the emitter and collector are patterned with the same mask.

These transistors also have quite high current gains B. Values of $B = 50$ are typically reached; and higher values can be obtained, as well, i.e., 98 to 99 holes out of 100 can reach the collector ring from the emitter. This is extraordinary when we look at the cross-section: The emitter injects its charge carriers across the entire interface of the base-emitter transition. A large proportion of the carriers pass into the base at the bottom, in fact. These charge carriers have very little chance of reaching the collector ring by diffusion, one would think. The n-type buried layer (NBL) is key in this process. Although Nwell and NBL are two n-doped zones, a small space-charge region forms at the interface between them, because of the extreme difference in dopant concentrations in these two zones. The holes diffusing down from the top are repelled by this space-charge region (see Fig. 6.18, middle), so that most of them ultimately reach the collector.

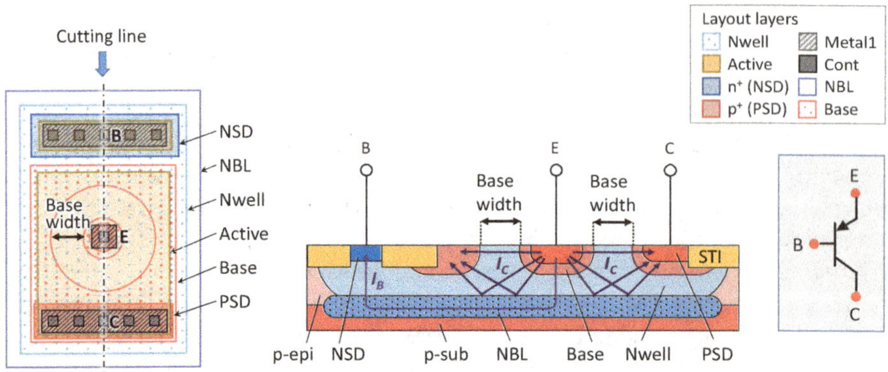

Fig. 6.18 PNP transistor; layout (left), sectional view showing the main current flow (middle) and schematic symbol (right)

Sizing. If the emitter or collector are resized, the electrical behavior (especially the current gain B) is extremely altered. Therefore, the PNP basic layout (ring radiuses) must remain unchanged. Hence, the transistor can only be sized for larger currents by replicating this basic layout.

6.4 Cell Generator: From Parameters to Layout

6.4.1 Overview

Integrated analog IC devices are typically individually sized. It was still common in the 1980s to manually lay out the devices by adapting the basic shapes from a library. This time-intensive layout work, often called "polygon pushing", was done manually in a graphics editor, and the circuit was built up this way.

In the 1990s, *cell generators* became commonplace. Cell generators are procedures that process the required electrical or geometrical device properties as input parameters and automatically create a correct layout cell for a specific technology (Fig. 6.19).

Cell generators are typically called directly from the circuit schematic ("Pick and Place"). They load the sizing parameters from the instance of the device's schematic symbol.

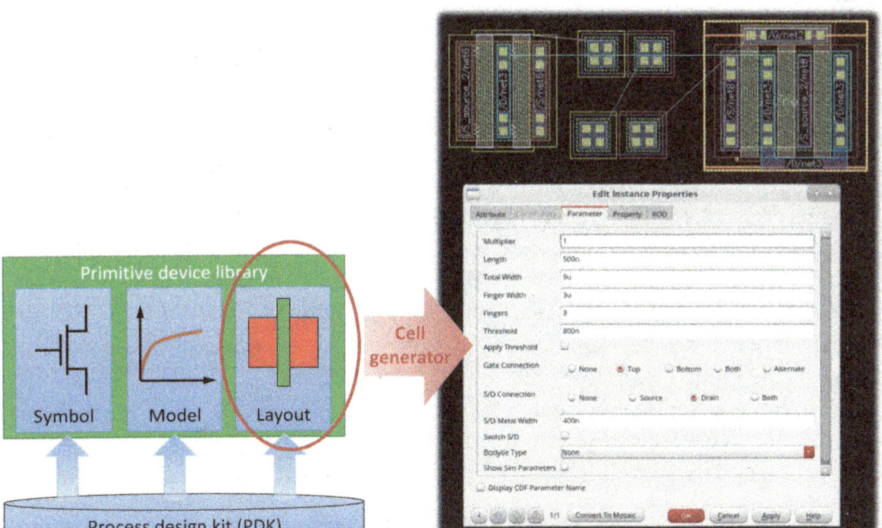

Fig. 6.19 Cell generators are part of a PDK (left) provided by the foundry. Analog devices, such as a transistor, can be laid out (right) with cell generators utilizing sizing parameters

Aside from parameters for electrical sizing, other parameters are used to define further layout attributes. They include number, shape and location of contact pins and splitting devices into smaller subdevices (e.g., folding a FET). These subdevices are then suitably placed and routed w.r.t. one another.

Cell generators are standard in every process design kit (PDK) today (Chap. 3, Sect. 3.5.1) and are supplied for most semiconductor processes. Their functionality differs from PDK to PDK, and they are closely connected with the design tools of leading EDA companies. Those cell generators are often referred to as *PCells*, which stands for *parameterized cells*.

Layout generators based on cell generators in the PDKs are developed by IC design teams in large corporations. Nowadays, whole circuits or at least standard subcircuits are often automatically created by these layout generators [6, 7]. These higher-level layout generators are sometimes also called *module generators*.

6.4.2 Example

We demonstrate how the cell generator of an NMOS-FET can be established with the aid of the Cadence® *PCell Designer* [4]. The PCell of the transistor we will build in this example shall get five parameters: "width", "length", "fingers" (finger count), as well as "leftBulk" and "rightBulk" (backgate contacts).

Figure 6.20 shows the command screen. The command tree is depicted in the "Command" column. Each line (numbered in the "Line" column) contains a command, whose parameter set is given in the "Parameters" column. To better understand the procedure, we have added some semantic notes in brown and some notes on the syntax in dark blue. The five parameters are highlighted in pink to illustrate where they appear in the code.

The layer names from the Cadence "GPDK180" are mapped to the layer names in our example in line 1. The remaining code consists of five blocks (named in lines 2, 18, 23, 30, 35), whose functions we shall explain briefly below.

The basic NMOS-FET layout (Fig. 6.21, left) is produced in the "Create initial shapes" block (lines 2–17). There are only rectangular shapes in this example; they are assigned to three groups "Core", "BulkLeft", and "BulkRight". The channel is stretched to the required measurements "width" and "length" in the "Stretch transistor channel" block (lines 18–22). All corners of the shapes in a group are stretched relative to the group center line with the Stretch command.

The folding operation is carried out in the next block (lines 23–29). The MOSFET core is duplicated here and placed at calculated intervals "fPitch". The right-hand bulk contact is only shifted here. Unused backgate contacts are deleted in the next block (lines 30–34).

The metals are filled with contacts in the last block (lines 35–38). The sample contact hole, generated in line 36, is deleted again at the end. The "Fill" command is a good example of the power of the functions provided.

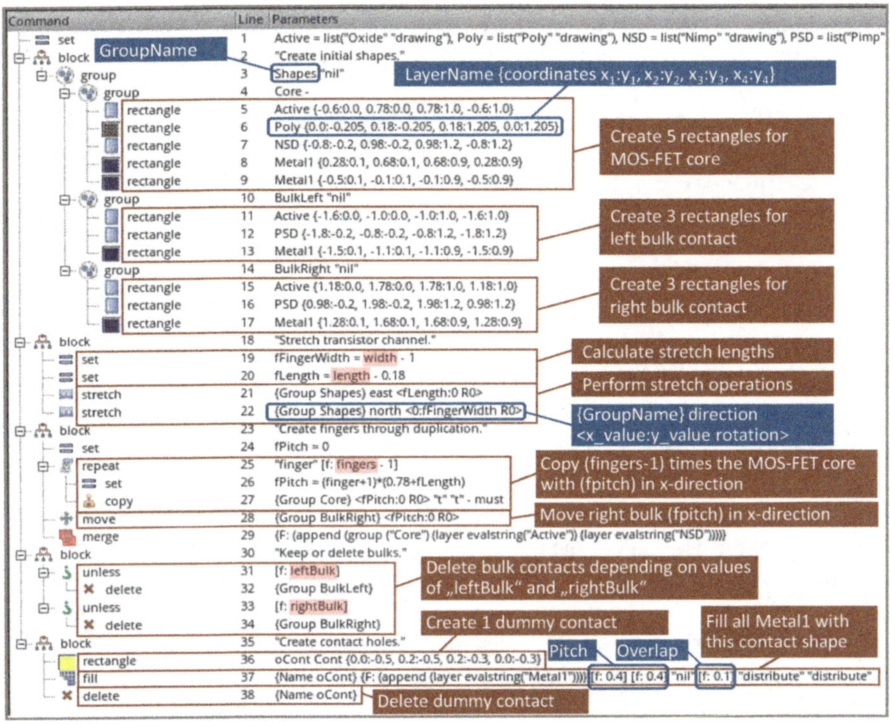

Fig. 6.20 Sample code for a cell generator for an NMOS-FET in the *PCell Designer* from Cadence with notes on syntax (dark blue) and semantics (brown). The PCell's parameters are highlighted in the code (pink background)

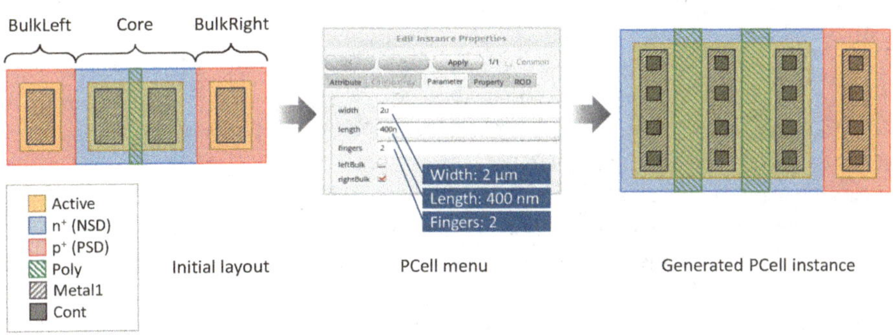

Fig. 6.21 Application of the PCell of an NMOS-FET shown in Fig. 6.20: initial layout (left), parameter settings in the PCell menu (middle), generated PCell instance (right)

The commands are executed in real time when the PCell is developed. The PCell designer can thus immediately see the layout produced by the code. The required shapes can be simply drawn in the graphics editor. If a sample layout is available, an experienced developer can create the PCell shown here in about one hour with this tool.

The PCell menu (Fig. 6.21, middle) is automatically generated based on the parameter definitions. The PCell instance created for the parameter settings in Fig. 6.21 (middle) is pictured in Fig. 6.21 (right).

6.5 The Importance of Symmetry

6.5.1 Absolute and Relative Accuracy: The Big Difference

If we measure the electrical parameters of individual devices in a completed chip, we find that their absolute values are often very different from their nominal values. The reason for this is the complexity of semiconductor manufacture, which comprises a very large number of fabrication steps (typically many hundreds). Although every fabrication step is carried out with the utmost precision, there always remains an unavoidable tolerance. As the final variations are caused by the sum of these tolerances, the final variations can be very large—typically in double-digit percentages.

If we compare the same device types, we find that the parameter differences between one device and another are always much smaller if the devices are from the same manufacturing cycle (production run). This is because these devices go through many common process steps. In other words: They have the same "fabrication history". The more precisely these fabrication histories coincide, the smaller the relative variations, or to put it another way, the higher the relative accuracy.

Take, for example, a poly resistor with a nominal resistance of 1 kΩ. This resistor type typically has a tolerance of ±30%. This means that the real resistance value of this device can lie between 0.7 and 1.3 kΩ. Two of these 1 kΩ resistors are now laid out in the same way and placed side by side on a chip. If, in a random sample, we measure a value of 1.16 kΩ, for example, for one of the two resistors, we will find that the other resistor also has a resistance value very close to 1.16 kΩ. The value will deviate by only a few Ω from this value, i.e., the two devices have a high relative accuracy.

The relative accuracies of two identically designed devices with different fabrication histories is reviewed in Table 6.1. The results are based on the above example with an absolute value tolerance of ±30%. This value is the same as the relative accuracy for two devices from different production runs and is therefore given in the top line in the table.

The values in the table are examples only and are intended to indicate the magnitudes of the relative parameter variations. They can change from one technology

Table 6.1 Relative parameter accuracy of two identically designed devices as a function of fabrication history

Manufacturing-specific "distance" (fabrication history)		Relative accuracy (%)
Within a fab	From lot to lot[a]	±30
Within a lot	From wafer to wafer	±20
Within a wafer	From reticle section to reticle section	±15
Within a reticle	From chip to chip	±10
Within a chip	Arbitrary	±5
Within a chip	With further layout measures	±1 to ±0.01

[a]Several wafers are processed in a production run in a wafer fabrication plant. The number of wafers is defined by the containers holding the wafers and in which they are transported from station to station. The wafers in a container make up a "lot"

to another. Capacitance values for capacitors normally fluctuate slightly less than resistance values for poly resistors. The current gains of bipolar transistors tend to fluctuate more.

The last row in the table suggests that the relative accuracy can be improved considerably by certain layout measures—which are called *matching*. The basic premises of matching are:

- The absolute accuracy of the devices is bad.
- The relative accuracy of the same type of devices on a chip is good.
- Only the same types of devices can have relative accuracy.
- The relative accuracy of the same type of devices can be optimized in the layout by matching measures.

The ability to match devices is one of the most important skills an analog layout designer can have. We shall therefore cover this topic thoroughly in Sect. 6.6. But first we want to explain why relative accuracy, and thus matching, is so important (Sect. 6.5.2).

6.5.2 Obtaining Symmetry by Matching Devices

Given that the absolute values of the electrical parameters of integrated devices are subject to such large variations, it is almost impossible to build good analog circuits whose quality depends on the accuracy of these absolute values. Instead special analog circuit technologies are deployed, whose performance depends on the symmetry of the electrical behavior of specific devices. This is the reason why integrated analog circuits, especially, benefit from the *relative* accuracy of the devices.

These relative accuracies are influenced to a large extent by the layout. The relative accuracy of two or more devices can be greatly impacted by matching, and improved by about two orders of magnitude, in fact. Consequently, the following knowledge and abilities are mandatory for an analog IC layout designer:

(1) The designer must identify which devices in a circuit should behave symmetrically. These devices must be “matched” in the layout.
(2) Several matching methodologies are available and they differ greatly in the amount of work and chip area they require. Hence, he/she must be able to select a sensible matching level.
(3) He/she must choose and implement the appropriate matching measures: how the devices are laid out, placed and routed.

It should be clear at this point why the layout designer of an analog circuit needs a schematic as input data instead of a netlist. It is much easier to understand a circuit topology and how the electronic circuit works from a graphical circuit diagram. This is especially important for picking out subcircuits to be matched. This is made all the more easier because the elements of these subcircuits are usually placed symmetrically and closely to each other in the circuit schematic. Appropriate guidance should be given where this is not possible. Furthermore, advanced design environments allow symmetry constraints to be defined where necessary.

Some standard subcircuits based on the symmetry of devices are [8]:

- *Current mirror*: The current flowing in a reference transistor is copied in one or more transistors in the current mirror. The ratio of the copied currents to the reference current can be set for MOS-FETs by the channel widths, and by the emitter area or the number of emitters for bipolar junction transistors.
- *Differential pair*: The differential pair evaluates the voltage difference of two paired transistors at the control electrodes (gates or bases). The differential pair is an input driver for all operational amplifiers.
- The *IPTAT* circuit: This circuit produces a temperature-proportional current (IPTAT: current *I* proportional to absolute temperature). It can only be created with bipolar transistors and is deployed in temperature-compensated and supply-voltage-independent voltage reference circuits and current reference circuits.
- Many circuits are based on the synchronicity of the electrical behavior of passive devices, i.e., networks of resistors or capacitors. A simple example is the *voltage divider*.

We described a bandgap circuit as a typical representative of an integrated analog circuit in Chap. 3 (Sect. 3.1.3). This is quite a small circuit that contains many of the above subcircuits. We have depicted these two schematics in Fig. 6.22 again and enclosed the devices to be matched in blue boxes. Boxes (d) and (f) contain current mirrors, box (e) contains a differential pair, and boxes (b) and (c) together make up an IPTAT.

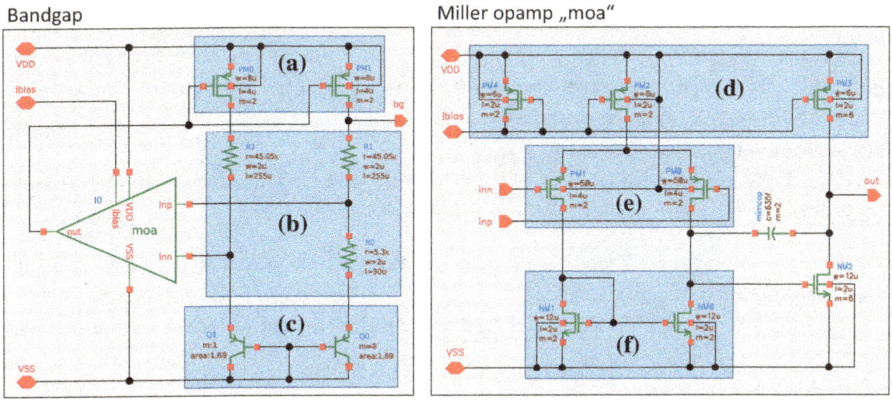

Fig. 6.22 Schematic of a bandgap circuit with marked matching groups of devices. The Miller opamp (right) is a function block in the bandgap schematic (left), where it is represented by a triangular schematic symbol labeled "moa". Matching devices are surrounded by blue boxes (**a**) to (**f**). Boxes (**b**) and (**c**) together make up an IPTAT, (**d**) and (**f**) contain current mirrors and (**e**) contains a differential pair

6.6 Layout Matching Concepts

A circuit should be laid out such that the electrical parameters of to-be-matched elements are the same, as far as possible. As electrical parameter deviations cannot be avoided, matching aims for identical parameter deviations. In other words: the purpose of matching is to eliminate *relative* parameter deviations.

Further, we need to know the possible causes of parameter variations so that we can choose suitable layout measures. Parameter variations can be caused by (i) manufacturing tolerances, (ii) nonlinearities in the device layouts, and (iii) effects of the application of an IC.

Manufacturing tolerances. All fabrication steps (Chap. 2) have tolerances, which are often stochastic. Typical issues here include inhomogeneous photoresist coating on the wafer; variations in the dimensional stability of masks, such as due to thermal expansion; mask alignment tolerances, distortions in optical mapping (e.g., lens faults); and different inhomogeneities occurring across the wafer, e.g., during layer growth, doping, etching, and chemical-mechanical polishing (CMP).

Nonlinearities in device layouts. Devices in analog circuits are individually sized in the layout. This results in "non-ideal" behavior due to design nonlinearities. The issue is that the electrical parameters do not change in proportion to the parameters used in graphics operators (e.g., distance in a Stretch function).

Effects caused by IC application. When in use, chips are exposed to various physical effects, such as heating and mechanical stress. These influences impact the electrical behavior of the devices.

The accumulated impact of these effects is very diverse and complex. Luckily, the appropriate matching measures can be selected without an exhaustive investigation of the physical relationships between all these effects. We are able to roughly classify their effect on parameter variations, and can then decide on appropriate methodologies based on these categories.

We shall now explain these issue categories and present general matching concepts based on them. The categories are defined as follows:

- *Fringe effects*, classified as internal device fringe effects (Sect. 6.6.1) and external device fringe effects (Sect. 6.6.3),
- *Location-dependent effects*, mostly called *gradients*, subdivided into unknown (Sect. 6.6.2) and known gradients (Sect. 6.6.4), and
- *Orientation-dependent effects* (Sect. 6.6.5).

We partly use simplified representations in our layout examples, e.g., we use N-Active for NSD and Active, and P-Active for PSD and Active (cf. Fig. 3.18 in Chap. 3).

6.6.1 Matching Concepts for Internal Device Fringe Effects

We have already seen in Sect. 6.3 that electrical device parameters are proportional to the size of specific layout structures. Devices are sized by stretching their structural dimensions. However, effects occur at the structure peripherals, which also impact the electrical parameters, but which do not scale with these changes. These effects, which we call *fringe effects*, can be stochastic or deterministic. They cause the electrical parameters to scale nonlinearly with stretching.

We resolve this issue by constructing to-be-matched devices with absolutely identically patterned *basic devices*. In other words, we build the devices with several suitably sized basic devices. These basic devices are connected in series or in parallel, depending on the requirements. We thus achieve a parameter ratio that is different from 1:1 if required. We shall call this process *splitting devices* from now on. Here, the fringe effects are duplicated with the same ratio such that the required ratio is maintained in the final result.

This has been a brief introduction to splitting, a very important and powerful matching concept that is applicable to all device types. We shall now delve deeper into its inner workings with several examples.

Resistors

There are two fringe effects associated with resistors: at the resistor body and at the resistor heads. We will look at them more closely here in turn, beginning with the resistor body.

The resistance of a resistor body is proportional to the quotient of its length l and width w, according to Eq. (6.4). Both variables are subject to stochastic edge shifts. We only need to consider the width here, as the length is defined by the position of the resistor heads. Width variations are caused by tolerances during exposing and etching. This applies to poly resistors and to PSD and NSD resistors, as the width of PSD and NSD resistors is determined by the location of the STI (shallow trench isolation). These tolerances form additive contributions, which do not scale when the resistor is sized by stretching the resistor body.

With Nwell resistors (see Fig. 6.10), the width is affected by lateral outdiffusion. This is a deterministic effect and it is considered using a correction factor in simulation models. However, it is not dealt with by pre-sizing (Chap. 2, Sect. 2.4.2 and Chap. 3, Sect. 3.3.4) in the layout post process. It is therefore not visible in the layout, i.e., the layout measurements correspond with the mask openings. Thus, the outdiffusion causes a mismatch in the case of different resistor widths.

Let us assume, for example, that these additional amounts add up and that the effective resistor width increases by 50 nm over the nominal widths. Hence, the width ratio of $1/0.5 = 2$ changes to $(1 + 0.05)/(0.5 + 0.05) = 1.91$ for two resistors with the nominal widths of 1 μm and 0.5 μm. This is equivalent to a mismatch of $(2 - 1.91)/2 = 4.5\%$! This mismatch can be avoided if we lay out the resistors with the same width. The conclusion here is clear: matched resistors always have the same width.

The second fringe effect is the inhomogeneous current flow in the resistor heads, which we discussed at the end of Sect. 6.3.2. Let us assume we wish to design two resistors R_1 and R_2 with the ratio R_1:$R_2 = r > 1$, as a voltage divider, for instance. We lay out the two resistors with the same width $w_1 = w_2 = w$, based on the above insight into resistor widths. Thus, they differ only in their lengths l_1 and l_2. An example is given in Fig. 6.23 for $r = 3$.

The resistors can be sized with a cell generator (Sect. 6.4), as per Eq. (6.8). With the required ratio r between the two resistivities of the two resistors R_1 and R_2, their lengths l_1 and l_2 are governed by:

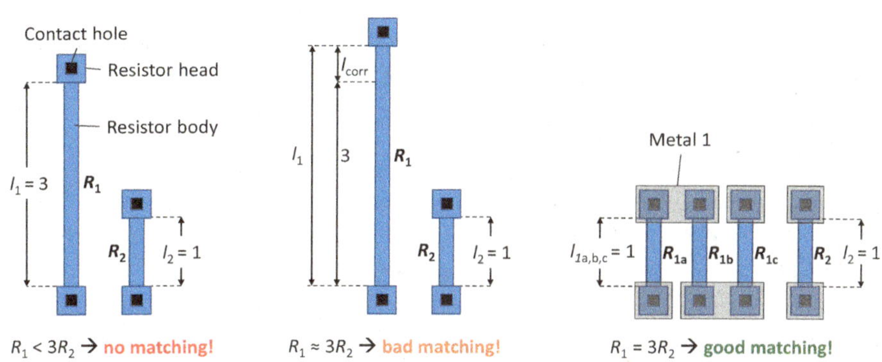

Fig. 6.23 Example of matching two resistors with the ratio R_1:$R_2 = r = 3$ (schematically). The resistors do not match if the head resistances are not considered (left). Generated resistors with different lengths are badly matched (middle). The resistors are well matched when split into identical resistors that are appropriately connected afterwards (right)

$$l_1 = r \cdot l_2 + l_{corr} \text{ with } l_{corr} = 2(r-1) \cdot \frac{R_H}{R_\square} \cdot w. \tag{6.10}$$

The resistor R_1 must be corrected by the distance l_{corr}. If this is not done, the ratio r would be between the two resistor bodies only. Since the two resistors have two (and thus the same number of) heads, the resistance of $2(r-1)$ heads would be missing from the resistor R_1 in this case in order to reach the required ratio r in the final result (see Fig. 6.23, left).

The contribution of the "missing" heads is replaced by the length l_{corr} (see Fig. 6.23, middle). The ratio of the two resistors is thus mathematically correct. However, recall that only the nominal values of R_H and $R_\square$ from the PDK can be used to calculate l_{corr}. The real values are subject to hefty process tolerances. We can, nonetheless, expect the R_H and $R_\square$ tolerances to be slightly correlated, as R_H is also impacted by the value of $R_\square$. Hence, the error is probably smaller in the $R_H/R_\square$ ratio. However, there is always a residual error, as R_H is affected by other factors too. This effect can only be eliminated by designing matching resistors from connected "basic resistors" and ensuring that the ratio of the resistor heads is the same as the required resistance ratio. This is precisely the case when all "basic resistors" are the same length (see Fig. 6.23, right).

Rule of thumb. Matching resistors must always have the same width and the same length. Use the splitting method to create resistance ratios $r \neq 1$. This is done by building the resistors from identical smaller resistors, which are preferably connected in series (see example in Fig. 6.23, right), or in parallel if this allows better layout solutions.

A final note on resistor sizes. Clearly, resistors can easily be "zoomed" as the number of squares and, thus, the resistance value itself (see Eq. 6.4) remains identical. At the same time, the relative impact of fringe effects on the absolute scale is reduced by enlarging devices this way, which, in turn, increases their relative precision. Consequently, we can improve resistor matching by stretching w and l evenly.

MOS-FETs

The size of the field-oxide opening and the width of the poly gate finger, which determine the channel width w and the channel length l, respectively, in MOS-FETs, are subject to stochastic edge shifts. Deterministic fringe effects are also at play. The channel length is shortened by a constant amount by the sub-diffusion of the source and drain zones under the poly gate.

In processes with LOCOS field oxide, there is a significant edge shift of the field-oxide opening (Chap. 2, Sect. 2.5.5) caused by the bird's beak effect (Chap. 2, Fig. 2.13). The effective electrical channel width is reduced by this edge shift.

These variations are handled in the same way as variations for resistors. This means that matching MOS-FETs should be split into "basic transistors" with the same channel lengths and widths. We would also like to point out that even without these effects, MOS-FETs can only be sensibly matched through uniform channel lengths

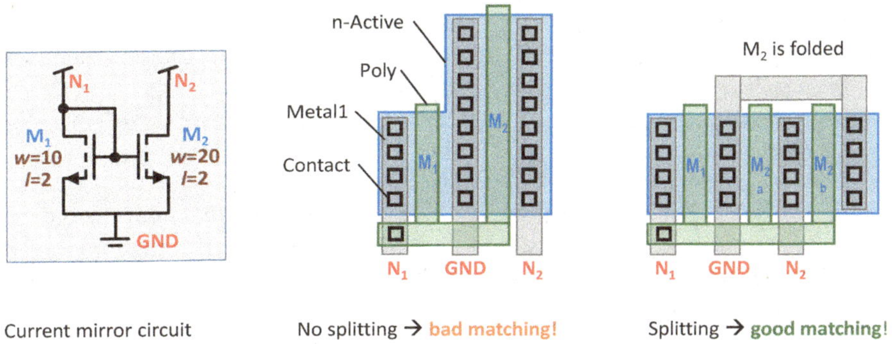

Fig. 6.24 Simple current mirror with current ratio 1:2 (left), layout without transistor folding (middle), layout with folded transistor M_2 (right) (layouts are shown simplified without bulk contacts)

because of circuit constraints. This is caused by the peculiarities of the response characteristic of MOS-FETs, e.g., the threshold voltage V_{th} is a function of the channel length in the case of "short" channels.

Rule of thumb. Matching MOS-FETs always have a uniform channel length. The w/l ratios are set by folding, such that the channel widths of the single transistors are all the same as far as possible. Hence, the gate fingers in all single transistors have the same length and width.

We illustrate this in Fig. 6.24 with an example of a simple current mirror, which mirrors the reference current through transistor M_1 in net N_1 in the ratio 1:2 in transistor M_2. The channel width of M_2 is twice that of M_1. As the fringe effects are always additive, the ratio of the channel widths is worsened by these fringe effects in the first layout variant (see Fig. 6.24, middle), where the transistors have different widths. The result is bad matching. The fringe effects are offset in the second layout variant (see Fig. 6.24, right) by splitting the larger transistor by simple folding. Matching is better in this case.

Capacitors

The calculation of the capacitance C of an ideal parallel-plate capacitor with Eq. (6.9) applies only to the proportion of the capacitance produced by the homogeneous field in the space between the plates. Fringe fields (Fig. 6.25), which also contribute

Fig. 6.25 Formation of the electrical field at a parallel-plate capacitor, including the fringe fields

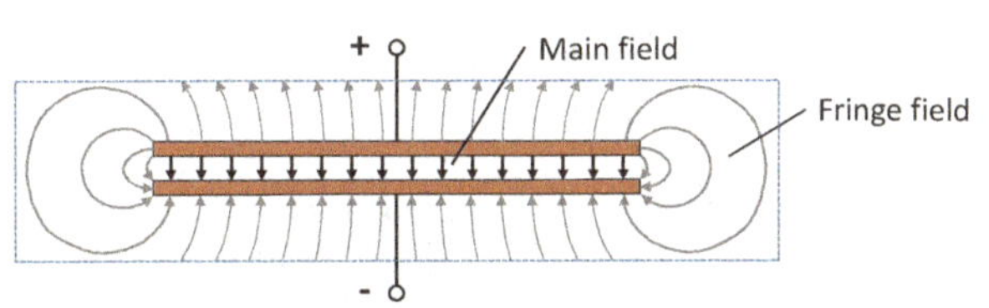

to the capacitance, occur outside this main field. Fringe fields cause significant fringe effects, given their contribution is not modeled in Eq. (6.9) that calculates the capacitance of an ideal parallel-plate capacitor.

Designs that come closest to the ideal parallel-plate capacitor are used for matching capacitors, as they have the smallest fringe fields. Lateral layers are used as electrodes, whose main field is aligned only in z-direction. The distance d in Eq. (6.9) is a layer thickness defined by the process, which is thus independent of the layout. The nominal capacitance value arising from the main field is therefore proportional to the plate surface A, which is set in layout design.

We assume rectangular capacitors in our deliberations below, without loss of generality. We use the width w and length l in the layout when sizing a capacitor as per Eq. (6.9). The surface area A is expressed as a product of these two parameters w and l as follows:

$$A = w \cdot l. \tag{6.11}$$

Similar to resistors and MOS-FETs, these two parameters are subject to manufacturing tolerances, which cause a fringe effect. What has been said on fringe effects in the context of resistors and MOS-FETs applies analogously here, as well.

Let us take a look at the fringe effect caused by fringe fields (see Fig. 6.25). One of the plates is often made slightly larger than the other(s)—thus reducing the fringe field to some extent (cf. Fig. 6.16, middle). Nonetheless, a fringe field that also contributes to the capacitance remains. This contribution is not proportional to the plate area A, but to the peripheral length of the capacitor. In the case of a rectangle, this is the perimeter P, which is calculated as follows:

$$P = 2w + 2l. \tag{6.12}$$

The surface areas of the capacitor plates, as well as their perimeters, must be in the required proportion to create matching geometries that represent as accurately as possible a required ratio C_1:C_2 of two capacitance values. This means the following relationships must be met:

$$P_1/P_2 = A_1/A_2 = C_1/C_2. \tag{6.13}$$

There are two possible approaches to meet this requirement, which we discuss below.

The first approach is similar to the options with resistors and transistors: the capacitors are split into identical single capacitors connected in parallel. Because these single capacitors are symmetrical, fringe fields and main fields are in the same proportions, i.e., Eq. (6.13) is satisfied. Fabrication tolerances, which modify the width and length of the plates, have the same effect on all single capacitors due to their identical sizes, that is, there is no mismatch. Two options for the ratio C_2:C_1 = 1.5 are pictured in Fig. 6.26 (middle, right). We shall review the additional positive effect of the symmetry axis in Fig. 6.26 (right) in Sect. 6.6.2.

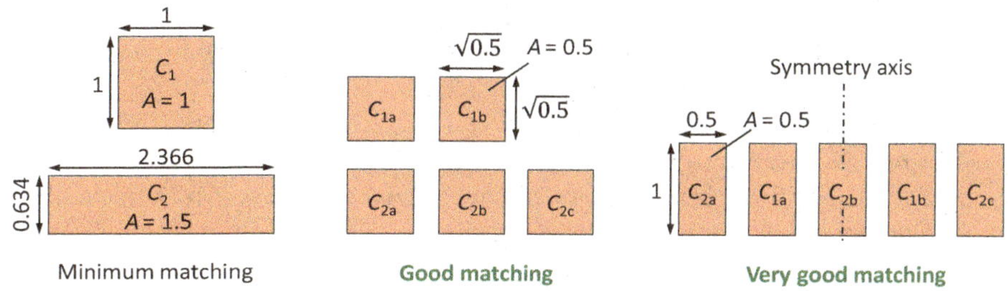

Fig. 6.26 Matching options for two capacitors with the ratio $C_2{:}C_1 = 1.5$ by stretching the capacitors (left) or by splitting them into identical basic capacitors (middle and right). All cases aim for maintaining the required ratio between main fields and fringe fields, thus minimizing the mismatch of the two capacitors C_1, C_2. The solutions in the middle and on the right also minimize mismatch due to fabrication tolerances

In the second approach, Eq. (6.13) is used as the conditional equation for the dimensions of two rectangular capacitors. We then apply Eqs. (6.11) and (6.12) for the two capacitors to obtain a set of equations for the unknowns w_1, l_1, w_2, l_2. Theoretically, there is an infinite number of solutions for this set of equations.

As a rule, capacitors should be laid out as compactly as possible. We aim for a solution where the ratio of the perimeter to the surface area P_1/A_1, P_2/A_2 is minimized; this is best obtained with a quadratic capacitor. We select capacitor C_1, as our "quadratic candidate", where $w_1 = l_1$. This value can easily be determined from Eq. (6.11) with Eq. (6.9).

A quadratic equation with the two solutions for the dimensions of the second capacitor C_2 is then obtained from the set of equations. These two solutions are

$$w_2 = \frac{C_2}{C_1}\left(1 + \sqrt{1 - \frac{C_1}{C_2}}\right) \cdot w_1 \tag{6.14}$$

and

$$l_2 = \frac{C_2}{C_1}\left(1 - \sqrt{1 - \frac{C_1}{C_2}}\right) \cdot w_1. \tag{6.15}$$

The capacitor with the smallest capacitance C_1 should always be chosen with a square shape for the equations to yield real solutions. This is clear from Eqs. (6.14) and (6.15). Figure 6.26 (left) shows the solution for the ratio $C_2{:}C_1 = 1.5$.

This methodology of individual stretching, i.e., the matching of area to perimeter ratio, works for more than two capacitors, as well. Its greatest advantage is the overall surface area needed is reduced for the configuration, as no splitting to multiple capacitors is required. No further routing is needed either.

Unfortunately, our second approach of individual stretching has some considerable drawbacks. It only makes sense when the capacitances of the matching capacitors do not differ too much from one another. The reason for this is that the aspect ratio for the larger capacitor is degraded as the ratio C_2:C_1 increases. The ratio is already $l_2/w_2 = 3.73$ for C_2:$C_1 = 1.5$ (see Fig. 6.26, left).

We should also remember that fabrication tolerances, which impact the dimensions w and l additively, impact the surface areas differently. Although the mismatch arising from fringe fields is mathematically minimized as per Eq. (6.13), the method of individual stretching causes an unavoidable remaining mismatch. This effect scales disproportionately with the increasing ratio C_2:C_1 and is caused by the fabrication tolerances described above. Hence, high-precision matching capacitors cannot be built with this stretching method. This can only be achieved with splitting (i.e., our first approach), as illustrated in Fig. 6.26 (middle and left).

NPN transistors

The total current of the NPN transistor is determined by its emitter area, as we described in Sect. 6.3.4. It is therefore sized based on its emitter area. Emitter outdiffusion occurred in older processes without STI (shallow trench isolation) whereby there was an additional lateral injection of electrons into the base, which caused a large fringe effect. Such NPN transistors can only be successfully matched by duplicating identical single emitters.

This known fringe effect has been eliminated in more advanced processes, where the outdiffusion is cut by the subsequently applied STI. However, stochastic edge shifts do occur in the STI opening, which defines the emitter area. A fringe effect occurs in this case too, as these edge shifts do not scale with larger emitter dimensions. Hence, the transistors can only be matched with high accuracy by creating all of them with uniform emitter structures, as shown in Fig. 6.17.

Another option is to stretch the single emitters. However, the emitters should only be stretched in the direction parallel to the base contact to avoid "current crowding", as described in Sect. 6.3.4. This ensures that the distance between all emitters and the base contact remains constant. The matching accuracy of this option, however, is lower than the one that can be achieved with the aforementioned matching by duplicating uniform emitter structures.

PNP transistors

We explained in Sect. 6.3.4 that this lateral transistor type can only be sized by replicating the basic structure, which is unchangeable. The device can also be matched in this way. In Fig. 6.27 (left), we show a simplified layout of a current mirror using two PNP transistors in the ratio 1:3. Please note that in this example the two transistors lie in a common n-well, which is the common base.

Collector rings of PNP transistors were cut radially in older processes with structure sizes larger than 1 micron, as shown in Fig. 6.27 (right). Area-saving current mirrors with shared base and emitter could thus be generated.

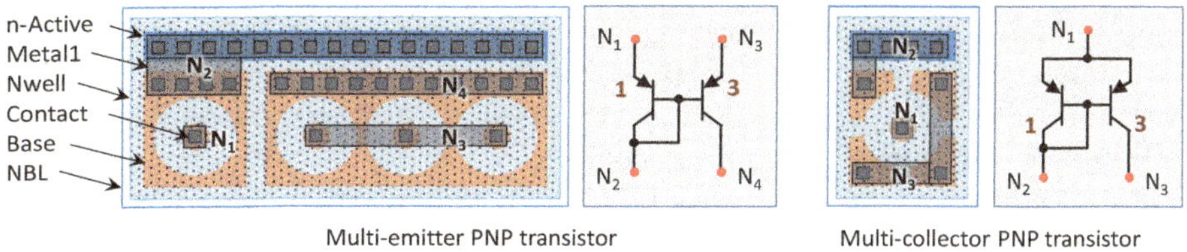

Fig. 6.27 Two versions of a current mirror of PNP transistors (layouts are shown without p-Active); version with multi-emitters (left), version with multi-collectors (right)

6.6.2 Matching Concepts for Unknown Gradients

As we mentioned at the beginning of Sect. 6.6, location-dependent variations in parameter settings on the wafer are caused by tolerances in the fabrication steps. These parameter variations impact the devices' electrical properties. An example of an Nwell sheet resistance across a wafer is visualized in a false-color image in Fig. 6.28 (top left). The data along the axis on the wafer is plotted underneath in the diagram. We could have similar charts for all other device parameters—for example, the gate oxide thickness or some other parameter could be represented by the *y* axis. These curves (and the parameters that determine them) are typically unknown, as fabrication tolerances are stochastic. How can we deal with these uncertainties?

Figure 6.28 (middle) shows an enlarged (extended in *x*- and *y*-directions) section of the curve across a die. The fictitious locations of two devices (MOS-FETs) are depicted above the curve. The two MOS-FETs are to be matched to one another in a A:B = 2:1 ratio. The devices are placed near the left-hand and right-hand die edges, respectively, and are separated in the layout by a very large distance of

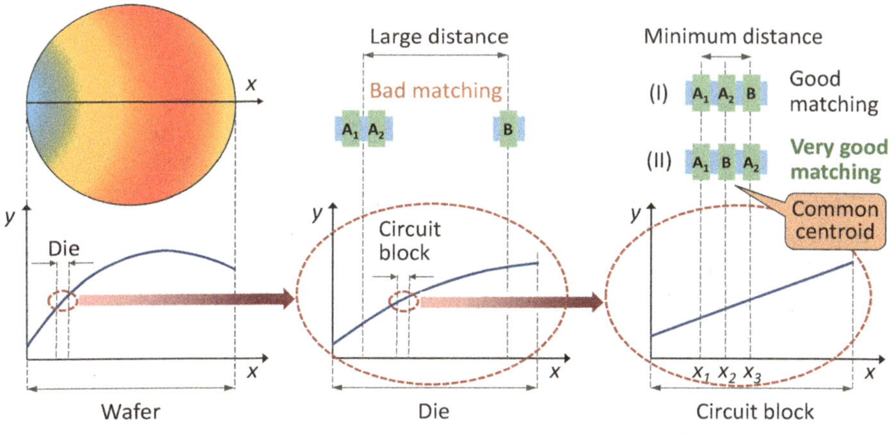

Fig. 6.28 Location-dependent gradients in device parameters caused by fabrication tolerances. As an example, the Nwell sheet resistance is considered (*y* axis) on the wafer (left), within one die on the wafer (middle), and within a die's circuit block (right)

several millimeters. The matching quality is clearly bad in this case, despite the consideration of fringe effects. This occurs because the parameter setting (here the Nwell sheet resistance) impacts devices A and B very differently, due to the large distance between them. We can now formulate another key matching concept based on this scenario.

Rule of thumb. Matching devices should always be placed as close as possible to each other in the layout. The smaller the layout distance, the less unknown gradients—regardless of what causes them and their magnitude—can contribute to a mismatch.

Another excerpt from the curve is enlarged (stretched in *x*- and *y*-directions) and plotted in Fig. 6.28 (right). Now the devices are placed at a minimal distance (in the μm scale) in layout variant (I). This scenario lends itself to good matching. Matching can be further improved if we split the devices and subsequently intertwine (or interdigitate) the subdevices between one another. In this case, the device focal points come closer together and the effective distance between the devices is further shortened.

Common Centroid Layout

The focal points are placed exactly on top of one another in the best-case scenario, as shown in Fig. 6.28 (right) layout option (II). In fact, an effective "zero distance" can be obtained with this approach. This type of constellation is known as a *common centroid layout*. We shall verify its effect now with a sample calculation.

We can quantify the effect of matching on the *y* value (sheet resistance $R_\square$ for the Nwell in our example of Fig. 6.28), given that it directly affects the electrical behavior of the circuit block. We can assume the parameter curve is *linear* within such a small circuit block. Thus, we can express the curve as a simple linear equation of the form $y = mx + b$. This enables us to carry out the following calculations:

$$\text{Layout variant (I)}: \ \mathrm{A}_1 \triangleq y(x_1) = mx_1 + b,\ \mathrm{A}_2 \triangleq y(x_2) = mx_2 + b,\ \mathrm{B} \triangleq y(x_3) = mx_3 + b$$

$$\text{and thus } \mathrm{A}/\mathrm{B} = \frac{\mathrm{A}_1 + A_2}{\mathrm{B}} = \frac{m(x_1 + x_2) + 2b}{mx_3 + b} \neq 2, \text{ because } x_1 + x_2 < 2x_3$$

$$\text{Layout variant (II)}: \ \mathrm{A}_1 \triangleq y(x_1) = mx_1 + b,\ \mathrm{A}_2 \triangleq y(x_3) = mx_3 + b,\ \mathrm{B} \triangleq y(x_2) = mx_2 + b$$

$$\text{and thus } \mathrm{A}/\mathrm{B} = \frac{\mathrm{A}_1 + A_2}{\mathrm{B}} = \frac{m(x_1 + x_3) + 2b}{mx_2 + b} = 2, \text{ because } x_1 + x_3 = 2x_2$$

The mathematics above show that layout variant (I) in Fig. 6.28 does not exactly produce the required ratio 2:1 due to the gradient. The focal points come together if we place element B exactly in the middle between A_1 and A_2 (layout variant II). The gradient is constant in this case so that the average for device A is exactly equal to the value for B. Hence, we obtain exactly the required ratio $A{:}B = 2$.

This effect is also used in the earlier discussed configuration in Fig. 6.26 (right) of two matching capacitors. Unknown gradients as well as fringe effects are offset in this layout solution. The layout in the example is also compact due to the 2:1 aspect ratio for the single elements. Compact layout segments minimize the effect of gradients.

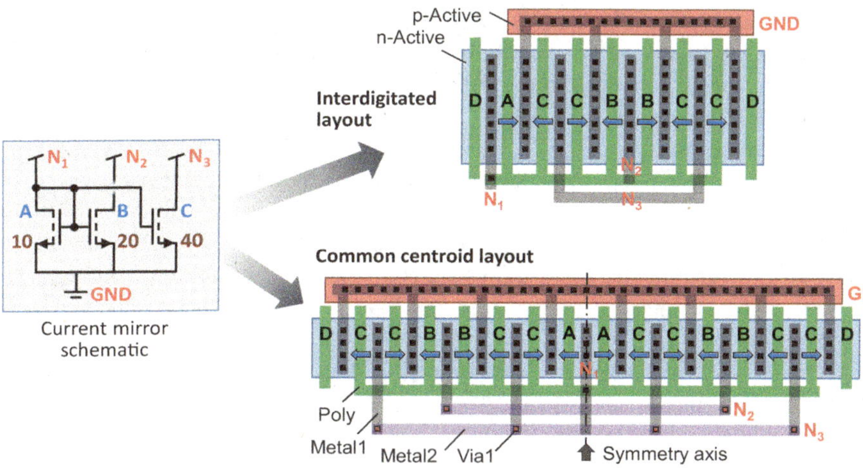

Fig. 6.29 Current-mirror circuit of NMOS-FETs with current ratio 1:2:4 (left); interdigitated layout (top), common centroid layout (bottom), (all layouts are drawn schematically, D dummy)

Two layout versions of a current-mirror circuit are depicted in Fig. 6.29. The net N_1 reference current is mirrored in nets N_2 and N_3 in a 1:2:4 ratio. The circuit is shown schematically on the left. In the first version (see Fig. 6.29, top), transistor B is folded in two fingers and transistor C in four fingers. This version is perfectly matched with regard to fringe effects. It is only when all single transistors are folded again that we can obtain a common centroid layout (see Fig. 6.29, bottom).

Rule of thumb. The effective distance can be further reduced by splitting the devices and interdigitating the subdevices. Unknown gradients can thus be more effectively offset. The ideal case is a common centroid configuration with no effective distance. These geometries always feature an axis or point symmetry.

6.6.3 Matching Concepts for External Device Fringe Effects

Good matching geometries comprise identical, closely packed single elements—as we now know. The elements of such a group that lie in the group center have the same surroundings. Unfortunately, this does not apply to the elements that lie at the group boundary. This discontinuity at the edge of a matching group can cause a mismatch, if local inhomogeneities arise in one (or multiple) fabrication steps due to the neighboring structures.

Features placed next to a matching group may require particularly strong (or weak) etching, for example. In this case, the properties of a device at the boundary of the matching group may be altered w.r.t. other devices in the group due to local excessive etchant saturation or insufficient etchant saturation. In order to avoid these effects, the matching group is surrounded by so-called *dummy elements* that are identical to the devices in the matching group but have no electrical function. Their purpose is to provide the same "neighborhood condition" for all the devices in the matching group.

Dummy elements are also placed at both sides in the two layout versions in Fig. 6.29. They are the poly gates labeled "D", which are not connected to any other elements.

Furthermore, dummies are placed at the beginning and end of any matched resistor series. Two-dimensional capacitor arrays are completely surrounded by dummies as well, which, unfortunately, take up significant space.

Well Proximity Effect (WPE). This is a well-known and dreaded effect. Higher dopant concentrations occur at approximately 1 μm from the boundary of wells as a result of scattering during the implantation of the wells [2]. This effect can severely alter the threshold voltage V_{th} of MOS-FETs and thus cause a large mismatch.

Trench-isolation stress. This effect is not caused by fabrication tolerances. Instead, it is a result of mechanical pressure that trench isolations exert on the directly adjacent silicon. The extent of the pressure that is exerted is temperature-dependent due to the different coefficients of thermal expansion (CTEs) of silicon and oxide. The pressure increases the carrier mobility of holes and lowers it for electrons. (Carrier mobility is a measure of how quickly a hole or an electron can move through a substance.) These mobility changes are also a much feared effect, which can cause large mismatches.

Rules of Thumb

- The same neighborhood conditions should prevail at both the edge of a matching group and inside, to prevent general (unforeseeable) external device fringe effects. This is done by surrounding the group with the same elements as inside. These dummy elements have no electrical function.
- The well proximity effect (WPE) can be avoided by placing matching elements at a distance from the well boundary.
- Dummies should also be used as a standard measure to counter stress caused by shallow and deep trench isolation (STI, DTI), as per [2].

6.6.4 Matching Concepts for Known Gradients

The conditions associated with a chip's use case produce known gradients. Heat distribution on the chip and mechanical stress are typical use-case conditions. Of course, the matching concepts for unknown gradients in Sect. 6.6.2 apply here too.

Given that the gradients are known, we can apply additional optimization options here as described below.

Heat Distribution on the Chip

Integrated circuits have extremely high power densities. Hence, they generate considerable thermal losses, which must be dissipated externally. The heat is not generated evenly on the chip, however, but in single devices. The heat is thus not uniformly distributed across the chip, especially in case of smart power ICs. Electrical properties, such as resistance, diode forward voltage drop, and the like, are key for device matching, as they can be very much a function of temperature.

Power transistors (DMOS-FETs) are significant heat sources: they can produce temperature differences of several 10 K on a chip. Given that power transistors are always placed at the chip boundary because of their high currents, we can estimate their heat distributions quite well. The matching process is simplified because it is the temperature difference that is key for matching and not the absolute temperature.

A chip with a DMOS-FET is pictured in Fig. 6.30 (left). Lines with the same temperature, so-called *isotherms*, are shown, and the direction of the temperature gradient perpendicular to them.

Rule of thumb. Matching devices should be placed along isotherms. Temperature-related mismatches can thus be minimized. The aforesaid applies also for common centroid layouts.

Mechanical Stress

When a chip is being packaged (cf. Fig. 3.24 in Chap. 3), the injected mold mass exerts pressure on the die. The die is permanently deformed as a result, similar to a rectangular sail on a boat when exposed to wind. This deformation causes mechanical stress in the silicon crystal. This mechanical stress has a significant effect on the mobility of the charge carriers and thus affects the devices' electrical properties. As

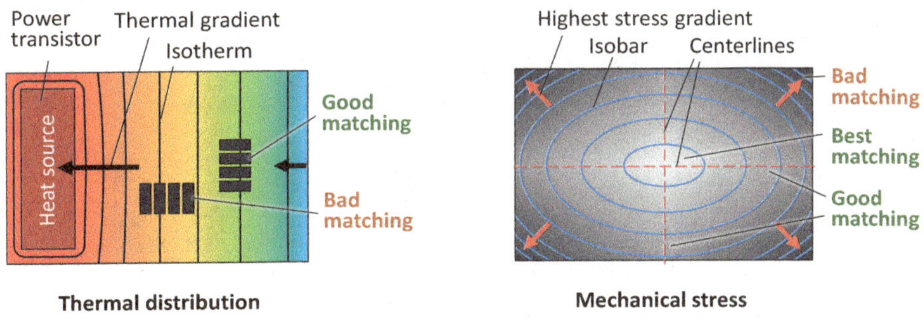

Fig. 6.30 Recommendations for matching devices on a chip with thermal gradients (left) and with mechanical stress from a chip package (right). While the former requires devices to be placed along isotherms, mechanical stress should be considered by avoiding high gradient areas, such as corners, for placing devices

the extent of the effects also depends on the crystal lattice alignment, we only know *in general* which regions on a chip are affected.

The lines of constant mechanical pressure in the crystal, the so-called *isobars*, which run perpendicular to the stress gradients are depicted in Fig. 6.30 (right). The distances between the isobars reflect the magnitude of the stress gradients similar to the slope of the contour lines on a map. The lowest stress is at the chip center (bright) and the highest in the corners (dark). The stress gradient is moderate in the middle of the chip's edges.

Rule of thumb. Chip regions with high stress gradients should be avoided if matching requirements are tight. The chip center is the ideal place for good matching results. Good results are also achieved along the two chip centerlines.

6.6.5 *Matching Concepts for Orientation-Dependent Effects*

Rule of thumb. Orientation is irrelevant in the case of capacitors. To match all other devices, they should always be aligned parallel to one another.

We shall explain below the reasons for this recommendation.

Alignment tolerances. Alignment tolerances are unavoidable. However, by aligning all matching elements in the same way, we can at least ensure that an alignment tolerance will impact all elements uniformly. The misalignment of resistor contacts is a case in point. Take the example in Fig. 6.31 where the contact holes are moved upwards w.r.t. the layer of resistance material (e.g., poly). While vertical resistors (or rather their resistances) are unaffected by this shift, the contact holes in horizontal resistors are off-center. The resistance in the resistor head will increase as a result. Parallel resistors (see Fig. 6.31, middle and right) are always matched regardless of such a shift.

Carrier mobility. We have already touched on this topic earlier. Charge carrier mobility is affected by the crystal lattice alignment when subject to mechanical forces. The same layout philosophy as above for alignment tolerances applies here

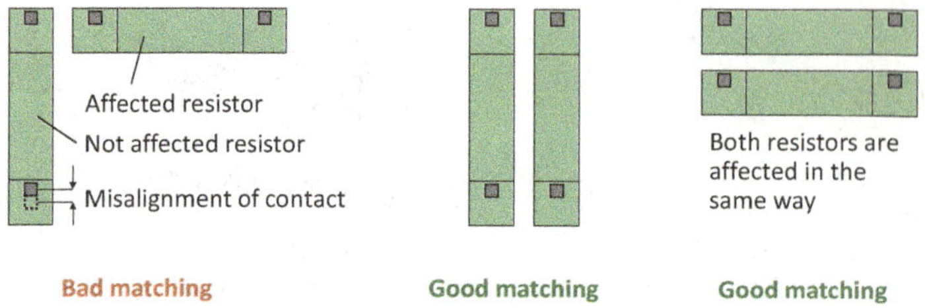

Fig. 6.31 Matching measure to compensate for misalignments

too. All elements in a matching group must be aligned to ensure that they are impacted in the same way.

Thermoelectric effect. Every contact surface between two different materials produces a potential difference. This applies to all interfaces: p–n transition, silicon-metal, or metal-metal. These voltages normally cancel each other out in a circuit. Given that this effect is a function of temperature, the said voltages do not cancel out when the contact boundaries are at different temperatures. The temperature dependency is defined by the material-specific *Seebeck coefficient* and causes so-called *Seebeck voltages* of between 0.1 and 1 mV/K at the interfaces [3]. These voltages greatly impact device matching as there are often large temperature differences on the chip.

Metal-silicon contacts are particularly relevant. For split devices, the current should flow in both directions (with half the current in each direction) as much as possible in a device's segments. The Seebeck voltages will then offset each other.

This approach was adopted in the common centroid layout in Fig. 6.29 (bottom). Figure 6.32 (left) depicts another example with split resistors. (Please note that Fig. 6.32, left, does not show a complete matching situation, but instead shows only one element of the matching group.)

If this measure cannot be implemented, or only partially, then a good approach is to make sure the currents in the matched devices flow in the same direction. This scenario is depicted in Fig. 6.32 (right).

In his/her quest for symmetry, a layout designer should not rely on the geometrical symmetry axis. As shown in Fig. 6.32 (right), this can mirror the currents in the associated devices which should be avoided.

Rule of thumb. Currents in the segments of split devices should be aligned antiparallel, if possible, to offset Seebeck voltages. If this cannot be realized, the currents in matched devices should flow in the same direction.

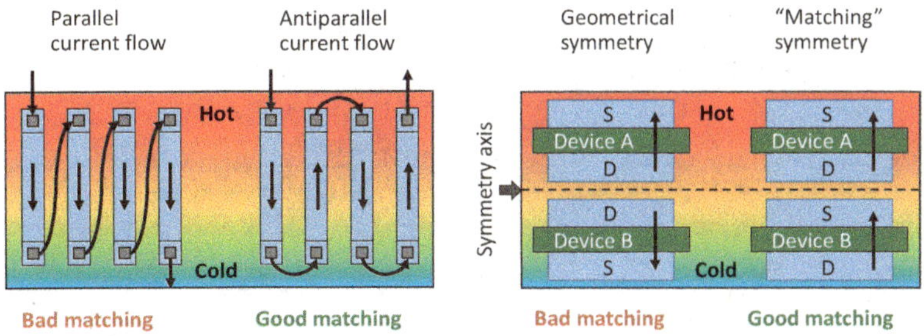

Fig. 6.32 Compensation of the Seebeck voltage within a split device (which is a member of a not-shown matching group) by using antiparallel currents (left). If this splitting within a device is not feasible, then the devices to be matched (here: A and B) should be placed such that the respective currents flow in the same direction (right)

6.6.6 Summary of Matching Concepts

Table 6.2 summarizes the aforementioned matching concepts of Sects. 6.6.1–6.6.5. We differentiate between matching for (a) normal, (b) higher, and (c) highest requirements. This classification should be beneficial for a layout designer, who should always be aware of "how much matching" is required given his/her design constraints. Note that measures listed under (a) are mandatory for any matching concept. Furthermore, depending on the circuit's requirements, it may be appropriate to apply the matching concepts also to the routing structure.

Table 6.2 Summary of matching concepts (T = transistor in general, M = MOS-FET, R = resistor, C = capacitor)

	Devices	Effect/Explanation
(a) Matching for normal requirements		
Same device type	All	Prerequisite for matching!
Same size and shape (splitting in identical basic elements)	All	Internal device fringe effects (Sect. 6.6.1)
Minimum distance	All	Unknown gradients (Sect. 6.6.2)
Same orientation	R, T	Alignment tolerances, carrier mobility (Sect. 6.6.5)
Same ratio of area to perimeter as an alternative to splitting	C	Internal device fringe effects (Sect. 6.6.1)
(b) Matching for higher requirements		
Interdigitation 1- or 2-dimensional	All	Unknown gradients (Sect. 6.6.2)
Same temperature (placing along isotherms)	All	Thermal gradient (known gradient) (Sect. 6.6.4)
Same environment (dummy elements)	All	External device fringe effects (Sect. 6.6.3)
Consider current flow direction	R, T	Thermoelectric effect (Sect. 6.6.5)
Increase dimensions	R, T	Internal device fringe effects (Sect. 6.6.1)
Distance to well border >1 μm	M	Well proximity effect (Sect. 6.6.3)
(c) Matching for highest requirements		
Common centroid layout	All	Unknown gradients (Sect. 6.6.2)
Placement in low stress chip regions	All	Carrier mobility (Sect. 6.6.5)
Symmetrical routing	All	Depending on circuit function

References

1. Cadence Design Systems. https://www.cadence.com. Accessed 1 Jan 2020
2. P.G. Drennan, M.L. Kniffin, D. R. Locascio, Implications of proximity effects for analog design, in *IEEE Custom Integrated Circuits Conf. (CICC)* (2006), pp. 169–176. https://doi.org/10.1109/CICC.2006.320869
3. A. Hastings, *The Art of Analog Layout*, 2nd edn (Pearson, 2005). ISBN 978-0131464100
4. G. Jerke, et al., Hierarchical module design with Cadence PCell designer, CDNLive! EMEA, 2015, Session CUS02. https://www.cadence.com/content/dam/cadence-www/global/en_US/documents/services/cadence-vcad-pcell-ds.pdf. Accessed 1 Jan 2020
5. O. Kononchuk, B.-Y. Nguyen, *Silicon-On-Insulator (SOI) Technology: Manufacture and Applications* (Woodhead Publishing Series in Electronic and Optical Materials, Vol. 58) (Woodland Publishing, 2014). ISBN 978-0857095268
6. D. Marolt, J. Scheible, G. Jerke, et al., SWARM: a self-organization approach for layout automation in analog IC design. *Int. J. Electron. Electr. Eng. (IJEEE)* **4**(5), 374–385. https://doi.org/10.18178/ijeee.4.5.374-385
7. B. Prautsch, U. Hatnik, U. Eichler, et al., Template-driven analog layout generators for improved technology independence, in *Proceedings of the ANALOG 2018*, pp. 156–161. https://ieeexplore.ieee.org/document/8576850
8. B. Razavi, *Design of Analog CMOS Integrated Circuits*, 2nd edn (McGraw-Hill Education, 2016). ISBN 978-1259255090
9. Y. Taur, T.H. Ning, *Fundamentals of Modern VLSI Devices*, 2nd edn (Cambridge University Press, 2013). ISBN 978-0-521-83294-6

Chapter 7
Addressing Reliability in Physical Design

Reliability of electronic circuits, which has always been an important issue, is becoming an increasing concern due to the ongoing downscaling of the structural dimensions and the continuous increase in performance requirements. This final chapter addresses the many options available to a layout designer, given the enormous influence of physical design on circuit reliability. Hence, the goal of this chapter is to summarize the state of the art in reliability-driven physical design and related mitigating measures.

We start by presenting reliability issues that can lead to temporary circuit malfunctions. We discuss in this context parasitic effects in the bulk of silicon (Sect. 7.1), at its surface (Sect. 7.2), and in the interconnect layers (Sect. 7.3). Our main goal is to show how these effects can be suppressed through appropriate layout measures.

After having presented temporarily-induced malfunctions and their mitigation options, we discuss the growing challenges of preventing ICs from irreversible damage. This requires the investigation of overvoltage events (Sect. 7.4) and migration processes, such as electromigration, thermal and stress migration (Sect. 7.5). Again, not only do we discuss the physical background to this damage, we also present appropriate mitigation measures.

7.1 Parasitic Effects in Silicon

When examining circuit schematics, it is very easy to think—incorrectly—that they represent actual physical circuits. However, a schematic is only an idealized model of the real world (Chap. 3, Sect. 3.1.2). There are always unwanted, but unavoidable, side effects—which do not appear in schematics. We call them *parasitic effects*. Every layout designer needs to be familiar with these effects and appropriate counter measures.

Parasitic effects can be modeled with the well-known lumped device types. These (virtual) devices are called *parasitics*. We show them as violet symbols in our images

J. Lienig and J. Scheible, *Fundamentals of Layout Design for Electronic Circuits*,
https://doi.org/10.1007/978-3-030-39284-0_7

throughout this chapter. Parasitic effects can be calculated when the parasitic electrical parameters are either known, are extractable from a layout, or can be estimated in some other way.

There are a multitude of differently n- and p-doped regions in a chip's bulk. These regions have parasitic track resistances; they also form parasitic diodes and bipolar transistors when combined. Focusing on the bulk of silicon, we present several important parasitics and respective counter measures in the following Sects. 7.1.1–7.1.4. We start with substrate debiasing, which occurs when parasitic substrate currents induce voltage drops in a resistive substrate.

7.1.1 Substrate Debiasing

The (typical p-doped) substrate—known as *ground* (GND or VSS)—is the reference potential for all circuits on a chip. The ground potential is raised locally by a current I_{sub} flowing in the substrate through the parasitic track resistor R_{sub} in the substrate. The ground potential increases up to the current source where its value reaches

$$\Delta V = R_{sub} \cdot I_{sub}. \tag{7.1}$$

As this is an unwanted effect, substrate currents are always parasitic.[1] Hence, one should always try to avoid current flows in the substrate if possible.

The current I_{sub} is injected into the substrate by a current source ("substrate current injector") and discharged using two contacts in the "SUB" net to the ground bond pad, as depicted schematically in the example in Fig. 7.1. The resistance R_{sub} in this context is composed approximately of the parallel track resistances R_{sub1}, R_{sub2} and the contact resistances R_{cont}.

A heavily doped (i.e., low-resistance) p-substrate is often used to minimize any possible voltage drop, aka *IR drop*, as per Eq. (7.1), as substrate currents cannot always be avoided. One always needs (at least) one additional lightly doped layer in this case, as the numerous redopings required to create devices can only be carried out in such an environment (Chap. 2, Sect. 2.6.4). This is one of the reasons for using p^+-substrate and p^--epi, as indicated in Fig. 7.2.

The causes of substrate currents are varied. They often arise when NPN transistors are operated in "saturation mode", i.e., with a collector potential that is as low as possible. This state is often desired with power transistors[2] in order to minimize the voltage drop in the base-collector diode and thus losses. If, in this case, the base-collector diode becomes forward biased, it acts as a base-emitter diode in a parasitic

[1]The so-called substrate PNP transistor, whose collector is formed by the substrate, is an exception to this rule. Its use should therefore be avoided if possible. For more information please refer to [7].

[2]Two lightly doped epi layers with the n-buried layer placed between them are used in Bipolar-CMOS-DMOS (BCD) processes for power electronics to increase the breakdown voltage of the NPN collector w.r.t. ground (Chap. 6, Sect. 6.2.2). Note that we have not shown this arrangement in the example in Fig. 7.2 to aid clarity.

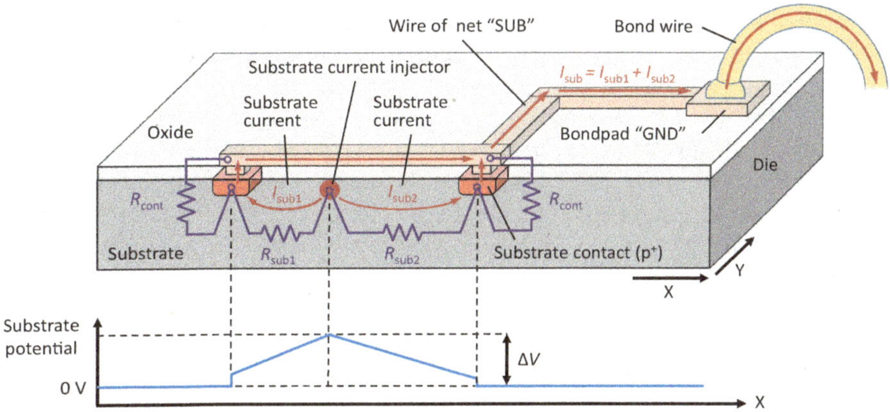

Fig. 7.1 Local increase in the ground potential due to the parasitic substrate current I_{sub}

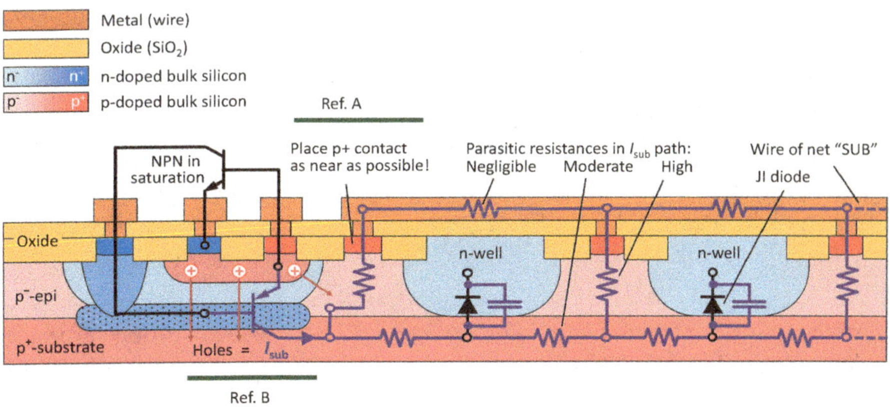

Fig. 7.2 Contacts for "tying down" the ground potential (Ref. A) should be placed as near as possible to the potential source of the substrate current I_{sub} (Ref. B). The parasitics' pins (violet symbols) are drawn as circles. They are placed on the doped regions that form the respective parts of the parasitic devices

PNP transistor. This transistor now emits holes into the n-well (parasitic PNP base). Some of these holes reach the p-substrate and the p-epi (parasitic PNP collector), where they produce the substrate current I_{sub} (see Fig. 7.2).

Subsequent analyses apply regardless of the source of the substrate current. The holes in a substrate current flow past neighboring n-wells because, as majority carriers, they are repulsed by the fields of the reverse-biased JI diodes (JI: junction isolation). The IR drop ΔV is thus increased because the carriers stay in the substrate where they can travel longer distances. The resulting potential can cause a well's JI to fail in the worst case. Hence, the proper functioning of the devices in the affected well is endangered. In addition, the JI diode may go forward biased and injects additional electrons in the p-substrate. These electrons are minority carriers there and can trigger further malfunctions. We shall discuss these effects in more detail in the next Sect. 7.1.2.

Besides this worst-case scenario, substrate currents can cause additional problems. If the JI remains effective, the depletion zones always form parasitic junction capacitances (see Fig. 7.2). The fluctuating ground causes interference through these parasitics (crosstalk) and, as a result, "modulates" adjacent circuitry. This unwanted voltage rise on the circuit ground is known as *ground bounce*.[3] The substrate current also causes more noise.

Counter Measures in Physical Design

Substrate contacts. To counter the negative effects of substrate currents, a goal in physical design is to allow these substrate currents to flow from the substrate in the most efficient manner. This is achieved by providing as many substrate contacts as possible for the current to flow into the metal, where it will have a very low-resistance path to the ground bond pad. The key here is that the track resistance to the nearest substrate contact should be small enough to allow most of the current to follow this route. The contact must be close to the current stream for this to happen, as indicated in Fig. 7.2. If this is not the case, the current will spread out evenly, i.e., it will also dive down into the depths of the substrate (chips can be up to about 1 mm thick), where it cannot be effectively "caught" again.

Design rules (spacings). A maximum spacing between neighboring substrate contacts is typically prescribed in the design rules contained in the PDK (process design kit). This requirement enforces a "minimum density" of substrate contacts across the entire chip.

Guard rings. If a potential substrate current source is known, as in our example above (see Fig. 7.2), substrate contacts are placed in the immediate vicinity of the source. Best results are achieved by completely surrounding the I_{sub} injector with these contacts. This arrangement is also called a *guard ring*.

"Currentless" ground. In Chap. 6 (Sect. 6.2.1) we recommended that the circuit ground net (GND or VSS) should not be used to connect substrate contacts to the ground bond pad. Instead, we said that a separate net, which is connected to the bond pad by a star point, should be used in the layout. We call this a *SUB net* throughout this chapter (Fig. 6.4 in Chap. 6). It is important to remember that no current-carrying device pins should be connected to this net, as these currents could otherwise contribute to a rise in the ground potential—something we want to avoid. This is why layout designers often call this a "currentless" SUB net. We know now what this means; namely, that no standard circuit currents flow through this net. Parasitic currents (which are orders of magnitudes smaller) do however pass from the substrate through this net. This is its purpose in fact.

[3]Ground bounce issues are of relevance especially in mixed-signal circuits. This is a topic that we shall not deal with in this book but refer the reader to [21] for further information.

7.1.2 Injection of Minority Carriers

As we have learned in Chap. 6 (Sect. 6.2.1), the task of the JI diodes is to electrically isolate the n-wells (cathode end) from the p-doped region (anode end) in order to block unwanted currents. The JI diode must be permanently reverse biased for this purpose. There are two ways by which this required polarity is suspended: (i) the ground potential rises inadvertently above the well potential, or (ii) the well potential drops inadvertently below the ground potential. The first of these two scenarios is caused by the substrate currents we mentioned above.

The second scenario is triggered by transients that are introduced by chip-external inductances into the interconnect that is connected to the well (Fig. 7.3, top left). The transients "pull" the well down under the ground potential; in this case the JI diode becomes forward biased and starts to conduct. It then acts as a switch that turns on an "invalid" current *I*, which flows from the chip ground contact through the nearest substrate contacts into the substrate and from there to the well. (This is the opposite direction to that depicted in Fig. 7.1 if we replace the substrate current injector there by the well.) This undesirable current *I* then leaves the chip through the affected interconnect and pad. The effect of this hole current is normally restricted to the affected well and therefore not shown in Fig. 7.3.

It is of much greater concern that a forward-biased JI diode emits electrons in the p-doped region. The injected electrons are minority carriers in this scenario that spread mainly by diffusion. If they reach neighboring n-wells, they are transported by the electric fields of these (correctly) reverse-biased JI diodes into the wells and form an unwanted leakage current there. The entire process is shown in five steps (1)–(5) in Fig. 7.3.

The effect can be characterized as the effect of a parasitic NPN transistor with the n-well (whose potential is pulled down below ground) as the emitter ("culprit"), the p-region as the base and all other n-wells ("victims") as collectors (see Fig. 7.3, shown in violet). We consider its "current gain" *B* to estimate its parasitic effect next.

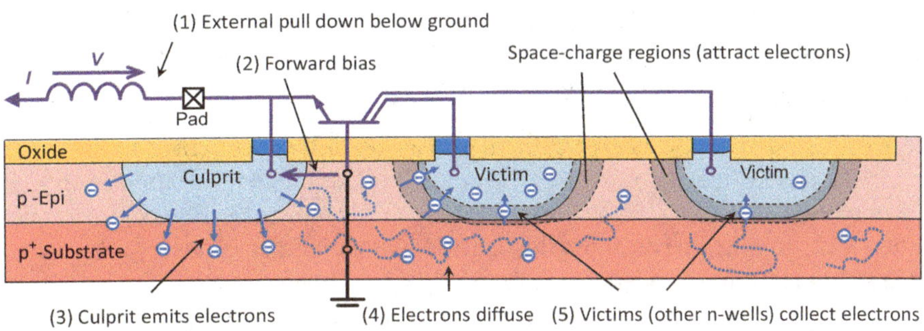

Fig. 7.3 The process of minority-carrier injection represented in five steps (1)–(5)

A heavily doped n-type buried layer (subsequently abbreviated with "NBL")[4] acts as a strong parasitic emitter in BiCMOS and BCD. The n-well in CMOS emits strongest near the surface, where it is most heavily doped and where n+ contacts are often located near the junction. The lightly doped p-epi is a "good" base, because it has very few holes to recombine with the injected electrons. The current gain B can be greater than 1 in the case of directly adjacent wells (i.e., small parasitic base width); it can even rise to 10 with injecting NBL [7]. Further, even if only a few parts per thousand of the emitter current arrive in a distant well (i.e., large parasitic base width) through a higher doped p-substrate, this can suffice to knock sensitive analog circuits operating at the μA scale out of kilter.

The best approach would be to eliminate the described causes during chip and system design. Unfortunately, this is almost impossible with the fast signal rise times in state-of-the-art applications that amplify the effects. The best option available then is to minimize the above effects by weakening the parasitic NPN transistor in the layout. We describe how this can be accomplished during physical design next.

Counter Measures in Physical Design

Identify possible injectors. Large inductances are usually only found outside the chip. Windings on electric motors are one example. All n-wells that are directly connected to a chip pad pose a high risk of becoming parasitic emitters (aka "aggressors" or "culprits"). Power transistors that drive chip-external active devices, so-called *actuators*, are typical culprits. The inductance of a bonding wire can also be enough to turn a well temporarily into a culprit in the presence of steep pulse edges.

Substrate contacts. These contacts, placed around the culprit, help to weaken the parasitic base by supplying holes for recombination with the diffusing electrons.

Spacing. Larger well spacings increase the parasitic base width, which increases the recombination rate, but at the cost of a bigger footprint. This remedial step is not very effective in a lightly doped p-epi. Furthermore, there is a small space-charge region at the boundary layer next to the higher doped p-sub. The electrons are repelled by the field in this space-charge region; as a result they stay in the p-epi, where they can travel greater distances than in the p-substrate.

Increase the p-doping. The parasitic base can be further weakened theoretically by locally increasing the p-doping concentration and thus the recombination rate. The only available option in standard CMOS are the substrate contacts mentioned earlier. Base doping is a possibility in the BiCMOS variant; it is not very penetrative, however, and thus not very effective. A heavily doped p-substrate is effective at least underneath the n-wells and, hence, promising in this regard—which is another good reason for its use in Bipolar-CMOS-DMOS (BCD) processes.

---

[4] An n-type buried layer (NBL) is mostly used to form an NPN collector in BICMOS (Chap. 6, Sect. 6.3.4) or a power MOSFET drain in BCD. Both applications are not shown in Fig. 7.3.

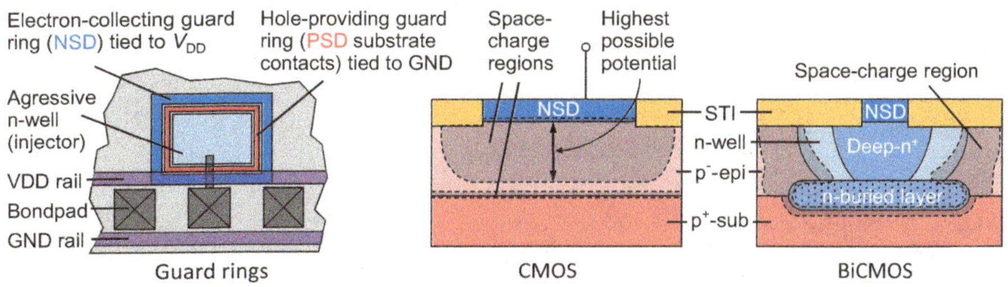

Fig. 7.4 Electron-collecting guard rings. Schematic layout instance (left) and examples of the internal structure showing the space-charge regions for CMOS (middle) and BiCMOS (right). The abbreviations NSD and PSD denote the n- and p-dopings used for source and drain regions of MOS-FETs

Guard rings. Offering the diffusion flow of electrons in the immediate vicinity of the culprit additional n-wells (with no function in the circuit) can be an effective counter measure. The field of their JI diodes attracts the electrons and so they act as "deputy collectors", which collect the electrons and render them harmless. Their effect increases with (i) the size of their electrical potential (this spreads the space-charge region and thus the field catchment area), (ii) their doping strength (enables a low-resistance current discharge into the metal), (iii) their width, and (iv) their depth (increases the catchment area, as well). We show several examples in Fig. 7.4.

All available n-doping variants are utilized in BiCMOS (see Fig. 7.4, right). Maximum depth is achieved here with NBL, and low-resistance discharge with Deep-n$^+$. There is a risk of saturation if the doping level is too low and the auxiliary collector itself becomes a parasitic emitter. This is why the n-well (whose track resistance is too high) is not used in standard CMOS (Fig. 7.4, middle) [7]. The NSD region should be at the maximum potential here. The space-charge region will then go very deep in a lightly doped p-layer. The small space-charge region at the p-epi/p-substrate interface (which is also depicted in Fig. 7.4) is beneficial here too: its field attracts electrons to the p-epi where they tend to stay.

The layout as a whole is guarded from the culprits by these deputy collectors, which is why they are also called *electron-collecting guard rings*. A schematic layout example is given in Fig. 7.4 (left). The inner (hole-providing) guard ring is made up of substrate contacts. It supplies holes to increase the recombination rate in the parasitic base.

Placement at the chip boundary. Culprits are typically placed very close to the bond pads as they are connected to the bond pads by (wide-sized) interconnects. Locating the culprits at the chip periphery makes it easier to shield them. The author of [7] recommends stretching the guard ring to the chip boundary, so that three—or in the chip corner only two—strips are sufficient.

7.1.3 Latchup

Latchup is an unwanted low-resistance path that can be triggered by the interaction between two complementary parasitic bipolar junction transistors Q_{NPN} and Q_{PNP}, which are built with an n-p-n-p doping sequence. A typical constellation in which they occur are adjacent NMOS- and PMOS-FETs, which operate with their backgates and sources at ground (VSS) and power supply (VDD), respectively. This is a very common scenario on chips. The "intended circuit" in Fig. 7.5 is a very simple and typical constellation often found in digital circuits (as an inverter) and in analog circuits (in amplifiers or as a circuit breaker).

The n-p-n-p sequence in this example (Fig. 7.5) is composed of NSD, p-substrate, n-well and PSD (NSD and PSD denote the n- and p-dopings used for source and drain regions of MOS-FETs). The "outer" regions are the sources of the NMOS- and PMOS-FETs. They act as the emitters for the parasitic bipolar junction transistors Q_{NPN} and Q_{PNP}. The "inner" regions in the sequence, that is, the chip's n-well and p-doped bulk, act reciprocally as their bases and collectors. The two parasitic bipolar junction transistors Q_{NPN} and Q_{PNP} are linked by these dual roles played by the two backgates. We label the parasitic track resistances of the two regions as R_{well} and R_{sub}. The parasitic circuit is drawn in the sectional view in Fig. 7.5 showing the location of the ring-shaped pins indicating the respective doped regions. The intended and parasitic circuits are also sketched in standard format on the right of the figure, for easier readability.

All p–n junctions are reverse biased during standard operation. The latchup effect is triggered if for any reason sufficient forward voltage is present at the Q_{NPN} (Q_{PNP})

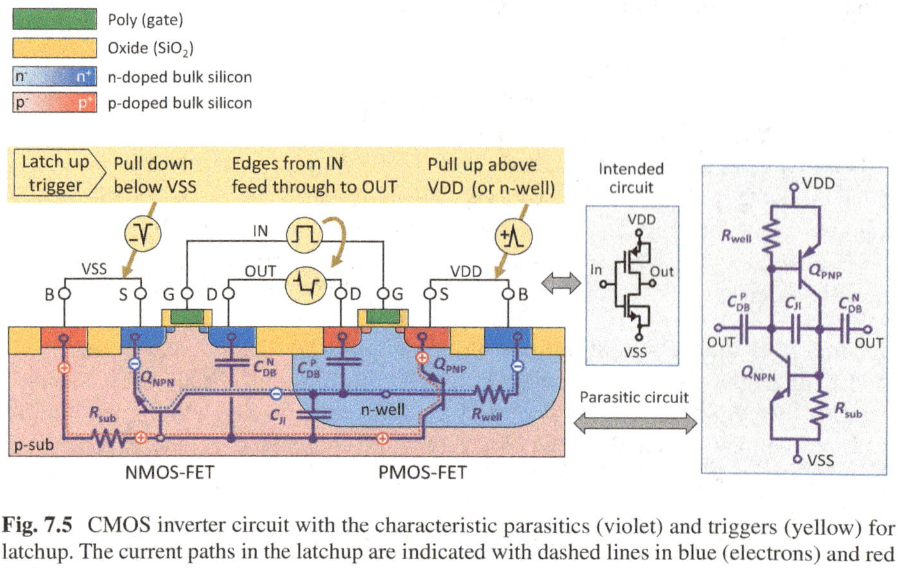

Fig. 7.5 CMOS inverter circuit with the characteristic parasitics (violet) and triggers (yellow) for latchup. The current paths in the latchup are indicated with dashed lines in blue (electrons) and red (holes)

base-emitter diode for carriers to be injected into the parasitic base formed by the p-substrate (the n-well) where they become minorities. This causes the transistor to be pulled up, and a parasitic collector current to flow in the n-well (in the p-substrate). The collector current comprising majority carriers leads to a voltage drop across R_{well} (R_{sub}), which forward biases the base-emitter diode of the complementary transistor Q_{PNP} (Q_{NPN}) to a greater degree. This is pulled up too as a result and produces a collector current, whose IR drop across R_{sub} (R_{well}) results in the first transistor being pulled up even more.

This is evidently a feedback loop in which Q_{NPN} and Q_{PNP} mutually amplify one another. Two parallel current paths of electrons and holes (drawn in red and blue in Fig. 7.5) are created between VDD and VSS. If the product of the two current gains $B_{NPN} \cdot B_{PNP} \geq 1$, the current flow stabilizes itself. This phenomenon is called a *latchup*. This state can only be eliminated if the external power supply is removed. When latchup occurs, it is possible for the current to destroy the circuit, in the worst case.

A latchup is often triggered by spikes on the supply lines. If the VDD potential is temporarily pulled up or the VSS potential is pulled down, a transient charging current flows across the junction capacitor C_{JI} at the JI diode between n-well and p-sub (see Fig. 7.5). The IR drop across R_{sub} (R_{well}) caused by this transient current flow activates the transistor Q_{NPN} (Q_{PNP}).

Coupling signals with steep edges can also trigger the latchup effect. The following mechanism is described in [1]: The edge at the inverter input "IN" feeds through directly to "OUT" through the parasitic gate-drain capacitors (not drawn in Fig. 7.5) and the drain-backgate capacitances ${C^N}_{DB}$, ${C^P}_{DB}$ (see Fig. 7.5). "OUT" thus overshoots initially in the same direction as "IN", before the inverter operates as required. This overshoot/undershoot can also produce sufficient IR drop to cause a latchup.

Counter Measures in Physical Design

Spacing. The parasitic base width for Q_{NPN} can be increased by enlarging the spacing between the two MOS-FETs, and the (above discussed) loop gain $B_{NPN} \cdot B_{PNP}$ becomes smaller as a result. (A larger spacing does not have the same effect for the vertically acting Q_{PNP}. Its base width is determined by the depth of the Nwell layer, which cannot be altered by the layout.) Obviously, this measure typically requires additional surface area. Other device types, which may be required in some analog circuits, can be placed between NMOS- and PMOS-FETs.

Higher levels of doping. The parasitic bases could in theory be further weakened by introducing heavily doped regions. This is not a practical option, though, because it is not very effective for lateral Q_{NPN} and suitable process options are often unavailable.

Low resistance backgate connection. The best approach is to reduce the parasitic resistances R_{well} and R_{sub} such that an IR drop large enough to cause latchup cannot occur under any circumstances. (The parasitic base-emitter diodes are then effectively short-circuited.) This is achieved by always contacting the backgates (i.e., the n-wells and p-sub) with metal as close as possible to the transistors. If we consider the current

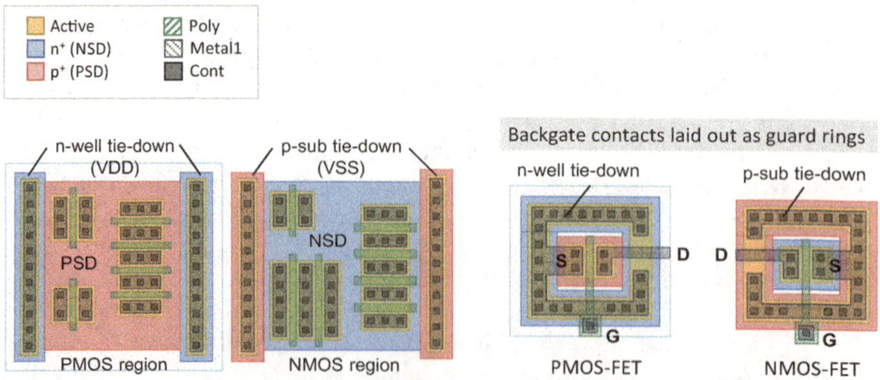

Fig. 7.6 Preventing a latchup. Isolating NMOS- and PMOS-FETs with additional backgate contacts between the regions (left), MOS-FETs with backgate contacts as guard rings (right)

paths in Fig. 7.5, we find that a very good option is to place the contacts directly at the interface between the MOS-FETs (Fig. 7.6, left).

Guard rings. The best way to inhibit a latchup is to ring-fence NMOS-FETs (PMOS-FETs) with n-well contacts (substrate contacts). These backgate contacts are also called *guard rings*. Many PDKs contain device generators that can automatically create such guard rings. Examples of guard rings for NMOS- and PMOS-FETs are shown in Fig. 7.6 (right). The closed NSD and PSD rings are fully contacted with Metal1 to minimize the resistance. The metal is only interrupted for the drain contact.

There is no risk of latchup severely damaging a chip in advanced CMOS processes anymore, due to the doping levels deployed. Hence, the latchup effect is not feared nowadays as it was in the past. Still, it is important to be familiar with the effect and to suppress it in analog circuit layouts, as small fault currents can cause malfunctions in these circuits.

7.1.4 Breakdown Voltage, aka Blocking Capability, of p–n Junctions

Most p–n junctions on a chip are reverse biased. Each of these junctions has a specific breakdown voltage V_{BD}, which should not be reached in standard operation. The circuit designer is responsible for compliance with this requirement by correctly designing the circuit and selecting and sizing its elements. The device properties are defined by the PDK elements, and the blocking capabilities between the devices are assured by the design rules. As a result, the layout designer is often left with little room to maneuver.

It may be possible and necessary in some specific cases to influence the breakdown voltage of a p–n junction by means of layout measures. Please refer to Sect. 6.2.2, where we have dealt in depth with the topic, for guidelines on how to select the right layout measures. We repeat the basic rules here:

- The breakdown voltage and, hence, the blocking capability of a p–n junction decreases with increasing dopant concentration.
- A high blocking capability requires that at least one side of the junction be lightly doped.

7.2 Surface Effects

Having presented parasitic effects in the bulk of silicon that can cause temporary circuit malfunctions, we now discuss parasitic effects that occur at the surface. Again, our goal is to show how these effects can be suppressed through appropriate layout measures.

7.2.1 Parasitic Channel Effects

MOS-FETs consist of lateral n–p–n or p–n–p sequences, over which a conductive structure separated by a dielectric is arranged. These types of constellations are also found on a chip in large numbers outside the intended MOS-FETs. They form parasitic MOS-FETs. Here, the region in the middle of these sequences plays the role of the parasitic backgate, the outer regions form the parasitic source and drain, and the conductor above them is the parasitic gate.

Figure 7.7 depicts two typical constellations with the necessary voltage conditions for channel formation. A fault current flows in each case if the parasitic sources and drains are at different potentials. The parasitic gate can also be created in metal

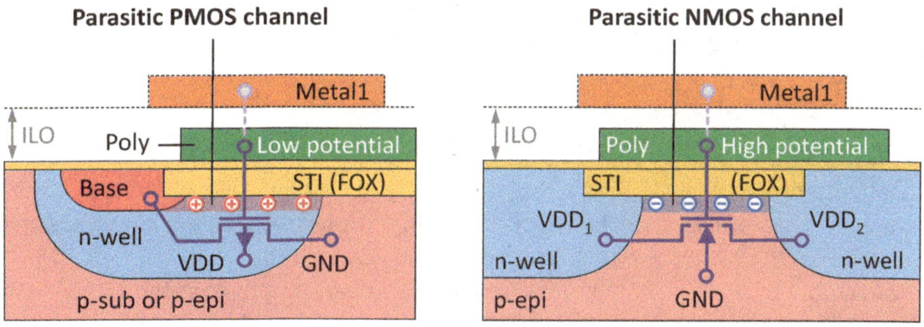

Fig. 7.7 Parasitic PMOS-FET (left) and NMOS-FET (right) with biasing (STI: shallow trench isolation; FOX: field oxide, i.e., FOX is formed by STI here; ILO: interlevel oxide)

instead of poly. This is indicated for Metal1 in Fig. 7.7 by dashed lines. The channel effect is then weaker due to the additional interlevel oxide (ILO).

Rule of thumb. We have learned in Chap. 2 (Sect. 2.9.1) that the field effect is greater if (i) the dielectric is thinner, and (ii) the backgate is more lightly (i.e., less) doped. We can make many critical inferences from these observations.

A first inference is that correspondingly higher voltages are needed to create a parasitic channel in the underlying silicon with oxide layers (FOX, field oxide, plus possible ILOs) that are about two orders of magnitude thicker than the GOX (gate oxide). This voltage is also known as the *thick-field threshold*. So-called *channel-stop implants* are inserted in many processes before the FOX is applied (this is always the shallow trench isolation, STI, in our examples). These implants ensure that the p-substrate or the n-well underneath the FOX has a higher p-doping or n-doping, respectively. This further raises the thick-field threshold.

Hence, there is no need to worry about parasitic channels underneath the FOX in chips operating at only a few volts. On the other hand, it is very important to consider these issues when dealing with higher operating voltages[5] as there is a limit to the doping strength of the channel-stop implants for supporting the breakdown voltages. We recommend to the reader to refer to the PDK for details on the thick-field thresholds of poly and metal layers.

Counter Measures in Physical Design

Active region. There is always a greater risk of parasitic channel formation in the active region because of the missing FOX. The lower interconnect layers should therefore only be used here to contact the devices. Hence, aside from blocking parasitic channel formation, one thereby minimizes the risk of unwanted capacitive coupling as well. There should be no routing apart from connecting devices in an active region of sensitive analog circuits.

Inactive region. If the voltages enable parasitic channels to form in the inactive region (i.e., above the FOX), the following methodologies are available to prevent them:

(1) Take care with endangered areas and avoid parasitic gates in the routing.
(2) Place interconnects in higher metal layers in order to thicken the parasitic gate oxide.
(3) Dope the parasitic backgate (locally) higher (insert "channel stoppers").
(4) Shield parasitic gates in metal with underlying interconnect layers.

Methodologies (2) and (3) are derived directly from our rule of thumb above. A parasitic channel needs to be interrupted only at one point—this simplifies the application of the methodologies. We provide several examples below.

A PMOS-FET, whose gate connection runs through poly to outside the n-well, is depicted in Fig. 7.8 (left). The applied control voltage V_{GB} could reach the

[5]Smart power chips in automobile electronics have a voltage capability of 60 V. Other Bipolar-CMOS-DMOS (BCD) chips even exceed 100 V.

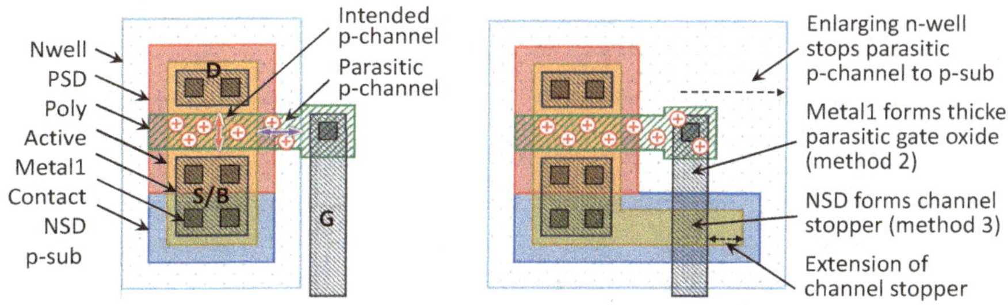

Fig. 7.8 Parasitic channel formation at the gate connection of a high-voltage PMOS-FET (left) and methodologies for interrupting the parasitic channel (right)

thick-field threshold of Poly to the Nwell layers in high-voltage applications. A parasitic p-channel is then formed that connects the intended channel between the PSD regions of the PMOS-FET (parasitic source) with the P-substrate (parasitic drain) [7].

The parasitic channel is interrupted before it reaches the p-substrate by enlarging the n-well (Fig. 7.8, right). If the thick-field threshold of Metal1 is greater than the NMOS-FET's V_{GB}, the parasitic channel is interrupted underneath the metal. This solution follows method (2). If this does not work, method (3) can be applied, whereby NSD is placed as a *channel stopper* underneath the metal. This is achieved by extending the n-well connection doping (i.e., NSD) in Fig. 7.8 (right). This thick-field threshold is typically higher than in the prior solution although there is no FOX now. There must be adequate extension of the channel stopper over the metal with this solution. The overlap should correspond to the sum of the mask alignment tolerance plus two times the oxide thickness in order to consider lateral fringe fields.

Method (3) is also a relatively simple approach of preventing parasitic n-channels between n-wells (as in Fig. 7.7 right). The method is applied by placing PSD between the wells. The stipulated well spacing is often so large that the PSD can be introduced with little or no additional surface area. The method is executed as standard in many processes, especially those with lightly doped p-epi, by enclosing all n-wells with PSD-channel stop rings (Fig. 7.9, left). The rings are then also ideally suited for substrate contacting.

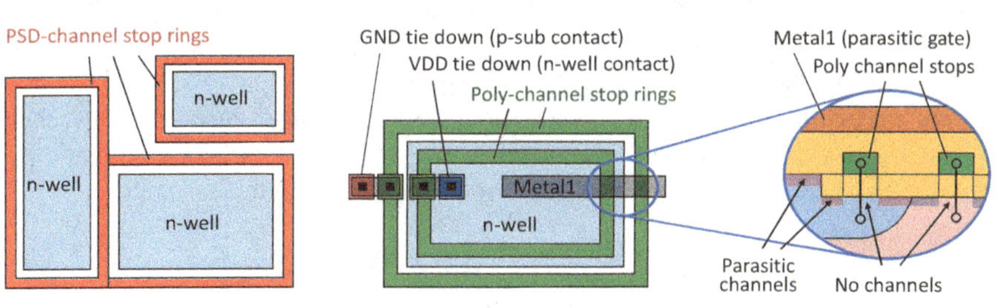

Fig. 7.9 Schematic layout structure of PSD-channel stop rings (left) and poly-channel stop rings (middle). The cross-sectional view (right) indicates the shielding effect of the poly structure

The geometries for the PSD-channel stop rings can be easily produced with graphics operations (Chap. 3, Sect. 3.2.3) from the structures in the Nwell layer. They could, in fact, be produced automatically in the "Layout to Mask Preparation" step (Chap. 3, Sect. 3.3.4).[6] In this case, however, the rings are not visible in the layout. The spacing rules must then ensure that they do not collide with active structures outside n-wells.

Our final example demonstrates method (4). In Fig. 7.9 (middle), *poly-channel stop rings* are in place at the inner and outer boundaries of the n-well, respectively. By placing the outer ring at the p-substrate potential (i.e., ground) and the inner ring at the n-well potential, the poly interconnects shield all fields emanating downwards from the metals. Hence, parasitic n- or parasitic p-channels as per Fig. 7.7 cannot then arise. Although only one case can occur electrically for a single interconnect at any given time, both cases are depicted in the cross-sectional view in Fig. 7.9 (right).

Poly-channel stop rings are widely deployed in high-voltage chips. Their use is normally restricted to wells at a high potential as they consume significant surface area. Suitable wells are selected on the basis of their voltage class assignments. The wells, including poly channel stops, are often created by layout generators. The poly channel stoppers can also be customized and applied selectively. The use of a combination of poly channel stoppers with NSD or PSD channel stoppers is not recommended, as there would be no FOX present then.

7.2.2 *Hot Carrier Injection*

We know that a conductive inversion channel is formed when an NMOS-FET is driven with a voltage $V_{GS} = V_{GB} > V_{th}$ (i.e., we assume $V_{BS} = 0$ here) between gate and source. If we apply a voltage V_{DS} between drain and source, a current will flow through the inversion channel. The MOS-FET saturates if $V_{DS} \geq V_{GS} - V_{th}$. In this scenario, the channel becomes increasingly pinched at the drain (Fig. 7.10, left) [21]. The electrical field, which produces the drift current in the channel, is at its maximum at this point.

The electrons reach a drift velocity of 10^5 m/s above a field strength of approximately 1 V/μm. This drift velocity cannot be further increased. The electrons are called *hot electrons*[7] in this state. They can then possess enough energy (i) to ionize silicon atoms and (ii) to overcome the energy barrier to the nearby GOX (gate oxide). The former produces more electron-hole pairs, which increases the drain current (electrons) and causes a noticeable substrate current (holes drift in the space charge region field to the backgate). The latter can cause charge carriers to enter the oxide and become trapped there. The same effect occurs in PMOS-FETs (where the carriers are holes), which is why we talk about *hot carriers* in this context.

[6] Advanced processes do not provide this option.

[7] The origin of this term is usually said to be that the drift velocity is greater than the velocity caused by thermal motion. Another explanation is that the effect of the observed "maximum velocity" resembles the effect of decreasing charge carrier mobility with increasing temperature.

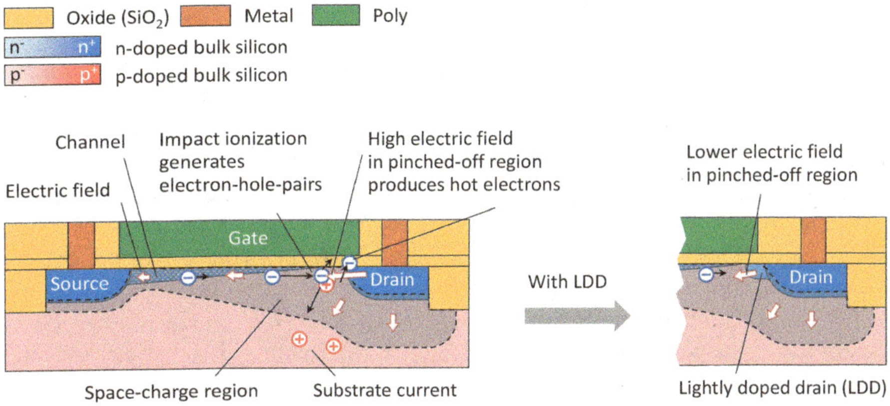

Fig. 7.10 Hot electron injection with NMOS-FET (left) and improvement with lightly doped drain (LDD) on the right

The charges trapped in the GOX can be negative or positive for both transistor types. They are impurities that alter the MOS-FET threshold voltages and degrade their gate breakdown voltages. If the transistor is regularly driven under these conditions, the charges accumulate in the GOX the longer it is in operation. An aging effect then sets in, which causes the threshold voltages V_{th} to drift.

Since the lateral field strengths increase with shorter channel lengths, the hot carrier injection also increases. The channel length should not be shorter than approximately 3 μm in older processes to inhibit the effect. This recommendation does not apply to the more advanced submicron processes. Here, methodologies have been developed to lower the field apexes that occur at the boundary to the drain zone (i.e., at the p–n junction).

The most important of these methodologies is the so-called *lightly doped drain* (*LDD*) concept. We learned in Chap. 6 (Sect. 6.2.2) that the field strength in a space-charge region is inversely proportional to the dopant concentration. This is leveraged in MOS-FET with LDD where the field can spread out further in the drain zone due to the lower local doping. The field apex is thus lowered at this critical point, which not only weakens the hot carrier effect, but also increases the breakdown voltage at the drain.

The disadvantage of LDDs is that an extra doped layer is required respectively for NMOS- and PMOS-FETs. LDDs also slightly increase the track resistance of the drain. Nonetheless, LDDs have become an absolute necessity with the very short channel lengths in advanced processes. Since they have become standard practice, we have included them in the process description in Chap. 2. LDDs are included in all our layout examples even when they are not required in the specific context. Our discussion here should provide the technical background for this process extension.

7.3 Interconnect Parasitics

Having presented parasitic effects in the bulk (Sect. 7.1) and on the surface of silicon (Sect. 7.2), we now discuss parasitic effects in the interconnect layers. Again, our goal is to show how these effects can be suppressed through appropriate layout measures.

We have already seen in Chap. 3 (Sect. 3.1.2) that real interconnects are not perfect short-circuits. In general, there is a *line resistance* per unit length R', a *line inductance* per unit length L', an *insulator capacitance* per unit length C', and an *insulator conductance* per unit length G' along a conductor. These primary line constants are illustrated as lumped parasitics R, L, C and G in the equivalent circuit diagram for a two-wire conductor in Fig. 7.11 (left). The circuit substrate can also assume the role of the second (bottom) conductor.

As long as the frequencies do not go too far into the GHz range, we can ignore the self-inductance on a chip due to the very small dimensions. In addition, the insulator conductance per unit length is typically negligible due to the excellent isolation properties of the oxides. However, the parasitics R and C play significant roles on a chip (Fig. 7.11, middle).

7.3.1 Line Losses

We shall first consider only the parasitic track resistance R (Fig. 7.11, right), which is calculated as follows:

$$R = R_{\square}\frac{l}{w} \tag{7.2}$$

with sheet resistance $R_{\square}$ (Chap. 6, Sect. 6.1), line length l and interconnect width w. This resistance R to the current I flowing in the conductor causes a thermal loss $I^2 \cdot R$. It also causes a potential drop $I \cdot R$ in the line—which is the difference between the voltage V_b at the beginning of the conductor and the voltage V_e at the end of the conductor. This *IR drop* always needs to be considered during the routing step.

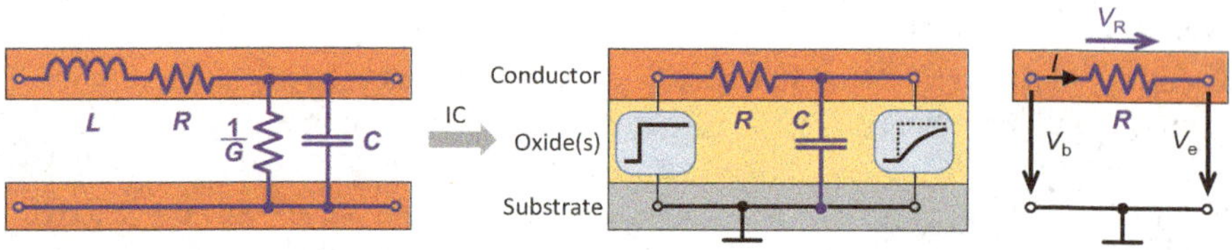

Fig. 7.11 Equivalent circuit diagrams for a standard two-wire conductor to illustrate conductor resistance and inductance, and insulator capacitance and conductance, labeled with the symbols R, L, C, and G, respectively (left). Also shown are an IC interconnect as RC element (middle) and with parasitic track resistance (right)

If the current I is known in layout design, the IR drop can always be altered by changing R. The quantities line length l, interconnect width w and sheet resistivity $R_{\square}$ (depending on the layer selected) can be tweaked by the layout designer, according to Eq. (7.2). As a rule of thumb, IR drops greater than 10 mV should always be checked.

Furthermore, the line length l depends to a large extent on the locations of the pins to be connected, so the placement of the devices should be optimized in this regard. Devices between which a high current flows should therefore be placed as close as possible to one another to avoid having to make the line width w too large. In general, the width w should be selected so that the current-carrying capacity of the line suffices (Sect. 7.5).

7.3.2 Signal Distortions

The thin deposited layers on a chip mean that every interconnect also has a significant parasitic capacitance per unit length C' to the chip substrate. A simplified model of this scenario with a lumped capacitance C is depicted in the equivalent circuit diagram in Fig. 7.11 (middle). The capacitor C must be charged by the resistor R (aka *RC element*) for every signal that is transmitted through the interconnect. This is shown in Fig. 7.11 (middle) for an ideal step function as input signal at the beginning of the line. The signal is delayed and distorted by the time it reaches the end of the line. The time delay is characterized by $R \cdot C$, which is known as the *time constant* of the RC element (67% of the final value is reached after this time). How can we alter the product $R \cdot C$ in the layout?

We can imagine that every interconnect is a capacitor electrode and the substrate is its counter-electrode. The well-known Eq. (7.3) for the parallel-plate capacitor:

$$C = \varepsilon_0 \varepsilon_\mathrm{r} \frac{A}{d} \tag{7.3}$$

(with $\varepsilon_0 \varepsilon_r$ the permittivity of the dielectric between the plates, d the distance between the plates and A the surface area of the plates) shows that C scales with the wire surface area A (i.e., with the width w and length l) and is inversely proportional to the oxide layer thickness d. Since the lower metal layers are nearer to the substrate, their value d is reduced; this means that the lower metal layers have greater parasitic capacitances than the upper ones.

The capacitance of a chip interconnect cannot be accurately calculated with Eq. (7.3), however, as this equation ignores the fringe fields. (The fringe fields on a regular parallel-plate capacitor are negligible because $w \gg d$ and $l \gg d$.) With chip interconnect, the lateral fringe fields contribute significantly to C. The capacitance per unit length C'_fringe arising from the fringe fields is greater than the plate capacitance per unit length C'_plate for typical narrow (i.e., small-width) interconnects (e.g., $w < 2$ μm for Metal1, $w < 4$ μm for Metal2) (Fig. 7.12a).

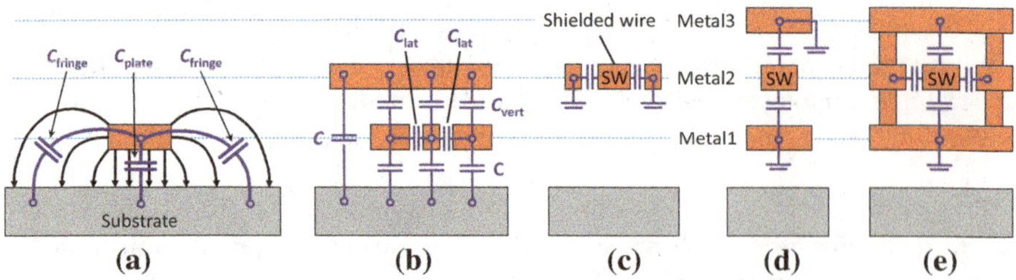

Fig. 7.12 Parasitic capacitances between interconnects and the substrate (**a**); coupling capacitances between interconnects (**b**); and lateral (**c**), vertical (**d**) and all-round (**e**) interconnect shielding

R·C optimization. The optimization of the product $R{\cdot}C$ impacts the layout directly. A reduction in the wire length l causes a proportional drop in C and R, i.e., $R{\cdot}C$ drops proportionally to $1/l^2$. A change to the interconnect width w has the opposite effect on C and R. The product $R{\cdot}C$ can nonetheless be reduced by widening an interconnect (especially a narrow-sized one), as the increase in C, which only impacts C_{plate}, is always less than the reduction in R, which is proportional to w. The product $R{\cdot}C$ can also be minimized by placing signals in upper interconnect layers.

It is a useful exercise to examine the C'_{fringe} and C'_{plate} data for the interconnect layers of a process in the PDK. Double-digit numbers in aF/μm^2 and aF/μm are typical for C'_{plate} and C'_{fringe}, respectively. A value table for a CMOS process is given in [1].

Differential-pair routing. External disturbances are another source of signal distortions on wires. Sensitive analog signals can be transported relatively securely over long distances by inverting the signal and carrying it in a second wire parallel to the original signal to the receiver (Fig. 7.13). The aim is that disturbances affect both interconnects equally by routing them both close together (Chap. 5, Sect. 5.3.3).

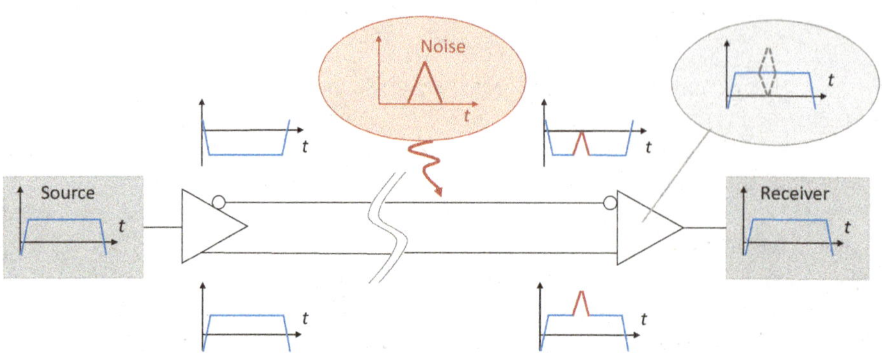

Fig. 7.13 Interference reduction using differential signaling, i.e., two adjacent wires, one with the original and one with the inverted signal. The noise is cancelled out when the signals are combined at the receiving end

The inverted signal is subtracted from the original at the receiver. The original is thus doubled, and all disturbances are neutralized. The effect can be increased by twisting the two interconnects.

7.3.3 Crosstalk

Aside from the capacitances to the substrate, many coupling capacitances are created between laterally adjacent interconnects (C_{lat}) and vertically adjacent interconnects (C_{vert}) (see Fig. 7.12b). These capacitances cause crosstalk between signals and thus disturb the electronic circuit. As IC downscaling of feature sizes progresses, the distances between interconnects are getting smaller, and these parasitics are increasing. Very narrow and high interconnect cross-sections can be produced with modern Damascene processes (Fig. 7.12c). Hence, the quantity C_{lat} is more significant than C_{vert} in state-of-the-art processes than in older processes. The most important guidelines for crosstalk reduction in physical design are summarized in the following.

Analog and digital isolation. Steep signal edges, which are typical for digital signals, contain high frequencies. These edges strongly disturb capacitively coupled interconnects (due to the AC reactance $R_C = |1/\omega C|$). Sensitive analog signals (e.g., sensor signals) should therefore be routed at a distance from digital signals. Clock nets are the biggest culprits in this regard.

Shielding. If lines cannot be physically isolated or this isolation is not sufficient, crosstalk can be minimized by enclosing interconnects with other interconnects that are grounded. The lines are thus shielded from disturbances. The three shielding options available are: lateral only, vertical only, or all-round (see Fig. 7.12c-e).

Avoiding minimum spacings. While the ILO (interlevel oxide) thickness cannot be altered in the layout, the *lateral* interconnect spacing within a layer can. It is a good idea to refrain from using minimum spacings for critical interconnects (culprits and victims alike), if the footprint allows it. After having routed all nets, it is good practice to use available whitespace to extend or adjust the wire spacing.

7.4 Overvoltage Protection

After having presented temporarily induced malfunctions and mitigation options in Sects. 7.1–7.3, we now discuss the growing challenges of preventing ICs from irreversible damage. This Sect. 7.4 describes the most common overvoltage effects, namely, electrostatic discharge (ESD) and the antenna effect. Both can potentially cause catastrophic damage in chips, which can nonetheless be prevented with specific layout measures.

7.4.1 Electrostatic Discharge (ESD)

In the early years of microelectronics, chips were often destroyed by electrical overstress before they were installed. Investigations showed that the damage was caused by undesired electrical discharge events, which occurred during fabrication, transportation or installation. How does this happen?

Everyone knows the electrical shock you can get if you touch a metallic object such as a door knob, which often happens when you have just been charged with static electricity, such as by walking across a carpet-covered floor. An electrical potential that can reach some 10 kilovolts (kV) builds up on your skin. This potential difference is caused by the *triboelectric effect*; electrostatics can also be triggered by a process known as *induction*. (We refer the reader to the literature, such as Sect. 6.5 in [14], for an explanation of these causes.)

Electrostatic discharge (*ESD*) is defined as the sudden flow of electricity between two electrically charged objects. More precisely, it is an electrical discharge from an electrically isolated material with a high potential difference that causes a very short and high electrical current pulse. An electrostatic discharge event can be felt at about 3 kV and higher. However, only a fraction of this voltage is needed to damage or destroy electronic circuits.

If someone who is statically charged touches a chip, this discharging current can flow through the chip. This is called an *ESD event*. ESD events on a chip can also be caused by statically charged equipment or packaging materials. The triggering voltages (V_{zap}) can be in the 100 V to many kV scale. However, only some few μJ are produced by the discharge. But substantial currents on the order of amperes flow due to the very short discharge time, normally around 100 ns. To make matters worse, the current flows in a very small area in the chip (typically $100 \times 100\ \mu m^2$), so that the temperature rises dramatically in this region.

Different damage mechanisms are described in [24]. They include oxide breaches caused by excessive field strengths as well as thermal damage caused by local current spikes. The reason for the first case is that gate oxides are especially vulnerable, because they are very thin and therefore very sensitive. In the second case, low-resistance short-circuits can form in p–n junctions and breaks in metal. Having said that, ESD damage does not always result in malfunctions. Sometimes, only circuit parameters, e.g., V_{th}, are shifted; the damage is irreversible, nonetheless.

ESD-Suppression Measures

A circuit must always be guarded against ESD events by allowing the discharging current to pass through a *shunt path* parallel to the circuit, so that it can (almost fully) discharge (Fig. 7.14, right). This is the same principle employed to protect a building from lightning strikes with a lightning rod (Fig. 7.14, left). An *ESD protection circuit* acts like a “lightning rod” on a chip. The many different pins on a chip complicate matters as the ESD pulse can enter and leave through any of these pins and the level of protection needed is not the same for all pins.

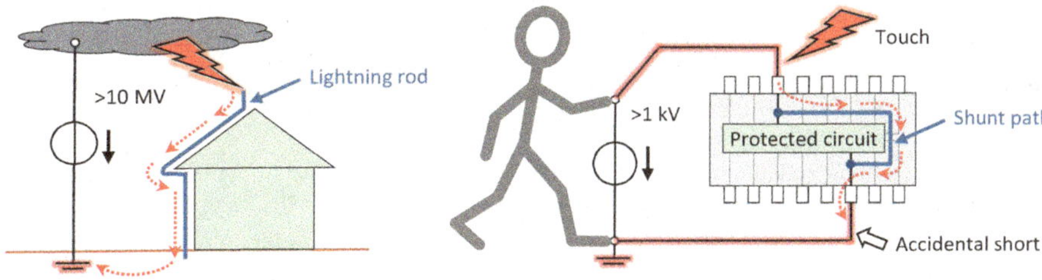

Fig. 7.14 Guarding a building against surge discharges with a lightning rod (left) and a chip's circuit against ESD events with a shunt path (right)

The ESD protection circuit on a chip must meet the following requirements:

- It should not interfere with the normal circuit functions.
- It must respond sufficiently fast to an ESD event.
- It must have a path for discharging current for every pin that leads to circuitry requiring protection.

The shunt path must have the following properties:

- It runs outside the protected circuit.
- Its resistance is sufficiently low.
- It is sufficiently robust against the (thermal) dissipated energy.
- It is triggered at an overvoltage level that is still non-critical for the functional parts of the circuit connected to the pin.

We can draw a few important conclusions from these four conditions that impact physical design.

The first three properties of a shunt path are achieved by designing the devices for the ESD protection circuitry sufficiently "large" and placing them as near as possible to the pads. Enlarging the devices (perpendicular to the direction of current) reduces the track resistance and thus the thermal losses. The peak temperature is reduced further as the thermal losses are distributed across a greater surface area, and robustness is considerably enhanced as a result. The size of the ESD protection devices should only be as large as necessary to avoid wasting surface area. Interconnections should be low-resistance. This can be easily achieved with the ESD devices placed near the pads. We shall outline further instructions on routing ESD protection devices at the end of this section.

All ESD protection devices have to fulfill certain conditions concerning their current–voltage (I–V) characteristics, which are constrained by the *ESD design window* shown in Fig. 7.15. This window is derived from both (i) the chip's non-influenceability (the ESD protection circuit should not interfere with the normal circuit functions) and (ii) the property of a shunt path that it is triggered at an overvoltage level that is still non-critical for the protected circuitry.

Let us first consider the protected device's operating range, which is generally defined by the supply voltage VDD. Here, the ESD protection device must be inactive, that is, it should have a very low conductance (flat part of the blue I–V curve in

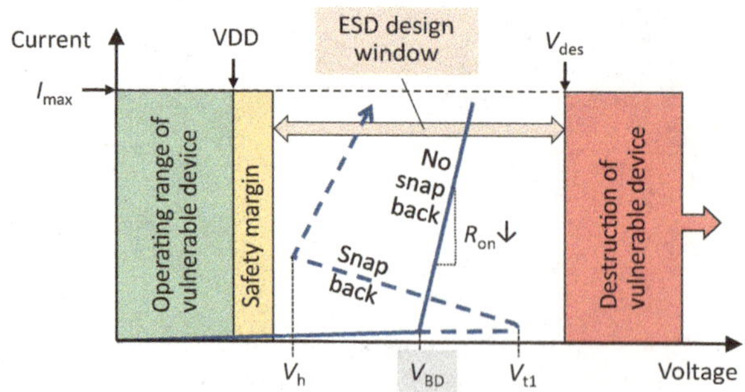

Fig. 7.15 The ESD design window visualizes the required *I–V* characteristics of ESD protection devices with and without snap back characteristics (dashed and solid curves, respectively)

Fig. 7.15). If the voltage exceeds this threshold by a set safety margin (typically 10–20% above VDD), the ESD device is triggered (breakdown voltage V_{BD}) and the desired shunt path is made available (steep part of the blue *I–V* curve with small R_{on}). The voltage is then "clamped" to just above V_{BD}.

Obviously, the ESD device's breakdown voltage V_{BD} and the entire curve must be below the voltage V_{des}, at which the protected device/circuitry would be damaged. Furthermore, the ESD protection device itself must be able to take the maximum current I_{max} occurring in the ESD event without failing.

Standardized models are available for determining the values V_{BD} and I_{max} for the ESD protection device and the destruction limit V_{des} for the protected circuitry. Typical ESD events are represented in simple equivalent circuits in these models. Among the most commonly used models for chips are the *human-body model* (*HBM*), the *machine model* (*MM*) and the *charged-device model* (*CDM*) [24]. HBM and MM cover the discharge of statically charged persons and machine elements through a chip. Applying the CDM, there is a static charge on the packaged chip, which is discharged through a pin. This condition can often only be met by placing ESD protection devices in the proximity of the protected circuitry, as the point where the charge enters the circuit is not known in this case [7].

The solid *I–V* curve in Fig. 7.15 is a typical curve for diodes. Simple ESD protection circuits are therefore constructed with diodes. These diodes are normally operated in reverse bias and their breakdown voltage V_{BD} is used as a clamp voltage, as the value required for V_{BD} is typically greater than the diode's forward voltage V_{on}. The ESD guard must be designed so that the breakdown, often caused by the *avalanche effect* [22], is reversible, i.e., it does not destroy the ESD guard.

Besides diodes, other special devices and circuits are deployed as ESD protection devices depending on the requirements. A so-called *snap back* effect is utilized by some of the guard devices (see dashed *I–V* curve in Fig. 7.15) to produce a clamp at a so-called *holding voltage* $V_h < V_{BD}$. We shall come back to these special devices later.

Basic Structure of an ESD Protection Circuit

An ESD protection circuit consists of multiple ESD protection devices, e.g., diodes, as illustrated in Fig. 7.16. The protected circuit's power supply pins—the supply voltage VDD and the ground connection VSS—are also shown. Any two I/O pins

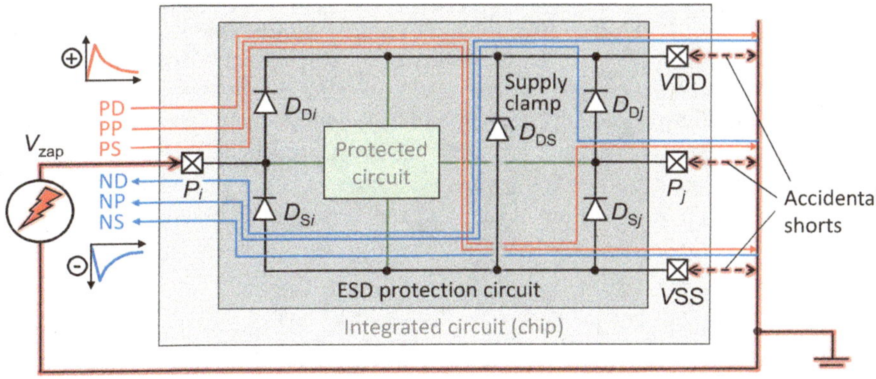

Fig. 7.16 Basic topology of an ESD protection circuit on a chip, consisting of multiple diodes that become forward biased ($D_{D,i}$ and $D_{S,i}$) or reverse biased (D_{DS}) in the case of an ESD event. Six different zapping modes (PD, PP, PS, ND, NP, NS) with their respective shunt paths are also shown

P_i and P_j represent all other pins. The diode D_{DS} designed in compliance with the ESD design window is in place between VDD and VSS. It is called a *(power) supply clamp*, as it "clamps" the voltage difference to a value within the ESD design window.

Every I/O pin P_i is connected to VDD by a diode D_{Di} and to VSS by a diode D_{Si}. These two diodes are reverse biased as long as the voltage V_i at pin P_i is between VDD and VSS (in standard mode). If V_i rises above VDD or drops below VSS, one of the two diodes becomes forward biased ("turns on") and clamps the voltage V_i to a value that is (slightly) above VDD or below VSS by the (small) diode forward voltage V_{on}. The forward-biased diodes D_{Di} and D_{Si} also ensure that the circuit section at P_i is not exposed to any voltage that is significantly outside the supply voltage range. The aforesaid applies to standard mode, i.e., when the chip is installed in the system.

To understand how the ESD guard works, we need to look at the disconnected chip, where all pins are usually floating (the "worst-case ESD scenario"). If a pin (pin P_i in Fig. 7.16) is exposed to a voltage impulse V_{zap} and another pin (inadvertently) comes into contact with the reference potential of the power source V_{zap} (we say: "the pin is grounded"), the circuit closes (this is shown as "accidental shorts" in Fig. 7.16), an ESD event is triggered, and a shunt path from P_i to the ground is established.

The voltage V_{zap}, which is either positive or negative, can occur between any two chip pins. There can be a total of six different zapping modes, which we refer to as PS, NS, PD, ND, PP and NP modes, in accordance with [24].[8] The zapping polarity is represented by the first letter (P = positive, N = negative). The "grounded" pin (D = VDD, S = VSS, P = signal pin) is represented by the second letter.

A specific shunt path, through which the ESD pulse signal can discharge, is available in every zapping mode. The current paths are drawn in red in Fig. 7.16 for the three Px modes and in blue for the three Nx modes. The schematic shows that the

[8]Only four of these modes are cited in [24]. We have added the PP and NP modes.

ESD diodes for the signal pins (D_D and D_S) are forward biased in all cases in which a current flows through them, while the clamp diode for the power supply net (D_{DS}) is reverse biased. The zapping voltage V_{zap} can of course also appear between the two power supply pins (although this case is not shown in the figure). In this case, the discharge current only flows (forward or backward) through D_{DS}.

This circuit design ensures that the maximum voltage between two arbitrary chip pins is clamped to the pin-to-pin voltage V_{pp}. This voltage V_{pp} depends on the type of ESD protection device used and on the zapping mode, as follows:

$$V_{pp} \leq \begin{cases} V_{BD} + 2V_{on} & \text{without snap back,} \\ V_h + 2V_{on} & \text{with snap back.} \end{cases}$$

As before, V_{BD} denotes the breakdown voltage, V_{on} the diode's forward voltage, and V_h the holding voltage. The maximum value is obtained in each of the PP and NP modes.

ESD protection devices are only custom designed in exceptional cases with a specific project in mind. Instead, they are normally offered as library elements in PDKs. ESD protection devices are often included in pad cells, which contain the bond pad including ESD protection as a combined layout element.

We show a sample pad cell layout in Fig. 7.17 from the "0.35 µm modular CMOS" process from the X-FAB Foundry® [27]. It contains the bond pad, diodes D_D and

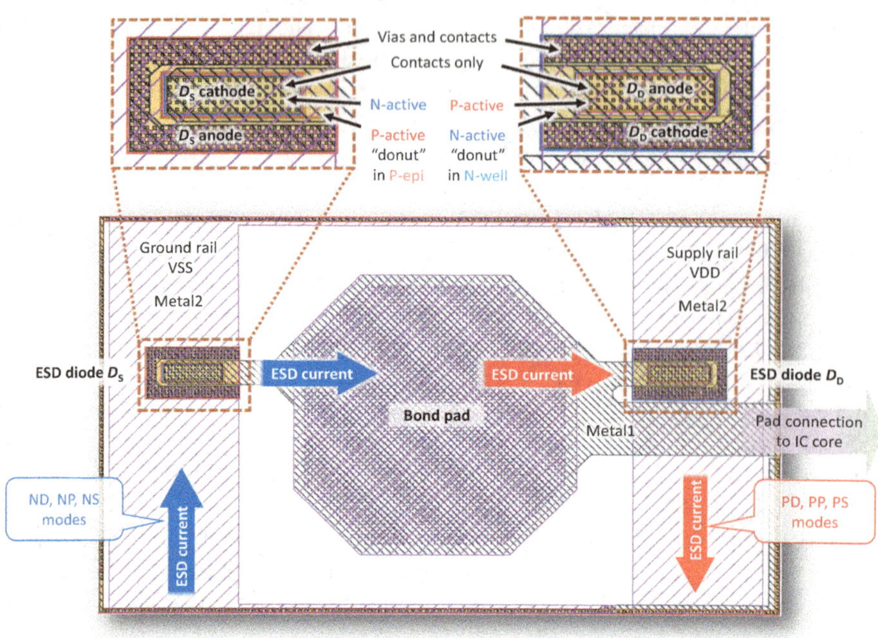

Fig. 7.17 A pad cell layout consisting of bond pad, power rails, and two ESD diodes. The latter short-circuit the bond pad to the vertical power rails in an ESD event, thus protecting the pad connection to the IC core from any ESD current

D_S as ESD protection devices, and power rails VDD and VSS for the chip power supply. We only show metal layers up to Metal2 so as not to complicate the example. The diodes are placed under the power rails to save space and are connected there by means of vias (D_S anode, D_D cathode).

The ESD diodes D_D and D_S appear very "small" in Fig. 7.17. These diodes are forward biased in an ESD event, which means that only the small forward voltage V_{on} (approximately 0.7 V) drops and the power loss remains small. Hence, the small surface area is sufficient. Since the power loss is mostly converted to heat in the supply clamp, the respective diode D_{DS} (see Fig. 7.16) should be given a suitably large surface area.

Two-Stage ESD Protection for Sensitive Inputs

Input pins are often directly connected to MOS-FET gates. Gate oxides can withstand a field strength of approximately 1 V/nm, according to [22]. They are vulnerable at a few volts (in advanced nm-CMOS processes <20 nm) as they are only a few nm thick. Unfortunately, the ESD measures introduced so far are not good enough to protect gate oxides, given that gate oxides are by far the most sensitive features on a chip.

The CMOS circuit operating voltages are often only very slightly below the limit noted above. Is there such a thing as a useful ESD window at all in the light of this? Fortunately, the gate oxide breakdown voltage V_{or} ("or" stands for oxide rupture) depends primarily on how long the field is present. This breakdown voltage V_{or} is approximately a factor of 3 greater than the supply voltage VDD for a typical ESD event duration of 100 ns [6]. For example, an IC core in a 100 nm process operates typically at VDD $=$ 1.1 V. The V_{or} for ESD events is approximately 4.5 V here. Consequently, the available ESD design window is still approximately 3 V wide [6].

A two-stage ESD protection circuit, consisting of two ESD devices for voltage clamping and one resistor, is used to guard the gate oxides. This circuit with two Zener diodes as ESD protection devices is depicted in Fig. 7.18 (left). The forward-biased primary ESD device D_1 can assume the function of diode D_S from Fig. 7.16 in a negative ESD event (Nx modes). In the case of a positive ESD strike (Px modes), it clamps the input to an "intermediate value" V_1. This voltage V_1 can then be clamped by the secondary ESD device D_2 to a sufficiently small value V_2 in the ESD window for the gate oxides.

The resistor R has two functions. First, it sufficiently limits the current I_2 to approximately two orders of magnitude less than I_1. The Zener diode D_2 can therefore be designed much smaller than D_1. Second, the value of R must be at least one order of magnitude greater than the D_2 track resistance, given that the resistor must take most of the voltage V_1.

The voltage V_1 can rise well above 100 V for a high ESD current in this arrangement (>10 A in the charged-device model, CDM), given that "large" Zener diodes can have very high track resistances (>10 Ω). The resistor R must have very high values (>10 kΩ) in this case to more effectively protect gate oxides in advanced

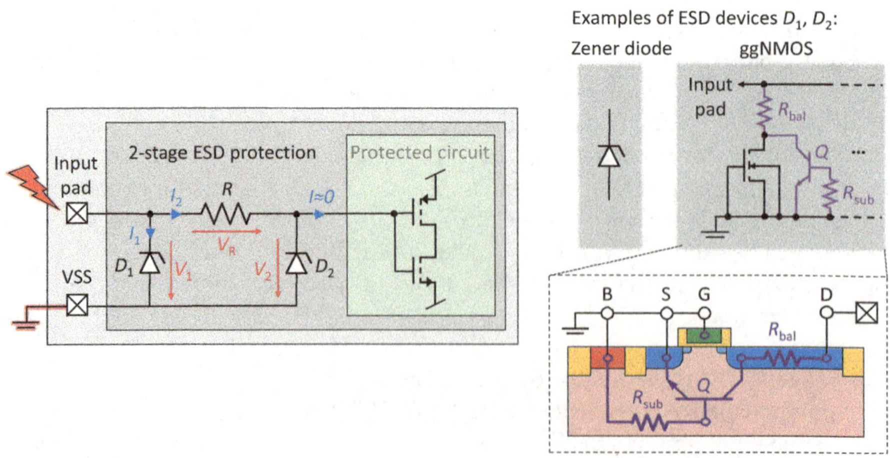

Fig. 7.18 Two-stage ESD protection circuit for sensitive inputs (left) to guard vulnerable gate oxides. Two potential ESD protection devices for this concept, a Zener diode and a grounded gate NMOS-FET (ggNMOS, operating as diode with snap-back characteristic), are shown on the right

nanometer (nm) processes. Given that R is in the signal path, this value is often too high for standard mode, as it results in too large an increase in the gate voltage slew rate and thus the response times.

ESD protection devices with a *snap-back* characteristic are therefore also deployed instead of Zener diodes in state-of-the-art processes (see dashed I–V curve in Fig. 7.15). The voltage drops from the so-called *trigger voltage* V_{t1} as the current rises until it reaches the *holding voltage* V_h. The curve rises again from there with a positive but small R_{on}. The clamp is improved by this characteristic: the voltages are lower and thus there is less thermal power dissipation.

An NMOS-FET is a widely used ESD protection device with this characteristic. This ESD device operates as a diode by connecting its source (S), backgate (B) and gate (G) to ground and the drain (D) to the input pad. This configuration, depicted in Fig. 7.18 (right), is known as a *grounded gate NMOS-FET* (*ggNMOS*). Key parasitic elements that are useful to explain the operation of the element are also shown (violet). The source and drain (heavily n-doped NSD) of the ggNMOS together with the p-substrate (or p-epi) form an n–p–n sequence that manifests itself as a parasitic NPN transistor (labeled Q in Fig. 7.18, right).

Let us now briefly discuss how this ggNMOS operates (Fig. 7.18, right). The p–n junction between drain and backgate (forming the so-called "DB diode") is reverse biased until avalanche breakdown. The breakdown current (holes) flows in the backgate to the backgate pin. It flows past the source and causes an IR drop in the parasitic (lateral) backgate resistor R_{sub}. This IR drop forward biases the source-backgate diode ("SB diode"). Since the SB diode is also the base-emitter diode for the parasitic NPN transistor Q, this transistor Q switches on and thus the snap back is triggered.

It is important that the current can spread across the entire surface area of the protection device before local hotspots cause thermal instability (so-called *second breakdown*, see e.g. [22]), leading to severe damage of the device. This is facilitated by additional ballasting resistors R_{bal} at the drain pins of every finger. These resistors are easily created by placing the drain contacts at a distance to the gate. This forces the current to travel some distance in the drain. The voltage at the DB diode of a finger is restricted by the ballasting resistor R_{bal}, which enables other fingers to take up current as well.

An example of this type of pad cell arrangement in the "0.35 μm modular CMOS" process from the X-FAB Foundry [27] is pictured in Fig. 7.19. Metal layers only go up to Metal2 in this layout view to aid clarity. The difference in size of the two ESD protection devices D_1 and D_2 is worth noting. The series resistor R is produced in poly here. Given that this resistor must carry many 10 milliamps in an ESD event, it must be designed wide enough and connected with a sufficient number of vias to assure ESD robustness. This resistor should not be bent either to avoid high field strengths at corners. The outer edges of Metal1 interconnects are chamfered for the same reason.

Given that strong substrate currents could be produced during ESD events, there is a risk that the voltage at the gate oxide could rise despite this protection because of substrate debiasing (Sect. 7.1.1). Hence, if the protected device is not situated in the

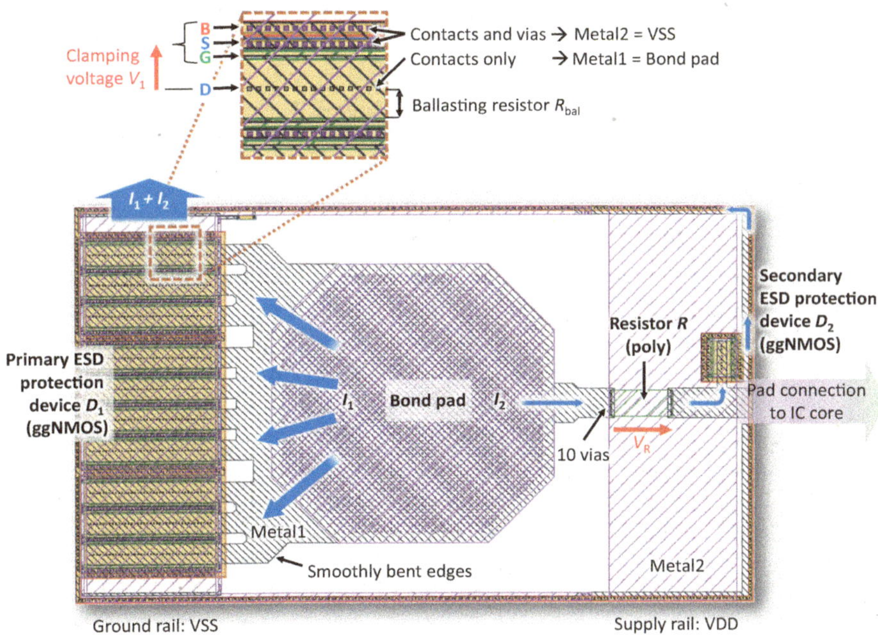

Fig. 7.19 Layout of a pad cell for a two-stage ESD protection circuit using a grounded gate NMOS-FET (ggNMOS) configuration as per Fig. 7.18

proximity of the bond pads, it may be advisable to place diode D_2 near the device.[9] In this case its connecting conductors need to be sized for I_2.

There are many more ESD devices available besides the ones described above. We kindly refer the reader who is interested in researching ESD devices further to two resources on this topic [7, 24]. Because of their size, ESD protection devices have significant parasitic capacitances, which impact the circuit and are also susceptible to latchup [21]. They are normally repurposed and characterized for every process. Hence, one should only use these known ESD protection devices and carefully select them for the specific application at hand, which is a decision for the circuit designer. The placing and routing of these elements is of no less importance, though. This task is in the remit of the layout designer. We shall outline some guidelines in this regard below.

Placing and Routing ESD Protection Circuits
The PDKs usually contain constraints for the interconnects, which carry the ESD current from the input pin to the output pin on a chip. Should no such constraints be available, we recommend the following guidelines:

- The total resistance should not exceed approximately 2 Ω. This must be fulfilled between two arbitrary points of the chip, for example between two opposing chip corners.
- The interconnect width should be designed for at least 100 mA permanent current. Computation rules are available in this regard. The ESD current is high, but of very short duration.
- The outer corners should be chamfered to minimize local field apexes.
- The lateral spacing to adjacent metals should be at least as large as the oxide layer thickness between the metal layers. We recommend 2 μm minimum.

Wide interconnects are needed from the pads to the chip's circuit blocks to power the chip. Since the ESD current also flows predominantly through these supply paths, it is advisable to run them at the chip periphery. This makes them directly accessible from all bond pads, which are always at the periphery. The connections to the pad and to the supply lines are typically very short, as most ESD protection devices are electrically assigned to bond pads. The contacts can thus be easily made ESD robust (i.e., wide), while keeping their contribution to the total resistance negligible.

The resistance of the connecting conductors to the ESD protection devices is therefore typically exclusively determined by the supply lines. These are the black subnets in Fig. 7.16. ESD protection devices that are shared by many bond pads are exceptions. This scenario is not covered in our examples; it can occur depending on the circuit, and is normally used as a space-saver. Note that these cases require special care during IC routing. Numerous supply clamps can be distributed throughout the chip, as well.

[9] Given that the ESD protection circuit is part of a library pad cell, this modification must be done in a local copy of the pad cell to prevent other instances of this cell from unintended changes.

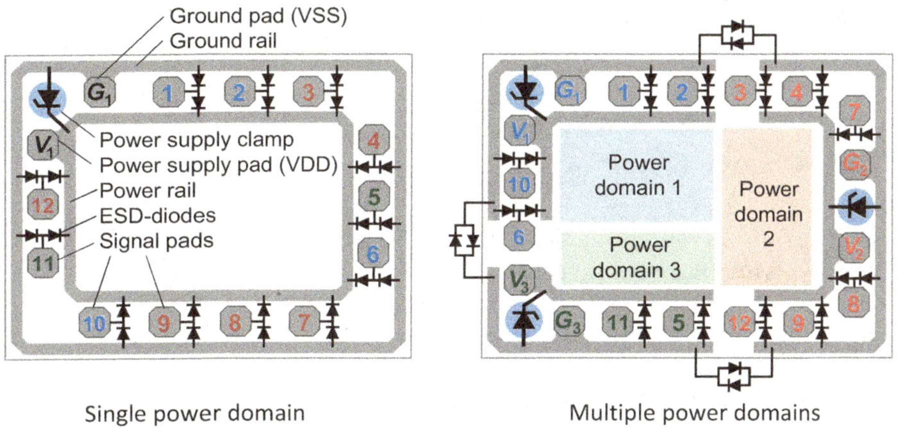

Fig. 7.20 Arrangement of power rail and ESD protection devices on a chip with one power domain (left) and a number of power domains (right)

A popular methodology is to place the ground line outside the bond pads and the supply line inside, both as closed rings. The pad cells shown in Figs. 7.17 and 7.19 are based on this principle. This routing concept is shown schematically in Fig. 7.20 (left). The seal ring (Chap. 3, Sect. 3.3.2, Fig. 3.15) can be used as a ground line in this concept too (if allowed by the manufacturer). This ring typically contains all metal layers and therefore enables a very low-resistance contact.

Modern mixed signal and smart power IC chips often contain different operating voltages and thus many separate ESD paths. Sometimes separate ground lines are laid to decouple circuit blocks in order to prevent crosstalk. This is not only a logical isolation, as described in Chap. 6 (Sect. 6.2.1), but an electrical isolation as well. It is a good idea to couple these ground lines by two antiparallel diodes (sometimes many stacked diodes are the method of choice) to enable the ESD current to use all ground lines. The floorplan for this type of constellation is depicted in Fig. 7.20 (right). (Caution: This solution is not possible if a seal ring is used.) Other concepts for the design of multi-power domains are to be found in [23].

Our final example in Fig. 7.20 illustrates the importance of setting out in detail the power supply concept and the associated ESD concept at the beginning of the layout phase. Why? Let's assume, for example, that the 12 signal pads on the chip in Fig. 7.20 should be assigned to three different power domains as indicated by the red, blue and green colors on the right. If the pins were assigned as per Fig. 7.20 (left), the supply lines would have to be routed almost three times around the chip. However, a pin assignment as per Fig. 7.20 (right) would enable a much better solution with locally separated supply lines. This type of decision needs to be taken at the beginning of the physical design phase, as it impacts the entire floorplan and thus the design of all blocks.

This leads us to our final piece of advice in this section: Do not only consider signal flows during IC pin assignment (Chap. 5, Sect. 5.3.1), but always think about the supply lines for the blocks and the ESD concept as well. If this is not done, considerably more effort will be needed for the layout design; more chip area will be required; and the layout quality will suffer, as well.

7.4.2 Antenna Effect

A so-called *antenna* is a conductor, such as a polysilicon or metal line, that belongs to a net which is only partially complete during the manufacturing process. As the layers above it have not yet been processed at specific wafer processing steps, the conductor is neither electrically connected to silicon nor grounded. Charge can accumulate on these (temporary dead-end) connections during the manufacturing process to the point at which leakage currents are generated and permanent physical damage can be caused to thin transistor gate oxide that lead to immediate or delayed failure (Fig. 7.21).

Gate dielectrics (GOX, gate oxide) can be damaged by impulse voltages in process steps in which the wafer comes in contact with charged particles. Dry etching by RIE (reactive ion etching, Chap. 2, Sect. 2.5.3) is a particular case in point. The etching medium applied is a plasma that typically contains different charges. These charges are partially absorbed by the wafer surface and can flow through conductive layers (metal, poly). They create a potential difference to the wafer substrate that can damage and even destroy the vulnerable GOX.

Poly layers are directly structured through RIE, which results in a strong antenna effect (see Fig. 7.21, left). In contrast, the antenna effect of metals (see Fig. 7.21, right) in the typical Damascene process (Chap. 2, Sect. 2.8.3) is reduced nowadays, as the RIE structuring takes place in the isolating oxide layer. Hence, the charge is only absorbed through the surfaces that form the vias to the next higher metal. Therefore, we focus our following considerations on the structuring of the Poly layer.

The structuring of poly with RIE in four phases (1)–(4) is outlined in Fig. 7.22. The process is marked by large fluctuations in the charge distribution in the plasma. This leads to charge differentials on the wafer. These charge differentials balance each other out as long as the poly has not yet been separated into sections. In other words, as long as the charges in the poly can flow freely across the entire wafer, no critical voltages occur (phase 1).

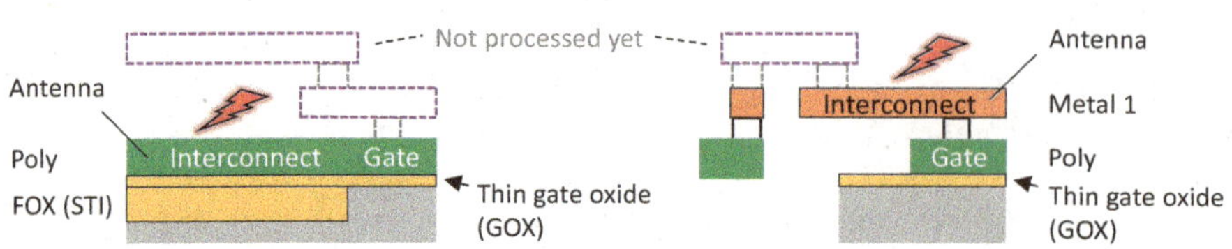

Fig. 7.21 Illustration of the antenna effect where charges, accumulated on temporary dead-end connections during the manufacturing process, damage the thin transistor gate oxide (side view). Poly lines (left) and metal lines (right) can act as antennas

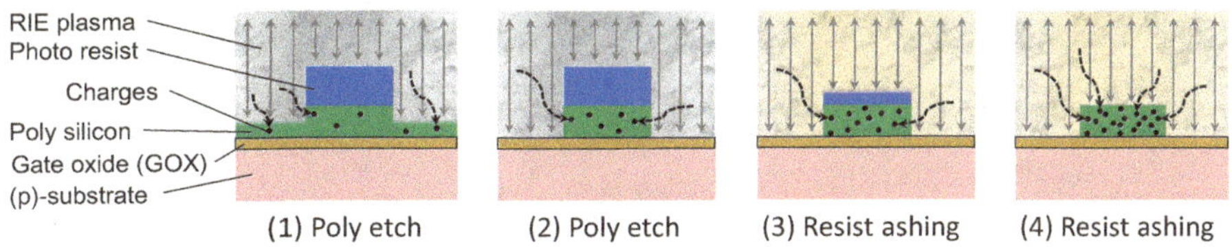

Fig. 7.22 Static charging of poly structures during RIE (reactive ion etching) in phases (1)–(4). Despite phases (2) and (4) showing the poly/resist removed, etching/ashing continues for some time until the etching/ashing depth is reached everywhere on the wafer. Separated poly structures collect charges by their side walls in phases (2) and (3) and by their entire surface in phase (4)

After the Poly layer has been etched through (phase 2), the charges are restricted to move within the created structures. These structures absorb further charges at their lateral edges in this critical phase. This phase lasts a set period, as the required etching time is selected such that the required etching depth is reached everywhere on the wafer. The charges therefore accumulate on poly elements in plasma areas of high charge density.

Given that the photoresist is hardened during the poly etching process, another RIE step is needed to remove the resist (phase 3); in this step the resist is destroyed in an oxygen environment by *ashing*. More charges, which accumulate on the poly, appear in phase (4) especially, since the photoresist has been removed at this point.

Overall, the poly structures (mostly paths) thus act as charge-receiving "antennas" in phases (2)–(4). This is where the effect got its somewhat misleading name. The effect is also referred to as *plasma-induced damage* (*PID*) in the technical literature.

If the poly is part of a gate, we have the situation depicted in Fig. 7.23 (left) with two parasitic capacitors (shown in violet) with the corresponding capacitances C_{GB} and C_{PS} between gate and bulk (backgate) and between poly and substrate, respectively. After all potential differences have been balanced, the same voltage V is present at both parasitic capacitors (Fig. 7.23, middle).

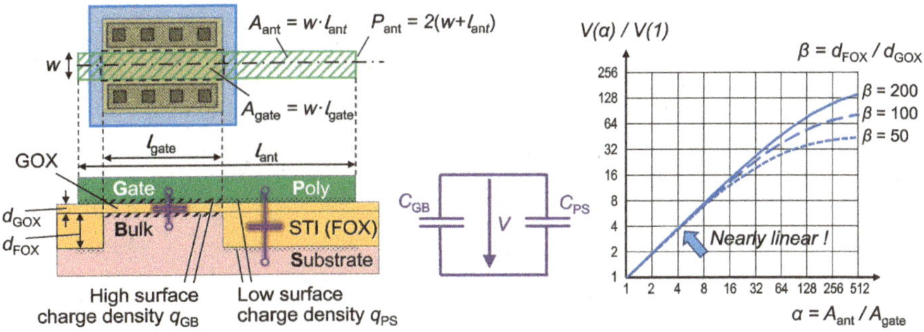

Fig. 7.23 Charge and voltage ratios as a function of structural ratios in the antenna effect (GOX: gate oxide, FOX: field oxide)

The following expressions can be easily derived from the known relation $Q = C{\cdot}V$ as well as the known Eq. (7.3) for the capacitance of a surface capacitor for the quantity of charge per unit area q_{GB} and q_{PS} on the capacitors:

$$q_{\mathrm{GB}} = \beta \cdot q_{\mathrm{PS}} \text{ with } \beta = \frac{d_{\mathrm{FOX}}}{d_{\mathrm{GOX}}}. \tag{7.4}$$

Assuming shallow trench isolation (STI) in our examples, the layer thickness d_{FOX} of the field oxide (FOX) typically is many 100 nm, while the layer thickness d_{GOX} of the gate oxide (GOX) is only a few nanometer. This mismatch explains the critical accumulation of charge Q on the gate. It is the resulting voltage V, however, that gives rise to the risk of damage to the gate oxide. This voltage V can be expressed with the above formulas as a function of the antenna and gate areas A_{ant}, A_{gate} and the layer thicknesses of the dielectrics d_{FOX}, d_{GOX}:

$$V = \frac{e_{\mathrm{ant}}}{\varepsilon_0 \varepsilon_{\mathrm{r}}} \cdot \frac{\alpha}{\alpha + \beta - 1} \cdot d_{\mathrm{FOX}} \text{ with } \alpha = \frac{A_{\mathrm{ant}}}{A_{\mathrm{gate}}} \text{ and } e_{\mathrm{ant}} = \frac{Q}{A_{\mathrm{ant}}}. \tag{7.5}$$

The assumption is made here that the charge Q absorbed by the antenna grows proportionally to its surface area A_{ant}. We call this proportionality factor the *antenna efficiency* e_{ant}. The term $\varepsilon_0 \varepsilon_r$ is the permittivity of the GOX and FOX dielectrics, which are assumed equal in this context.

The curves in Fig. 7.23 (right) show that the voltage and thus the risk of damage to the GOX scale almost linearly with the ratio α of the antenna area A_{ant} to the gate area A_{gate}. This is what makes the antenna effect so dangerous! It is only when the surface ratio α (antenna area vs gate area) approaches the thickness ratio β (FOX thickness vs GOX thickness) that the rise decreases.

We finally note that the assumption that Q scales with A_{ant} is a simplification. It is, in fact, more the length of the antenna's outer edge, i.e., its perimeter P_{ant} (shown in Fig. 7.23, upper left), that determines the amount of charge absorbed in phases (2) and (3) of Fig. 7.22. Its entire surface area then acts in phase (4). The aforesaid applies also to the ion implantation of the source and drain zones (Chap. 2, Sect. 2.6.3, not shown in Fig. 7.22) following the reactive ion etching (RIE). Consequently, the process rule sets contain design rules that prescribe maximum values for the ratios $\alpha = A_{\mathrm{ant}}/A_{\mathrm{gate}}$ and $P_{\mathrm{ant}}/A_{\mathrm{gate}}$ (or only one of them) to limit the voltages at the GOX. In general, the boundary values for the antenna rules depend greatly on the process steps in the back-end-of-line (BEOL) layers.

Counter Measures in Physical Design: Jumpers and Leakers

Based on the discussed causes of the antenna effect and its electrical and physical parameters, we are now able to derive appropriate counter measures in physical design.

As a matter of principle, gates should be connected as closely as possible by metallic interconnects to minimize the parasitic contact resistance. This will also prevent any dangerous antennas in poly. If this is impossible, poly conductors should

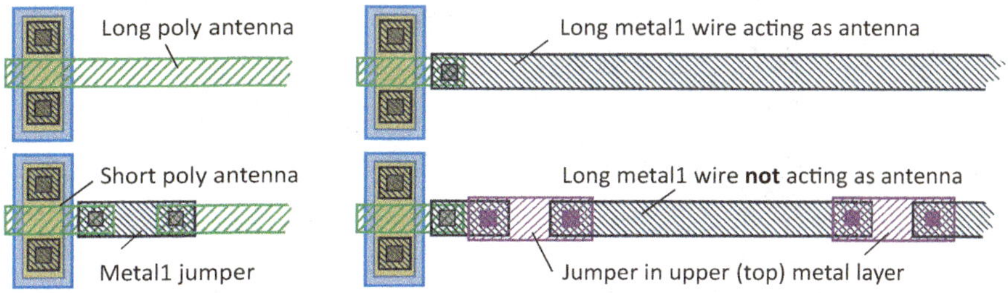

Fig. 7.24 Using jumpers to protect gate oxides endangered by the antenna effect. Protection against poly antennas with jumpers in Metall1 (left) and against metal antennas with jumpers in higher metal layers (right)

at least be isolated by metal bridges, so-called *jumpers*, in the proximity of the gate pins. This interrupts the antenna as the metals do not yet exist when the poly is being structured. An example of this scenario is given in Fig. 7.24 (left).

It should be noted that the antenna effect described earlier for poly also occurs with all metallic interconnects connected to gates. The contact can be made by other interconnects in lower layers (only lower layers exist at the specific fabrication step). As with Poly layers, we can work around the problem by interrupting these interconnects at both ends with metal jumpers in a higher metal layer. The (long) piece of interconnect is thus isolated during etching, with only a short antenna in the higher metal layer (Fig. 7.24, lower right).

There is another option for mitigating the antenna effect, which resembles the methodology adopted for countering the ESD effect. In this option, the charges are discharged as soon as they are created. This allows the charges to be simply discharged into the p-substrate, as the currents in question are very small. Additional p–n junctions are introduced and their diode characteristics are used for this purpose. These antenna diodes should not impair standard operation. They are also called *leakers* as they always cause small leakage currents. Two versions of this second protection option are depicted in Fig. 7.25.

The NSD/p-sub diode (a) is reverse biased (acting as insulator) in standard mode. It discharges the charge in forward-biased mode if the antenna is negatively charged (the blue current). The same thing happens if the antenna is connected to another NMOS transistor's source or drain. No diode is needed in this case (a), as the current is discharged by this source or drain in the substrate.

This diode's breakdown voltage V_{BD} (a) can be used in older processes if the antenna is positively charged. Its V_{BD} is much too high, however, for the thin GOX in state-of-the-art processes.

Consequently, the author of [7] recommends that this diode should be extended by the structure as shown in Fig. 7.25b. The Nwell/p-sub diode is reverse biased in standard mode. The diode PSD/Nwell becomes forward biased when the antenna is positively charged; the Nwell/p-sub diode is reverse biased in this case. Additional charge carriers are excited in the diode by the photoelectric effect as the plasma is excited in the RIE process such that it glows. The leakage current is sufficiently

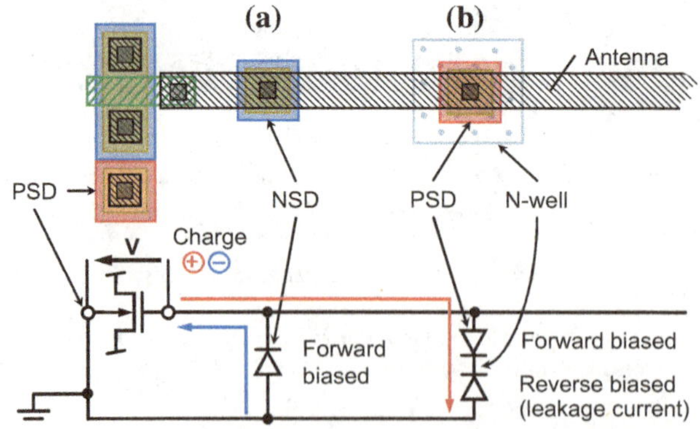

Fig. 7.25 Protection against the antenna effect with leaker diodes for negative (**a**) and positive (**b**) antenna charges

increased by these extra charge carriers that the antenna is discharged (current shown in red). The main thing to remember here is that a sufficiently large portion of the Nwell/p-sub diode must not be covered by metals, so that it receives adequate exposure to light.

The second method with leakers works with metal layers only. It is generally easier to use and is therefore preferred by many layout designers. However, the leakers must also be entered in the circuit schematic, otherwise the LVS check (Chap. 5, Sect. 5.4.6) will not produce an error-free result. This means that their behavior will also be considered in the simulation.

Some designers integrate the leakers in the layout cells of the digital libraries. Unfortunately, such cell extensions typically require additional cell area. The method with jumpers is, however, becoming increasingly popular again in advanced processes. The main reason for this is that the leakage currents of leakers increase the overall power consumption of the IC, whereas the method with jumpers does not.

Numerous algorithms have been proposed to minimize the manual layout effort to counter antenna effects (e.g. [18]), some of which have found their way into commercial routing tools.

7.5 Migration Effects in Metal

Besides overvoltage events, there is another effect that can cause irreversible damage in ICs: material transport, also known as *migration*. We differentiate between three types of material transport in metallic connectivity architectures that can significantly impact circuit reliability: electromigration, thermal migration, and stress migration.

We first describe each migration type (Sects. 7.5.1–7.5.3) before presenting mitigation options in physical design (Sects. 7.5.4, 7.5.5). Although we cover these types of migration separately (electromigration, thermal migration, and stress migration), it is important to point out that they are in fact closely coupled processes, as their driving forces are linked with each other and with the resultant change in migration. We refer the reader to the literature, such as Sect. 2.5 in [19], for further elaboration on these mutual interactions.

7.5.1 Electromigration

Current flow through a conductor produces two forces, which act on the individual metal ions in the conductor. The first of these forces is an electrostatic force F_{field} caused by the electric field strength in the metallic interconnect. Since the positive metal ions are shielded to some extent by the negative electrons in the conductor, this force can safely be ignored in most cases. The second force F_{wind} is generated by the momentum transfer between conduction electrons and metal ions in the crystal lattice. This force acts in the direction of the current flow and is the primary cause of *electromigration* (*EM*, Fig. 7.26).

If the resulting force in the direction of the electron wind (which also corresponds to the energy transmitted to the ions) exceeds a given trigger known as the activation energy E_a, a directed migration process starts. The resulting material transport takes place in the direction of the electron motion, that is, from the cathode (−) to the anode (+).

The actual migration paths are material-dependent and are primarily determined by the size of their respective activation energies. Every material has multiple, different activation energies for migration, namely for migration (i) within the crystal, (ii) along grain boundaries, and (iii) on surfaces. The relationships between the individual energy levels determine which of the migration mechanisms (i)–(iii) dominates, as well as the composition of the entire migration flux [19].

The interconnects of a fabricated IC chip contain numerous required features that result in inhomogeneities; as a result, the migration is also inhomogeneous. Subsequently, divergences occur in the migration flow, leading to tensile and compressive

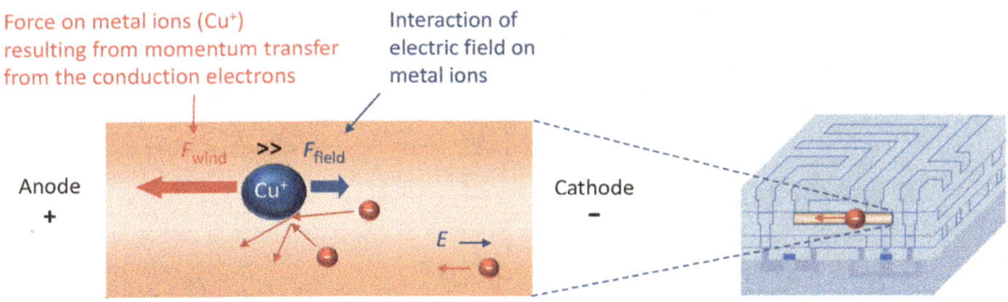

Fig. 7.26 Two forces act on metal ions (Cu) that make up the lattice of the interconnect material [19]. Electromigration is the result of the dominant force, that is, the momentum transfer from the electrons that move in the applied electric field

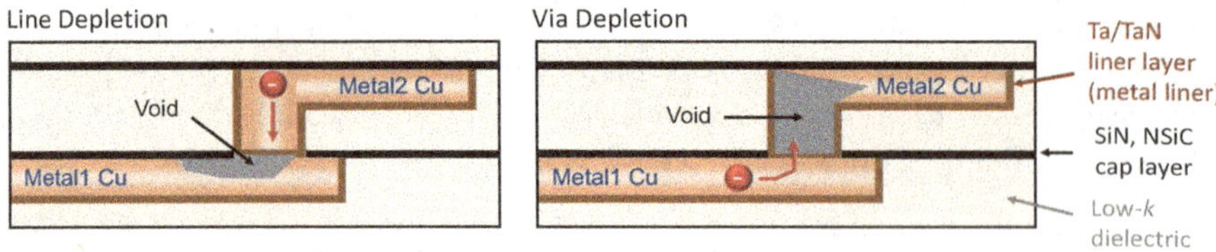

Fig. 7.27 Line depletion (left) and via depletion (right) are common failure mechanisms due to EM in integrated circuits

stresses in the vicinity of such inhomogeneities. Tensile stress can result in metal depletions, so-called *voids*, while compressive stress can lead to metal accumulations, so-called *hillocks*. Another sign of EM in wires is *whiskering*, which is a crystalline metallurgical phenomenon involving the spontaneous growth of tiny, filiform hairs from a metallic surface.

EM failures in modern chip manufacturing processes are mostly due to voids [5]. They result from the building up of tensile stress with two stages of EM degradation: In the *void nucleation phase*, the tensile stress increases over time but no void has nucleated. When the stress reaches a critical threshold, the void nucleates and the void *growth phase* begins. This can be observed by an increase in line resistance. Hence, the conductance of the line decreases but will not reach zero (in most cases) due to the remaining conduction in the surrounding metal liner [5].

Void damages in ICs are differentiated in *line depletion* and *via depletion* (Fig. 7.27). Electron flow from a via to an attached interconnect line (below or above the via) can cause line depletion due to obstructed material flow through the cap and liner layers. Reversing the electron flow, i.e., electron flow from a line to a via, may result in via depletion, sometimes also called *via voiding*. Here too, its causes are a combination of geometry and process. As with line depletion, the material migration is hindered by the surrounding cap and liner layers. In addition, as the ratio of the line width to the via width increases, the via must carry more current for the same wire-current density, making the via more susceptible to the voiding process.

Wires with unidirectional currents, such as power and analog lines, are most susceptible to EM failures. In contrast, digital signal and clock lines carry bidirectional currents and benefit from a so-called "self-healing" process due to bidirectional, and thus reversed and compensatory material migration.

7.5.2 Thermal Migration

Thermal migration (*TM*), sometimes also referred to as *thermomigration*, is produced by temperature gradients. Here, high temperatures cause an increase in the average speeds of atomic movements. Atoms in regions of higher temperature have a greater probability of dislocation than in colder regions due to their temperature-related activation. This causes a larger number of atoms diffusing from regions of higher

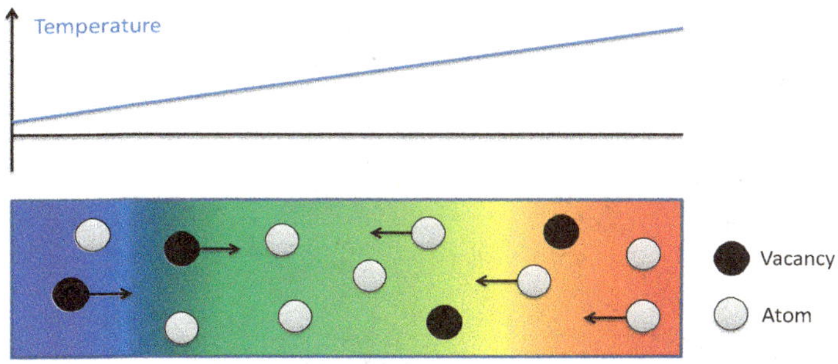

Fig. 7.28 Thermal migration (TM) is expressed by atomic and vacancy movement. It consists of mass transport from one local region to another, much like EM, with the difference that TM is driven by a thermal gradient rather than an electrical potential gradient [19]

temperature to regions of lower temperature than atoms in the opposite direction. The result is net diffusion (mass transport) in the direction of the negative temperature gradients (Fig. 7.28).

The main reasons for temperature gradients in metal wires are

- Joule heating inside the wire caused by high currents,
- External heating of the wire, e.g., by highly performant transistors nearby, and
- External cooling of the wire, which may result from through-silicon vias (TSV) connected to a heat sink, in connection with low thermal conduction of the wire and its surrounding, such as through narrow wires embedded in a thermally insulating dielectric [19].

Interestingly, thermal migration also contributes to thermal transport, as heat is coupled to the transported atoms. This means that thermal migration directly moderates its own driving force, which contrasts with EM, where current density is only indirectly reduced by increased resistance in some cases.

7.5.3 Stress Migration

Stress migration (*SM*), sometimes also referred to as *stress voiding* or *stress induced voiding* (*SIV*), is atomic diffusion that causes a balancing of mechanical stress. There is a net atomic flow into regions where tensile forces are acting, whereas metal atoms flow out of regions under compressive stress. Similar to thermal migration, this leads to diffusion in the direction of the negative mechanical tension gradient (Fig. 7.29). As a result, the vacancy concentration is balanced to match the mechanical tension.

The main reasons for mechanical stress as the driving force behind SM in metal wires are thermal expansion, electromigration, and deformation through packaging. A mismatch of the coefficients of thermal expansion (CTE) between metal, dielectric,

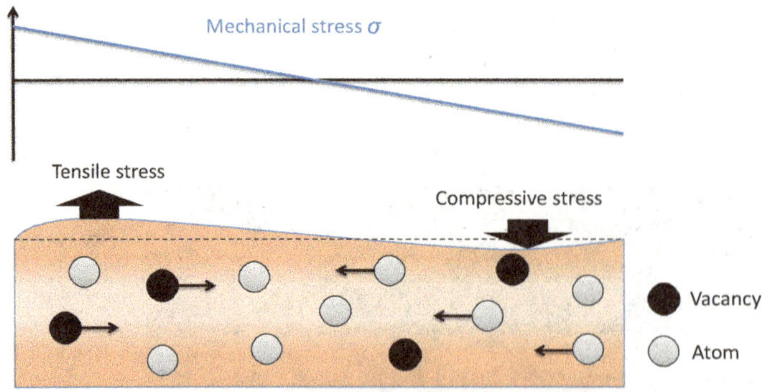

Fig. 7.29 Stress migration is the result of a mechanical stress gradient, either from external forces or from internal processes, such as electromigration or thermal expansion [19]. Voids form as a consequence of vacancy migration driven by the hydrostatic stress gradient

and die material, and the temperature change from fabrication to storage, as well as the working conditions, cause most of the stress.

Metal lattices usually contain vacancies, i.e., some of the atomic positions in the lattice are unoccupied. Although they are aligned with the lattice grid, vacancies consume less space than atoms at the same positions. Therefore, the volume of a crystal that contains vacancies is to some extent smaller than the volume of the same crystal with atoms in the place of former vacancies. Hence, vacancies play a major role in stress migration.

The stress gradient drives atoms from high pressure regions to regions with tensile stress and pushes vacancies in the opposite direction. This effect is equivalent to a highly viscous fluid that reacts slowly to an external pressure gradient. The external stress gradient is minimized in this case by structural deformation. Initially, microscopic atomic or vacancy motion facilitate this process. Temperature has a critical effect on the process, as it enables the "place-changing" of atoms, which, in turn, causes vacancies to move.

In the case of *external* mechanical stress, the crystal lattice is stretched or compressed depending on the type of stress. While there is an increased likelihood of atoms migrating to the stretched regions, atoms in the compressed regions are "pushed" outwards to increase the number of vacancies; the required volume and the stress are thus reduced (Fig. 7.30). The result is an atomic flux from regions of compressive stress to regions of tensile stress, until a static state with no stress gradient is reached.

If the stress is exerted *internally* by migration processes, e.g., by EM, there will be a greater concentration of vacancies in regions of tensile stress. This concentration will be balanced by stress migration to a steady state, where the atomic flux due to EM is compensated by SM.

If the number of vacancies induced by external or internal (EM) stress exceeds a threshold, the vacancies unite to form a void due to vacancy supersaturation

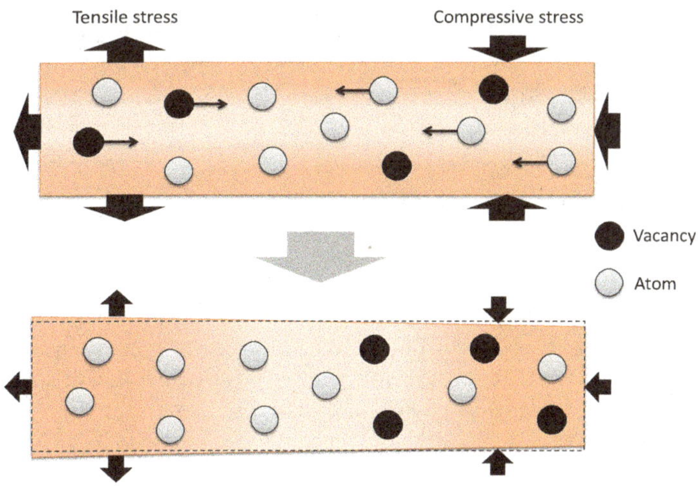

Fig. 7.30 Stress migration leads to diffusion of atoms and vacancies (top) to eliminate the origin of this migration (bottom) [19]. Atoms migrate into the stretched regions (left-hand side, outward facing stress arrows), whereas atoms in the compressed regions are "pushed" out of these regions (right-hand side, inward facing arrows). Note that this material flow from compressive to tensile stress is in the opposite direction to the EM flow

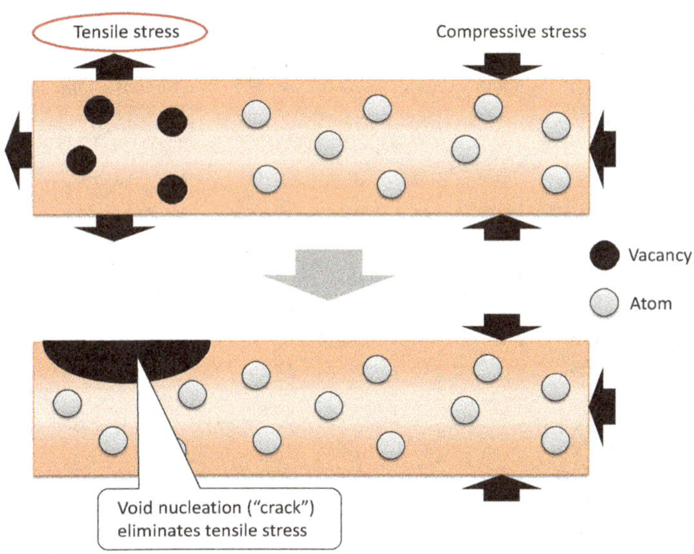

Fig. 7.31 Vacancy supersaturation (top) leads to the formation of voids (bottom), also called void nucleation [19]. Note that the resulting crack eliminates the (external) tensile stress

(Fig. 7.31). This phenomenon is often called *void nucleation*, the results of which are illustrated in Fig. 7.31 (bottom). Subsequently, the tensile stress is reduced to zero by the resulting crack [8].

7.5.4 Mitigating Electromigration

As introduced in Sect. 7.5.1, electromigration (EM) is caused by the momentum transfer between conduction electrons and metal ions in the crystal lattice. This process is often caused by excessive current densities, which makes current density J a fundamental parameter of interest in addressing EM. Current density is calculated from the quotient of the flowing current I and the cross-sectional area of the interconnect A as follows:

$$J = \frac{I}{A}. \tag{7.6}$$

How currents will grow in the future (which will increase I in the numerator of Eq. (7.6), thereby increasing J), and interconnect track parameters, such as the cross-sectional area (which will decrease A in the denominator of Eq. (7.6), thereby also increasing J), are clearly critical in the context of electromigration and its mitigation measures [16].

Besides using current density J as the decisive parameter for describing the EM risk in IC interconnect (Black's model [2]), EM-induced mechanical stress is increasingly applied for this purpose (Korhonen model [12]).[10] Here, the hydrostatic stress σ arising under the influence of EM is used to characterize EM risks and guide mitigation measures. For example, a void is said to nucleate once the hydrostatic stress exceeds a predefined threshold value $\sigma_{\text{threshold}}$.

Electromigration cannot be prevented from occurring in metallic routing wires; it can only be offset or restricted in its effect. This is done by (i) reducing the wire's material transport or by (ii) raising the tolerated boundary values for this transport. The latter, in turn, means that we increase the permissible current density, i.e., raising its allowed limits. Finally (iii), we can also reduce our layout configuration's required current density.

Based on these insights, we summarize next the most important remedial measures applicable in physical design. While our first measure (length limitations) counteracts the material transport (i), reservoirs and via configurations seek to raise the permissible current density (ii). The last three measures (via arrays, corner bends, and net topology) reduce the peak values of the required current density (iii).

[10] While Black's model calculates the reliability due to EM of a single wire segment, the Korhonen model and its subsequent extensions, e.g. by Chatterjee et al. [5], track the material flow in all branches of a net located within one layer of metallization.

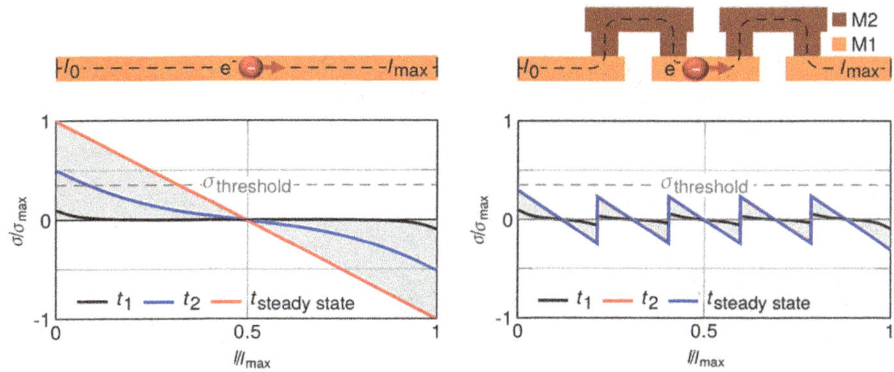

Fig. 7.32 Development of EM-induced mechanical stress over time in a long (left) and short-segmented interconnect (right), which keeps hydrostatic stress levels σ of the short segments below the critical EM threshold [4]

Length limitations. Any interconnect length below a threshold length ("Blech length") will not fail by EM [15]. Here, mechanical stress buildup causes reverse stress migration (SM), which compensates for the EM flow. An effective EM-suppression measure is to divide a long interconnect into several short segments below the Blech length as shown in Fig. 7.32 [4]. All segments then become "EM immortal," thus, preventing the formation of voids. The disadvantage is that much more routing resources are needed due to the additional layer changes. An alternative approach is to balance the individual segment lengths of a net. This saves routing resources because there are no further vias required (vias are only shifted to adjust the lengths of the segments). Specifically, this "balancing method" reduces the hydrostatic stress in the highest risk segment (i.e., longest segment) as this segment is shortened.

Reservoirs. Reservoirs, such as enlarged via overlaps, increase the maximum permissible current density by providing material for migration, thus preventing void growth from damaging the interconnect. The key enabler of the reservoir effect is the shifting of the prevailing equilibrium during void growth saturation [19]. Reservoirs can, however, have an adverse effect on reliability in nets with current-flow reversals, as the stress migration is reduced in this case [16, 19].

Via configurations. The robustness of Cu interconnects fabricated with dual-Damascene technology depends on whether contact is made through vias from "above" (via-above) or "below" (via-below) [19]. Via-below configurations are better from an EM-avoidance perspective than their via-above counterparts, as the higher permissible void volumes associated with via-below configurations allow higher current densities (Fig. 7.33).

Via arrays. Multiple (redundant) vias increase robustness against EM damage [19]. They should be placed "in line" with the current direction so that all possible current paths have the same length. This ensures a uniform current distribution, and, hence, avoids a local detrimental increase in current density between the vias (Fig. 7.34).

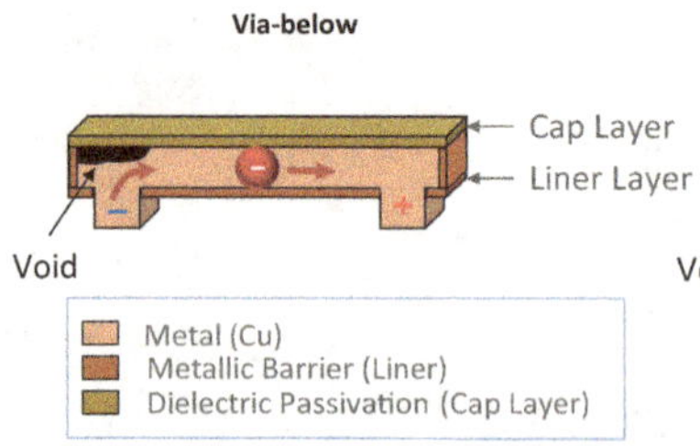

Fig. 7.33 As voids nucleate on the top surface of the Cu interconnects, segments with via-below configurations (left) are more EM robust due to higher permissible void volumes compared to via-above configurations (right)

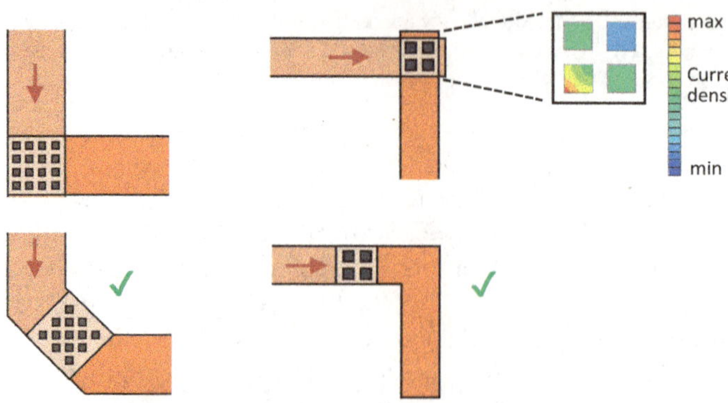

Fig. 7.34 Via arrays connecting two neighboring metal layers. If this array connects perpendicularly arranged interconnects, the "inner" via(s) can be affected by EM through current overload (top). EM robustness can be achieved by placing the via array such that the current does not change the lateral direction (bottom)

Corner bends. Attention needs to be paid as well to bends in interconnects. In particular, 90-degree corner bends should be avoided, since the current density in such bends is significantly higher than that in oblique angles of 135 degrees, for example (Fig. 7.35) [15].

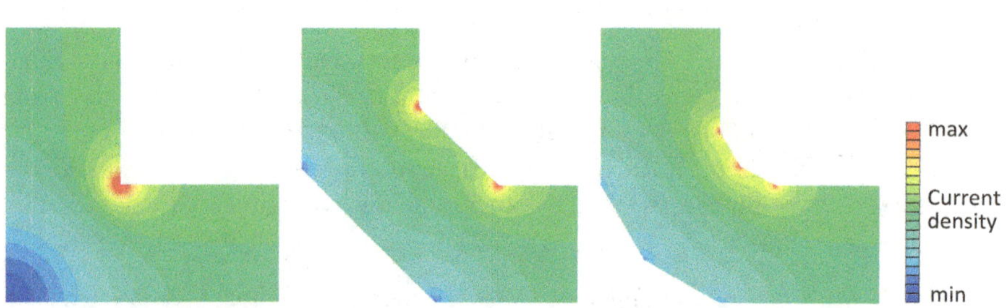

Fig. 7.35 Current-density visualization of different corner bend angles in a routing structure

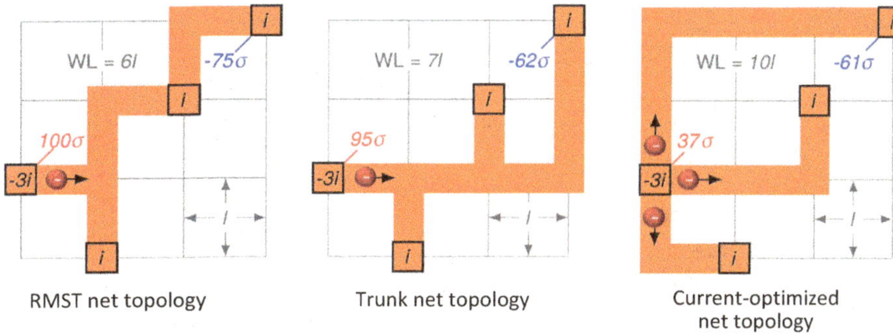

Fig. 7.36 Routing results of a four-pin net containing one output driving three inputs with current i [4]. The RSMT net topology on the left yields the lowest wire length (WL), but highest EM-induced hydrostatic stress (σ). The trunk net topology in the middle generates slightly less stress but increases the WL. The net topology on the right is the most EM robust of the three nets

Current-optimized net topology. Rectilinear Steiner minimum tree (RSMT) and trunk tree are the main net topologies used in today's routers to minimize both the wire length and routing congestion. However, the net routing procedure can also be tailored towards EM robustness by optimizing the resulting net topologies such that currents are split into different paths and so reduce EM-induced mechanical stress. Figure 7.36 contains examples of three different net topologies, each characterized with EM-induced hydrostatic stress (σ) and wire length (WL) [4].

A universal scheme for preventing EM damage can only be developed by (i) implementing the measures mentioned above with appropriate EM-aware design tools—such as current-driven routing [10, 17], wire resizing [20, 26], or stress-aware routing methodologies [3, 4]—and by (ii) evaluating these measures with analysis tools.

Regarding the latter, today's EDA tools offer some EM-analysis functionalities based on simplified models. Accurate results, as needed in analog and power circuits, can only be achieved with sophisticated numerical simulations. The finite element method (FEM) can represent spatially the impact of current density, hydrostatic stress, and other design parameters on the migration processes—allowing the effects and measures associated with EM to be analyzed by simulation. We refer the reader to the literature, such as [9] or Sects. 2.6 and 3.5 in [19], for a detailed discussion of these simulation and verification techniques.

7.5.5 Mitigating Thermal and Stress Migration

Thermal migration (TM) and stress migration (SM) are closely coupled through thermal expansion and therefore often act in the same direction. Hence, mitigating SM and TM are closely related.

The methods in reducing EM (outlined in Sect. 7.5.4) should be considered as a first step. Consequently, as a counteracting process to EM, SM will be reduced as well. TM will also be lowered at the same time if local current densities (and Joule heating) reduce as a result of these EM-mitigating measures. It is known that reducing the current density of alternating currents in signal nets may help prevent thermal migration.

To further mitigate TM and SM directly in physical design, special care needs to be given to the placement of high-power transistors. The overall goal is to avoid local hot spots by distributing transistors with high loads over a large die area.

Additionally, the routing should be organized to avoid high current densities and hot spots in the interconnect through Joule heating: (i) Implement wide wires for high currents, e.g., by using a higher layer in the metal stack, and (ii) avoid adjacent wires carrying high current at the same time.

TM can be accessorily mitigated by lowering thermal gradients. These gradients arise from external heating or cooling and internal heating of circuit elements, in particular transistors. Reducing the power consumption of transistors is one option here.

Experienced layout designers often include additional wires and vias as heat conductors (so-called *thermal wires* and *thermal vias*) regardless of their current-carrying capabilities. These additional wires and vias can significantly improve heat conductivity and reduce thermal hot spots and gradients. Good thermal conductivity throughout the chip is always beneficial, while the low power consumption mentioned above also reduces temperature gradients.

Another option is to route along smaller temperature gradients, e.g., mainly along isothermal lines (if feasible), until the region containing large temperature gradients is left, and only then lead the wire away from the thermal hotspot (cf. Fig. 6.30, left, in Chap. 6).

SM can be tackled by including additional assistant features in the metal stack to normalize the material distribution and to create a uniform coefficient of thermal expansion (CTE). However, these features may not always comply with thermal wires and vias used for mitigating TM as the latter tend to accumulate metal in one place.

Using "soft" dielectrics in terms of low stiffness is another promising SM suppression measure as it enables unhindered thermal expansion of the interconnect. Subsequently, almost no mechanical stress is introduced in the wires. Fully "relaxed" wires will suffer almost no SM. In physical design, the "softness" of the wire's dielectric environment can be further improved by using as few as possible of the available interconnect materials (which are "stiffer" than the dielectric) or by introducing less dense regions around otherwise mechanically stressed wires. However, this contradicts EM-damage counter measures because SM is often used to counteract EM in order to increase the wires' lifetime [25].

Three-dimensional circuits, i.e., 3D-ICs with multiple active layers, require special attention when it comes to reducing SM and TM. Through-silicon vias (TSVs) in these circuits are rather large obstacles [13] where material that is different from bulk substrate is added. This leads to mismatches due to different CTEs. Hence, production and usage of the circuit (i.e., its thermal loading) introduce mechanical stress around TSV locations. Wires near TSVs are especially affected by SM.

Keep-out zones are created around TSVs as a precaution. They not only avoid SM in wires but also minimize stress-related mobility changes in active devices [11, 19]. TM in 3D-ICs can be further mitigated by introducing the aforementioned thermal vias (which are called *thermal TSVs* here) to lower the thermal gradients.

References

1. R.J. Baker, *CMOS—Circuit Design, Layout, and Simulation*, 3rd edn. (Wiley, 2010). ISBN 978-0-470-88132-3
2. J.R. Black, Electromigration—a brief survey and some recent results. *IEEE Trans. Electron. Dev.* **16**(4), 338–347 (1969). https://doi.org/10.1109/T-ED.1969.16754
3. S. Bigalke, J. Lienig, FLUTE-EM: electromigration-optimized net topology considering currents and mechanical stress, in *Proceedings of 26th IFIP/IEEE International Conference on Very Large Scale Integration (VLSI-SoC)*. https://doi.org/10.1109/VLSI-SoC.2018.8644965
4. S. Bigalke, J. Lienig, G. Jerke et al., The need and opportunities of electromigration-aware integrated circuit design, in *Proceedings of the IEEE/ACM International Conference on Computer-Aided Design (ICCAD)* (2018). https://doi.org/10.1145/3240765.3265971
5. S. Chatterjee, V. Sukharev, F.N. Najm, Power grid electromigration checking using physics-based models, *IEEE Trans. Comput. Aided Design Integ. Circuits Syst.* **37**(7), 317–1330 (2017). https://doi.org/10.1109/TCAD.2017.2666723
6. H. Gossner, ESD protection for the deep sub-micron regime—a challenge for design methodology, in *Proceedings of the International Conference on VLSI Design (VLSID)* (2004), pp. 809–818. https://doi.org/10.1109/ICVD.2004.1261032
7. A. Hastings, *The Art of Analog Layout*, 2nd edn (Pearson, 2005). ISBN 978-0131464100
8. A. Heryanto, K.L. Pey, Y. Lim et al., Study of stress migration and electromigration interaction in copper/low-k interconnects, in *IEEE International Reliability Physics Symposium (IRPS)* (2010), pp. 586–590. https://doi.org/10.1109/IRPS.2010.5488767
9. G. Jerke, J. Lienig, Hierarchical current-density verification in arbitrarily shaped metallization patterns of analog circuits, *IEEE Trans. CAD Integ. Circuits Syst.***23**(1), 80–90 (2004). https://doi.org/10.1109/TCAD.2003.819899
10. G. Jerke, J. Lienig, J. Scheible, Reliability-driven layout decompaction for electromigration failure avoidance in complex mixed-signal IC designs, in *Proceedings of the Design Automation Conference (DAC)* (2004), pp. 181–184. https://doi.org/10.1145/996566.996618
11. J. Knechtel, I.L. Markov, J. Lienig, Assembling 2-D blocks into 3-D chips, *IEEE Trans. Comput. Aided Design Integ. Circuits Syst.***31**(2), 228–241 (2012). https://doi.org/10.1109/TCAD.2011.2174640
12. M.A. Korhonen, P. Borgesen, K.N. Tu et al., Stress evolution due to electromigration in confined metal lines. *J. App. Phys.* **73**(8), 3790–3799 (1993). https://doi.org/10.1063/1.354073
13. J. Knechtel, E.F.Y. Young, J. Lienig, Planning massive interconnects in 3D chips, *IEEE Trans. Comput. Aided Design Integ. Circuits Syst.***34**(11), 1808–1821 (2015). https://doi.org/10.1109/TCAD.2015.2432141
14. J. Lienig, H. Bruemmer, *Fundamentals of Electronic Systems Design* (Springer, 2017). ISBN 978-3-319-55839-4. https://doi.org/10.1007/978-3-319-55840-0
15. J. Lienig, Introduction to electromigration-aware physical design, in *Proceedings of the International Symposium on Physical Design (ISPD)* (ACM, 2006), pp. 39–46. https://doi.org/10.1145/1123008.1123017
16. J. Lienig, Electromigration and its impact on physical design in future technologies, in *Proceedings of International Symposium on Physical Design (ISPD)* (ACM, 2013), pp. 33–40. https://doi.org/10.1145/2451916.2451925

17. J. Lienig, G. Jerke, Current-driven wire planning for electromigration avoidance in analog circuits, in *Proceedings of the ASP-DAC* (2003), pp. 783–788. https://doi.org/10.1109/ASPDAC.2003.1195125
18. C.-C. Lin, W.-H. Liu, Y.-L. Li, Skillfully diminishing antenna effect in layer assignment stage, in *International Symposium on VLSI Design, Automation and Test (VLSI-DAT)* (2014), pp. 1–4. https://doi.org/10.1109/VLSI-DAT.2014.6834859
19. J. Lienig, M. Thiele, *Fundamentals of Electromigration-Aware Integrated Circuit Design* (Springer, 2018). ISBN 978-3-319-73557-3. https://doi.org/10.1007/978-3-319-73558-0
20. Z. Moudallal, V. Sukharev, F.N. Najm, Power grid fixing for electromigration-induced voltage failures, in *Proceedings of the 2019 IEEE/ACM International Conference on Computer-Aided Design (ICCAD)* (2019), pp. 1–8. https://doi.org/10.1109/ICCAD45719.2019.8942141
21. B. Razavi, *Design of Analog CMOS Integrated Circuits*, 2nd edn. (McGraw-Hill, 2015). ISBN 987-0-07252493-2
22. S.M. Sze, K.K. Ng, *Physics of Semiconductor Devices and Technology* (Wiley, 2007). ISBN 978-0-471-14323-9
23. C. Saint, J. Saint, *IC Mask Design, Essential Layout Techniques* (McGraw-Hill Education, 2002). ISBN 978-0-07-138996-9
24. O. Semenov, H. Sarbishaei, M. Sachdev, *ESD Protection Device and Circuit Design for Advanced CMOS Technologies* (Springer, 2008). ISBN 978-1-4020-8300-6. https://doi.org/10.1007/978-1-4020-8301-3
25. C. Thompson, Using line-length effects to optimize circuit-level reliability, in *15th International Symposium on the Physical and Failure Analysis of Integrated Circuits (IPFA)* (2008), pp. 1–4. https://doi.org/10.1109/IPFA.2008.4588155
26. A. Todri, M. Marek-Sadowska, Reliability analysis and optimization of power-gated ICs, *IEEE Trans. Very Large Scale Integration (VLSI)Syst.* **19**, 457–468 (2011). https://doi.org/10.1109/TVLSI.2009.2036267
27. https://www.xfab.com/home/. Accessed Jan 1 2020

Index

A
Acceptors, 10, 52
Alignment marks, 38
Analog design gap, 20
Antenna effect, 286
Antenna-rule check, 204
Area routing, 185
Assertion-Based Verification (ABV), 198

B
Back annotation, 173
Back-End-of-Line (BEOL), 82
Band gap, 8
BCD, 16
BICMOS, 16
Blind via (PCB), 4
Bottom-up design style, 145
Breakdown voltage, 218
Buried via (PCB), 4

C
Calculated layer, 115
Capacitors, 227
Carrier substrate (PCB), 3
Cell generator, 233
Channel-stop implant, 268
Channel stopper, 269
Charged-Device Model (CDM), 278
Chemical-Mechanical Polishing (CMP), 67
Circuit design, 22, 130
Circuit diagram, 87, 169
Circuit schematic, 87
Circuit topology, 25
Clock planning, 179
Clock tree synthesis, 131
CMOS process steps, 73
Compaction, 187
Conduction band, 8
Congestion-driven placement, 182
Conics, 93
Constraint
  design-methodology, 150
  electrical, 149
  functional, 149
  technological, 149
Contact holes (IC), 13
Contacts (IC), 13
Crosstalk, 275
Current mirror, 238
Czochralski process, 33

D
Damascene technique, 67
Deep Trench Isolation (DTI), 59
Defect electrons, 9
Depletion region. *See* space-charge region
Derived layer, 107
Design freedom, 153
Design gap, 20
Design-methodology constraints, 150
Design models, 133
Design Rule Check (DRC), 202
Design style, 139
Detailed routing, 184
Developer (PCB), 3
Device, 85
Differential pair, 238
Differential-pair routing, 183, 274

J. Lienig and J. Scheible, *Fundamentals of Layout Design for Electronic Circuits*,
https://doi.org/10.1007/978-3-030-39284-0

Diffusion, 52
Diffusion layer. *See* doping layer
Diffusion resistors, 224
Dishing, 104
Donors, 10, 52
Donuts, 92
Doping, 10, 52
Doping layer, 56
Drawn layer, 92
Dual-Damascene process, 69
Dummy error, 196
Dummy fill, 104

E
Electrical constraints, 149
Electrical Rule Check (ERC), 204
Electromagnetic Compatibility (EMC), 192
Electromigration (EM), 291
Electronic Design Automation (EDA), 20
Electrostatic Discharge (ESD), 276
EMC-compliant layout, 192
Epitaxy, 11, 58
Equivalence checking, 198
ESD design window, 277
ESD event (chip), 276
ESD protection circuit, 276
Extract, extraction file, 205

F
Fabricated layer, 92
False error, 196
Field-Effect Transistor (FET), 73
Field-Effect Transistor (MOS-FET), 220
Field-Programmable Gate Array (FPGA), 143
Floorplan, floorplanning, 175
Folding (transistors), 222
Footprint library (PCB), 124
Footprint (PCB), 190
Formal verification, 198
FPGA (design), 143
Fracturing, 110
Front-End-of-Line (FEOL), 78
Full-custom design, 140
Functional constraints, 149
Functional design, 130, 168
Functional verification, 199

G
Gajski-Kuhn Y-Chart, 136
Galvanic isolation, 218
Gate array (design), 142
Gcell, 184
Geometrical design rules, 111
Global routing, 184
Guard ring, 260, 263, 266

H
Hard block, 175
Hardware Description Language (HDL), 166
Heteroepitaxy, 58, 61
Hole conduction, 9
Holes, 9
Homoepitaxy, 58
Hot carriers, 270
Hot electrons, 270
Human-Body Model (HBM), 278

I
Identification letter (schematic), 170
Instance, 86
Instantiation (device), 169
Ion implantation, 54
IPTAT circuit, 238

J
Jumper, 289
Junction Isolation (JI), 217

K
Keepout area (PCB), 191
Keep-out zone (TSV), 301

L
Landpattern (PCB), 190
Latchup, 264
Layer, 24
  calculated, 115
  derived, 107
  diffusion, 56
  doping, 56
  drawn, 92
  fabricated, 92
  logical, 107, 108
  physical, 107
Layout design, 23
Layout generator, 234
Layout post processing, 102, 208
Layout Versus Schematic (LVS), 205

Leaker, 289
Legalization (placement), 180
Library, 119
  footprint (PCB), 124
  macro cell, 123
  model (PCB), 125
  pad cell, 123
  standard cell, 121
  symbol (PCB), 123
Lightly Doped Drain (LDD), 271
Line losses, 272
Local Oxidation of Silicon (LOCOS), 49
Logical layer, 107, 108
Logic design, 130
Logic synthesis, 168
Low Temperature Co-fired Ceramics (LTCC), 5
Lumped elements, 86

M
Machine Model (MM), 278
Macro cell (design), 142
Macro (cell) library, 123
Manufacturing documentation (PCB), 193
Mask (PCB), 3
Matching, 237
Matching concepts, 254
Meet-in-the-middle design style, 145
Mixed-signal IC, 158
Model-based OPC, 44
Model checking, 198
Model library (PCB), 125
Module generator, 234
Moore's law, 19
MOS-FET, 220

N
N-doped, 10, 52
Net, 84
Netlist, 87, 173
Node, 84
NPN transistor, 230

O
Optical Proximity Correction (OPC), 43, 110, 210
Oversizing, 100
Oxidation, 46
Oxide, 45

P
Pad, 117
Pad cell library, 123
Parasitic effects, 257
Parasitic Extraction (PEX), 207
Parasitics, 257
Partitioning, partition, 175
PCell, 120, 234
P-doped, 10, 52
Photolithography, 34
Photomask, 36
Photoresist, 34
Physical design, 23
Physical layer, 107
Physical synthesis, 169
Physical verification, 132
Pin, 84
Pin assignment, 178
Pinning, 118
Placement, 180
Plasma-Induced Damage (PID), 287
PNP transistor, 232
Poly resistors, 226
Port, 84
Power supply (floorplanning), 177
Printed Circuit Board (PCB), 2
Process Design Kit (PDK), 119
Process nodes, 17

R
RC element, 273
RC optimization, 274
Reactive Ion Etching (RIE), 48
Register Transfer Level (RTL), 166
Resist, 34
Resist mask, 35
Resistor head, 226
Resistors, 224
Resolution Enhancement Techniques (RET), 110, 209
Reticle, 37
Reticle layout, 105
Rip-up and reroute, 186
Robustness (layout, design rule), 111
Routing, 183
Routing layers, 63
Rule-based OPC, 44

S
Schematic, 169
Schematic entry, 169, 189
Schematic symbols, 172

Semiconductor, 8
Shallow Trench Isolation (STI), 59
Shapes, 91
Sheet resistance, 214
Shrinking, 18
Shunt path, 276
Signal distortions, 273
Silicon dioxide (SiO_2), 33, 45
Silicon (Si), 33
Simulation, 199
Simulation Program with Integrated Circuit Emphasis (SPICE), 90
Sinker, 217
Sizing, 100
Smart power IC, 16
Soft block, 175
Space-charge region, 218
Specification, 21, 129
Standard-cell design, 140
Standard cell library, 121
Standard cells, 141
Static Timing Analysis (STA), 201
Stick diagram, 187
Stimuli (simulation), 199
Stress Migration (SM), 293
Subcircuit, 85
Substrate currents, 258
Substrate debiasing, 258
Sub-wavelength lithography, 43
Surface-Mount(ed) Device (SMD), 4
Surface-Mount Technology (SMT), 4
Symbolic compaction, 187
Symbolic design entry, 169
Symbolic layout design, 187
Symbol library (PCB), 123
System specification, 129

T
Tank (well), 217
Technological constraints, 149
Technology nodes, 17
Thermal Migration (TM), 292
Thermal via, 300
Thermal wire, 300
Thick-field threshold, 268
Thick-film technology, 5
Thin-film technology, 5
Through contact (IC), 13
Through-Hole Device (THD), 4
Through-Hole Technology (THT), 4
Through-hole via (PCB), 4
Time constant (RC element), 273
Timing closure, 131, 202
Timing-driven placement, 182, 202
Timing-driven routing, 202
Timing verification, 200
Top-down design style, 145
Tub (well), 217

U
Undercut, 48
Undersizing, 100

V
Valence band, 8
Valence electrons, 8
Validation (vs verification), 147
Verilog, 166
Verilog-AMS, 166
VHDL-AMS, 166
VHSIC Hardware Description Language (VHDL), 166
Via, 13
Via doubling, 186
Via minimization, 186
Via (PCB), 4
Voltage-dependent spacing rules, 219

W
Well, 216

Y
Y-chart, 136

Z
Zone melting, 33